Performance and Durability of Bituminous Materials

Supporting organisations

Cleveland County Council
County Surveyors Society
Energy Efficiency Office
Institute of Asphalt Technology
Institution of Highways and Transportation
National Power
Tilcon Limited
The Department of Civil Engineering, University of Leeds

Technical Committee

J.G. Cabrera (Chairman)	University of Leeds
D. Bonner	University of Hertfordshire
D. Colwill	Transport Research Laboratory
H. Khalid	University of Liverpool
A. Nikolaides	University of Thessaloniki
D. Rockliff	Tilcon Limited
M. Sutcliffe	Cheshire County Council
A. Woodside	University of Ulster

Organising Committee

J.G. Cabrera (Joint Chairman)	University of Leeds
J.R. Dixon (Joint Chairman)	University of Leeds
J. Higgins	Cleveland County Council
B. Ley	National Power
P. MacDonald	ETSU on behalf of the EEU
G. Poad	Cleveland County Council
A. Ridley	Tilcon Limited
M. Sutcliffe	Cheshire County Council
Mandy Stamp (Secretary)	University of Leeds

Performance and Durability of Bituminous Materials

Proceedings of Symposium, University of Leeds, March 1994

Edited by

J.G. CABRERA

Professor of Civil Engineering Materials, Civil Engineering Materials Unit, University of Leeds, UK

and

J.R. DIXON

Lecturer in Construction Management, Department of Civil Engineering University of Leeds, UK

E & FN SPON
An Imprint of Chapman & Hall
London · Glasgow · Weinheim · New York · Tokyo · Melbourne · Madras

Published by E & FN Spon, an imprint of Chapman & Hall,
2–6 Boundary Row, London SE1 8HN

Chapman & Hall, 2–6 Boundary Row, London SE1 8HN, UK

Blackie Academic & Professional, Wester Cleddens Road,
Bishopbriggs, Glasgow G64 2NZ, UK

Chapman & Hall GmbH, Pappelallee 3, 69469 Weinheim, Germany

Chapman & Hall USA, 115 Fifth Avenue, New York, NY 10003, USA

Chapman & Hall Japan, ITP-Japan, Kyowa Building, 3F, 2-2-1 Hirakawacho,
Chiyoda-ku, Tokyo 102, Japan

Chapman & Hall Australia, 102 Dodds Street, South Melbourne,
Victoria 3205, Australia

Chapman & Hall India, R. Seshadri, 32 Second Main Road, CIT East,
Madras 600 035, India

First edition 1996

Printed in Great Britain by St Edmundsbury Press, Bury St Edmunds, Suffolk

ISBN 0 419 19730 3

Publisher's Note
This book has been produced from camera-ready-copy provided by the individual contributors

A catalogue record for this book is available from the British Library

∞ Printed on permanent acid-free text paper, manufactured in accordance with ANSI/NISO Z39.48-1992 (Permanence of Paper).

Contents

Preface

Heavier traffic loads, large growth of vehicles of high tonnage and the end of the service life of many motorways and trunk roads in the European Community, have taxed the ability, knowledge and experience of researchers and highway practitioners in the development and design of composite materials, especially bituminous composites of high performance characteristics and adequate structural properties for the construction of new road pavements, upgrading of existing ones and maintaining the in-service performance of road pavements.

The new CEN specifications will help to bring about improved methodologies for assessing the performance of bituminous materials and utilising the new technological advances in the production of improved materials and processes which demand lower levels of energy for their preparation and placement.

This book brings together contributions from researchers and engineers on new ideas and innovations on the properties, performance and durability of bituminous materials for the construction of pavements. It highlights particularly new research methodologies to design and construct bituminous composites which require low energy expenditure. It should be therefore of interest to all levels of researchers and highway practitioners.

The Civil Engineering Materials Unit (CEMU) of the Department of Civil Engineering, University of Leeds has for a long time pioneered developments in the field of high performance composites particularly bituminous composites. Its research group is active in the development of low energy bituminous composites and in the development of improved testing techniques and therefore is very pleased to have hosted the Symposium which has resulted in this book. CEMU is committed to an intensive programme of Continuous Postgraduate Education and the Symposium is a significant part of this programmc in this arca.

J.G. Cabrera
J.R. Dixon
Leeds

PART ONE
AGGREGATES AND FILLERS

1 THE ASSESSMENT OF THE SUITABILITY OF WASTE MATERIALS FOR USE IN A HIGHWAY STRUCTURE

A.R. WOODSIDE and D.J. McCUTCHEON
Department of Civil Engineering and Transport, University of Ulster, Carrickfergus, Northern Ireland
R.J. COLLINS
Building Research Establishment, Garston, Hertfordshire, UK

Abstract
An ever increasing amount of demolition debris is being produced and disposed of in landfill sites throughout the country. Research by the Institute of Demolition Engineers has estimated that the annual amount available for recycling is in the region of 25 million tonnes , (Lindsell 1990). At present little use is made of this material in the United Kingdom for recycling back into permanent road works, although it has been proven that the material can easily satisfy the 'Type 1' specification for granular sub-base materials. This lack of acceptance of recycled aggregates for use in highway construction is primarily due to the lack of British Standards which specifically consider recycled aggregates.
Keywords: Aggregate, Demolition Waste, Recycling, Sub-base.

1 Introduction

In recent years there has been a growth in the awareness of the need to conserve natural resources and recycle or reclaim those materials which are in short supply. It is generally agreed that the conservation of existing sources of construction materials is a major priority for future generations. The demand for raw aggregates has continued to grow over the last decade, and one third of demolition firms surveyed, (Lindsell 1990), expect the amount of demolition debris to increase by 50% to 100% in the next 10 years. This shall lead to an increase of dumping costs at landfill sites particularly in S.E. England were contractors may find it more expensive to dump demolition debris than it is to recycle it.

Research by the Institute of Demolition Engineers has estimated the annual amount of demolition debris to be in the region of 25 million tonnes. Given the correct conditions, efficient recycling of this material could reduce the demand on natural aggregates by 10%. In view of the increasing pressures

Performance and Durability of Bituminous Materials. Edited by J.G. Cabrera and J.R. Dixon. © E & FN Spon. Published in 1996 by E & FN Spon, 2–6 Boundary Row, London SE1 8HN. ISBN 0 419 20540 3.

on the supplies of natural aggregates and the decreasing availability of land fill sites more recycled materials should be used.

2 Reasons for Recycling

It is agreed that were possible the recycling of waste products is a desirable activity, whether it is motivated by either economic or environmental considerations, or more usually a combination of the two.

Whilst environmental factors may make the recycling of demolition debris an attractive option, it is the economic aspects which determine if recycling is to be implemented. The cost of dumping large amounts of debris in landfill sites in built up areas i.e. S.E. England, has become extremely expensive over recent years. Also licensed tips are often many miles from the areas of demolition activity, therefore the cost of transportation has become increasingly important. In such areas recycling of demolition debris is profit orientated, such that recycling only becomes an attractive option when the value of the recycled material returns a profit.

In contrast, areas with a ready supply of low cost natural aggregates and nearby landfill sites provide no opportunity for recycling to produce a profit. In such areas the debris will generally be dumped regardless of the quality of the material.

The use of recycled aggregates in highway construction would produce two major benefits. First, by supplementing the supply of natural aggregates the life of existing quarries can be extended and the opening of new quarries delayed. Second, the rate of consumption of space in landfill sites and the tipping of demolition debris on derelict or common land may be reduced. Taken together environmental and economic incentives present a powerful argument for the production and use recycled aggregates from demolition debris.

Several county councils have used crushed concrete successfully in their road building programmes, this includes Hampshire county council who were able to obtain a regular supply of high quality crushed concrete from a recycling firm based in Portsmouth. Previous research indicates that that suitably recycled aggregate could perform as well as natural aggregate in most cases. Therefore it could be considered for many potential uses however many people associate recycled materials from demolition debris with low quality and variable performance ,in practice this is not necessarily true.

At present there are 3 major barriers to the quality of recycled materials being improved i.e. cost, local authorities and lack of incentive. Most contractors can only afford the use of a basic recycling plant. Such an arrangement is limited in the quality of the aggregate it produces by the level of contamination in the original debris. The attitude of local authorities is very important as this will influence the type and size of recycling plant which can be set up. Lack of incentive is a major hindrance as natural aggregate may be more accessible and cheaper than recycled aggregate so contractors have little incentive to use recycled aggregates.

3 Use of recycled aggregate as a sub-base

The use of clean, graded brick or concrete aggregates in the construction of road sub-base appears to be accepted in several European countries including Belgium and Netherlands. The Netherlands has relatively poor reserves of natural aggregate and has become more dependent on recycled material for the construction of unbound aggregate road layers.

Subsequently it has become an important source of aggregate for Dutch construction industry. Research in the United Kingdom,(Mulheron & Mahony 1990), has proven that recycled concrete can be considered as a suitable granular sub-base and generally fell within the 'Type 1' specification. Despite this the use of recycled concrete aggregate in such applications often meets with resistance in the United Kingdom.

In the United Kingdom there are no Standards specifically covering the use of recycled aggregates, consequently recycled products can only be compared with existing specifications developed for natural aggregates. Such a comparison can be totally inappropriate and will tend to inhibit the future development of recycled materials. The largest potential use of many of the recycled materials has been identified in the lower specification materials for road construction (BRE Report 1993). There is evidence from the use of demolition debris in a few localised areas that these materials perform at least as well as high quality quarried aggregates, but there has been little research to substantiate this or to provide reasons for the difference.

In response to this lack of experience with the use of recycled materials, fundamental research is currently being carried out at the University of Ulster, Highway Engineering Research into determining the performance of such materials. This has initially involved comparitive testing of proven natural aggregate with those obtained from various types of low quality and recycled aggregate. The results given in this paper relate to their testing using a range of British Standard and proposed European CEN methods. The following tests were carried out:

- Ten Percent Fines Value
- Magnesium Sulphate Soundness Value
- Los Angeles Value
- Micro Deval Value

4 Discussion of results

The aggregates assessed were high quality Silurian greywacke; low quality Silurian shale, Magnesian limestone and Thames gravel, slate waste from North Wales and recycled concrete. A summary of the data obtained is shown in Table 1.

Table 1.

Rock type	TFV (dry)	MSSV	LA	MDE
Mag. Limestone A	65	75.9	47.7	39.2
Mag. Limestone B	70	79.5	52.6	39.3
Mag. Limestone C	170	95.9	26.7	10.8
Mag. Limestone D	20	66.3	89.8	73.2
Mag. Limestone E	150	77.6	28.1	32.0
Mag. Limestone F	85	79.8	47.4	34.8
Thames gravel	75	73.1	47.7	41.8
Recycled concrete A	170	99.9	28.2	20.5
Recycled concrete B	160	99.8	35.0	20.1
N. Wales slate	180	99.9	20.3	42.2
Silurian shale	155	76.0	25.0	33.0
Silurian greywacke	390	98.0	12.0	18.0

From the results given in Table 1 it is apparent that the marginal and recycled materials do not achieve the same high specifications as a high performance gritstone. The following figures show the variation in results obtained.

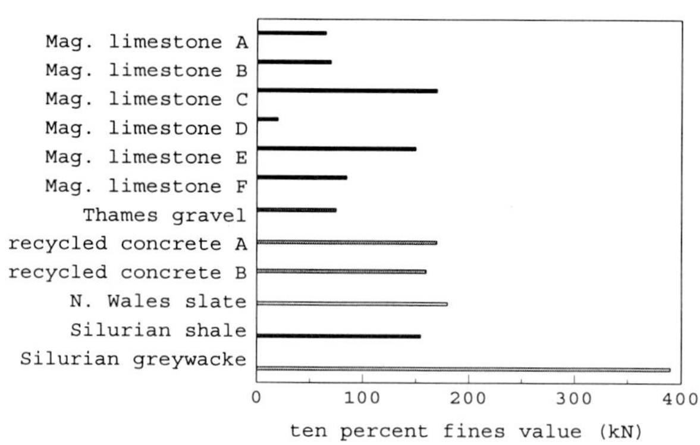

Fig. 1 Ten pecent fines value results

Fig.1 shows the data obtained for the dry Ten percent fines value test. It should be pointed out that most specifications require a value obtained from soaked aggregate and not the dry results shown. They indicate considerable variation within the Magnesian limestones assessed with the recycled conrete giving reasonably good results of 160 and 170 kN which were comparable with the slate waste and the Silurian slate.

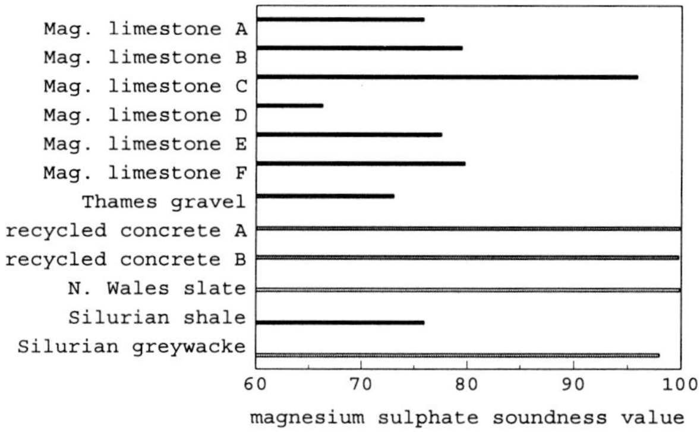

Fig. 2 Magnesium sulphate soundness value results

Fig.2. shows the data obtained for the magnesium sulphate soundness value test. Typically, a value in excess of 75% is required. Again considerable variation occurs for the Magnesian limestones in comparison to the recycled concrete which showed very little breakup during testing.

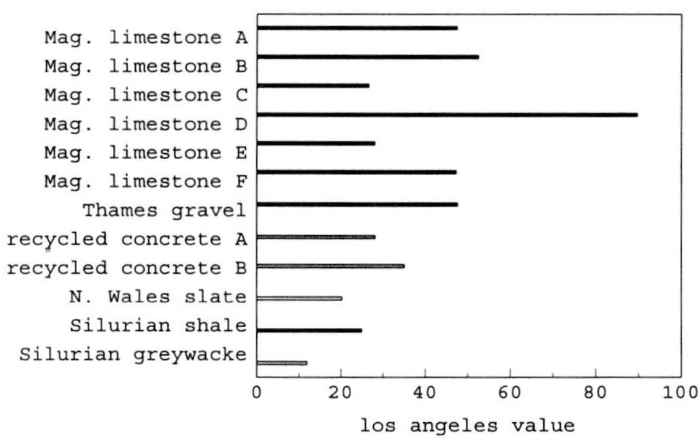

Fig.3. Los Angeles results

Fig. 3 shows the results obtained for the Los Angeles test. This has been included as it is currently proposed as a European CEN method for assessing the fragmentation properties of aggregate. Although no limits are as yet proposed, values of < 35 or < 45% are under review. Again with the exception of the variability of the Magnesian limestones and the Thames gravel, the remainder of the results for the low quality / marginal materials are all less than 35%.

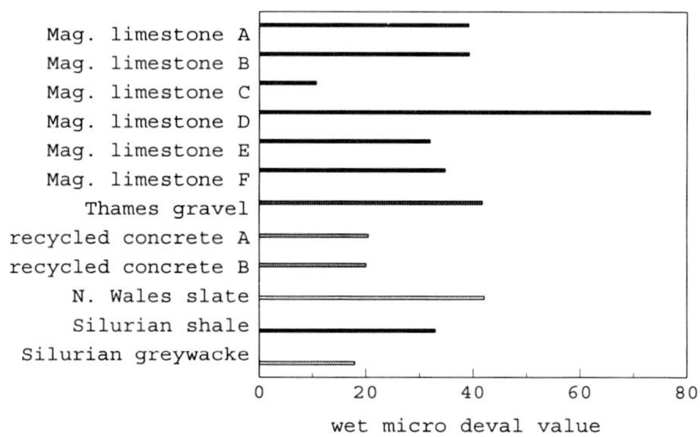

Fig.4. Wet Micro-deval results

The wet micro-deval test is also a proposed European CEN test method where it is felt that the particle to particle friction in the presence of moisture closely replicates the conditions in a suub-base layer. Again no limits have been set but it is likely that such a limit may be of the order of < 35 or 45%. With the exeption of the high quality greywacke and one of the Magnesian limestones, the values obtained for the recycled concrete were most encouraging at 20.5 and 20.1.

The initial conclusions from this programme of comparitive investigation using standard aggregate were the considerable variation of the Magnesian limestones and that the slate waste and recycled concrete could pass these test methods.

5 Performance Testing

Although specifications are required to ensure that unsuitable sources of material are not used, the presently used specification requirements, i.e. grading, mechanical strength and plasticity, do not consider the in-service performance characteristics of the material being used. With the trend towards the use of low grade and marginal materials, it may be argued that there is the requirement for a performance specification for lower grade material which could prove there suitablity for use.

A programme of performance based research is currently being carried out at the University of Ulster in conjunction with the Building Research Establishment. This is investigating the performance properties of low quality demolition material in an attempt to propose new standards for such traditionally recognised marginal materials. This involves the use of simulated trafficking conditions using a 'purpose designed segmental cell' and wheel track testing apparatus. Although at an early stage the initial results are proving satisfactory.

6 Conclusions

This paper has considered the ability of low grade and marginal sources of aggregate to meet current specification requirements for use as sub-base material. The results indicate the variability of Magnesian limestone. They also show that slate waste from North Wales and recycled concrete may offer an alternative to traditional sources of aggregate. Due to the availaibity of demolition concrete in urbanised areas were typically natural aggregate is in short supply, this type of material is currently being assesed to determine it in-service performance.

7 References

Ferguson,J., 1994. Waste from construction and duty of care. Proc. Instn Civ. Engrs Mun Engr,1994,**103**,Mar 23-29

Lindsell,P.,1990. Recycling of Construction Materials. A report for the Institute of Demolition Engineers

Mulheron,M. & O'Mahony,M.M.,1990. Properties and Performance of recycled aggregates . Highways and Transportation,1990,Feb 35-37.

2 AGGREGATE GRADING DESIGN FOR POROUS ASPHALT

J.G. CABRERA and M.O. HAMZAH
Civil Engineering Materials Unit, Department of Civil Engineering, University of Leeds, Leeds, UK

Abstract
Porous asphalt consists predominantly of coarse aggregate, its open porous structure is advantageous for eliminating splash and spray, improving skid resistance, reducing aquaplaning potential and lowering noise level and binder content requirement. A well designed porous asphalt exhibits high porosity where pores are continuous and form a network of drainage channels.

Current literature on the development of aggregate gradings show that the overriding design considerations are to produce blends that give maximum density or minimum porosity. Most of porous asphalt gradings currently specified are based on empirical studies and do not deal directly with the packing behaviour of the aggregate mass.

This paper proposes a grading for porous asphalt which is obtained by applying the theory of packing currently used to design dense bituminous mixtures. The packing behaviour of dry aggregate blends is studied using a vibratory compactor. The grading for porous asphalt is developed by applying the concept of designing to a target porosity starting with the minimum porosity of the coarse aggregate matrix. Fine aggregate gradings are varied to achieve the same target porosity. Mixes made with the gradation proposed exhibits slightly superior permeability and resistance to disintegration and overcompaction than the equivalent mix specified in BS 4987. Other gradations to achieve any desired porosity values can be easily established from a set of experimental curves developed using the vibratory compactor.
Keywords: Porous asphalt, aggregate grading, porosity.

1 Introduction

On high speed roads, surface water generates splash and spray, induces aquaplaning and reduces skid resistance. Porous asphalt was developed to facilitate the rapid removal of surface water during wet weather. In addition, traffic noise level is reduced and so is the binder content requirement. A well designed porous asphalt mix is one that exhibits high porosity where pores are continuous and form a network of drainage channels.

Performance and Durability of Bituminous Materials. Edited by J.G. Cabrera and J.R. Dixon. © E & FN Spon. Published in 1996 by E & FN Spon, 2–6 Boundary Row, London SE1 8HN. ISBN 0 419 20540 3.

2 Porous asphalt aggregate grading

Mix design of porous asphalt involves selection of an appropriate aggregate grading and blending it with an optimum binder content to produce a durable mix that can perform satisfactorily during the design life of the pavement structure.

Aggregate grading specifications for porous asphalts, over the world, differ widely. It appears that most porous asphalt aggregate gradings currently used, are based on empirical studies and do not deal directly with the packing behaviour of the aggregate mass. BS 4987 [1] which for the first time include porous asphalt gradation, specifies two porous asphalt mixes which provides two grading envelopes, amount and type of filler and binder grade and content for each grading envelop. The British Standard also defines fine aggregate as material passing the 3.35 mm sieve.

Field trials conducted by the TRL culminated in the publication of an Advice Note [2] detailing the design, use and properties of porous asphalt surface course. The aggregate grading proposed is similar, but not identical, to the BS 4987 gradation for 20 mm maximum size material.

3 Packing and void characteristics of aggregate blends

3.1 Design concept

Current literature on the development of aggregate grading for dense mixes shows that the overriding design considerations are to produce blends that give maximum density or minimum porosity. These range from the Fuller's curve due to Fuller and Thompson [3], it's modification by Brown et al. [4] and an extension of a two-component analysis into the realm of multi-component system by Lees [5].

A porous asphalt aggregate grading consists predominantly of coarse aggregate. The fine aggregate fractions are added so as not to bulk or interfere with the interlock of the coarse aggregate matrix but to leave enough voids to maintain a pervious structure.

In this paper, the grading for porous asphalt is developed by applying the concept of designing to a "target porosity" starting with the minimum porosity of the coarse aggregate matrix. A coarse aggregate matrix of minimum porosity provides the stable strong matrix required for adequate strength and minimum permanent deformation. The fine aggregate gradings are varied to achieve the target porosity. The gradation that exhibits potentially the highest permeability and stability characteristics is adopted as the proposed gradation.

3.2 Multi-component mix design for porous asphalt

When two aggregates of unequal sizes are blended, there is an optimum proportion of the two components which results in minimum porosity and this minimum value is always smaller than any of the porosities of the respective single components. Furnas [6] was the first to describe this behaviour. Accordingly, if a finer fraction is added incrementally to this coarser component, the resultant porosity reduces up to a limiting value beyond which the finer fraction no longer finds room in the pores of the larger aggregate. If the amount of fines increases further, the porosity of the finer fraction becomes more dominant and the character of the blend is governed by the finer component. At the minimum porosity, the proportions of the coarser and finer fractions are *balanced*. Blends that are not at the optimum are either *abundant* or *deficient* in fines.

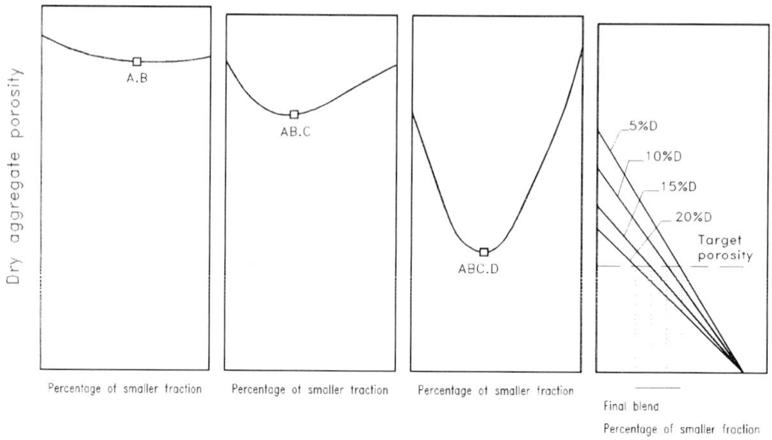

Fig. 1. Multi-component mix design for porous asphalt

The concept of blending to achieve minimum porosity is clearly seen by considering for example mixing five aggregates size fractions A B C D and E in decreasing size order. As seen schematically in Figure 1, the first step in the design of a multi-component system is to determine the proportions of A and B to achieve minimum porosity. The blend A.B is the balanced proportion between aggregate fractions A and B and for the next step, blend A.B then becomes a new coarser component into which the next finer component C is added incrementally to result in a new minimum porosity mix AB.C.

Mix AB.C represents the blend that gives the minimum porosity of the ABC aggregate matrix. If this matrix is then consider to be the stable matrix which will provide strength and resistance to deformation, the next step is to vary the fine aggregate gradings in order to achieve a target porosity that is considered suitable for a pervious mix. It is necessary to adjust the proportions of the components so that the blend will be deficient in fines but maintaining the balanced coarse aggregate matrix. This is done by building up the next curve, that is by considering AB.C as the new component into which the first finer aggregate fraction D is added incrementally. The final fine aggregate size E need to be combined with mix ABCD each starting with 5%, 10%, 15% and 20% of D to give curves shown in Fig. 1(d). Knowing the "target porosity", then various aggregate gradings could be obtained. The method enables the design of gradings with any desired porosity when the porosity of the coarse aggregate matrix is at its minimum.

4 Materials and their properties

Bitumen penetration grade 100 supplied by CRODA HYDROCARBONS was used as the binder. The penetration and softening point values are 98 dmm and 46°C respectively. Gritstone aggregates supplied by ARC Northern Ingleton Quarry was used in this investigation. The aggregates were washed, dried and then sieved into their respective size range or 'bins' as shown in Table 1.

Coarse aggregate is defined as material retained on the 2.36 mm sieve. From Table 1, the coarse and fine aggregate fractions were respectively separated into three and two bins. A 14 mm maximum aggregate size was chosen to make the proposed mix comparable to the

10 mm pervious wearing course aggregate grading specified in BS 4987. In addition, 2% hydrated lime filler was used as recommended by BS 4987.

5 Dry aggregate compaction

5.1 Experimental design

To assess the packing behaviour of the gritstone aggregates used, a vibratory method of compaction was utilised. The vibratory compactor consisted of a cement mortar vibrating table as described in BS4551 [7]. As shown in Fig. 2, the vibrating table was replaced with a standard Marshall mould with a collar. The amplitude and frequency of vibration were 12.5 mm and 50 Hz respectively. A 4 kg steel cylinder surcharge was placed on top of the sample to achieve a uniform compacted surface. After preliminary trials, a compaction time of 1 minute was chosen since a major proportion of compaction had occurred after such period had elapsed.

Table 1. Aggregate size ranges and their bin designations

Aggregate Size (mm)	14 - 10	10 - 6.3	6.3 - 2.36	2.36 - 0.60	0.60 - 0.075
Bin Designation	A	B	C	D	E

5.2 Blending and compacting

The laboratory process consisted of weighing (accurate to 0.1 g) the blend of fractions A plus xB (x varied in steps of 10%) aggregate fractions, mixing the aggregates in a bowl and pouring them into the Marshall mould from a constant height. After placing the surcharge on top of the aggregate specimen in the mould, the compactor was switched on for 1 minute after which the height of the sample at three equally spaced points to the nearest 0.02 mm was recorded. The results used for the design are the average of three tests. Aggregate degradation during vibratory compaction was negligible due to the nature of the aggregate which as indicated was gritstone.

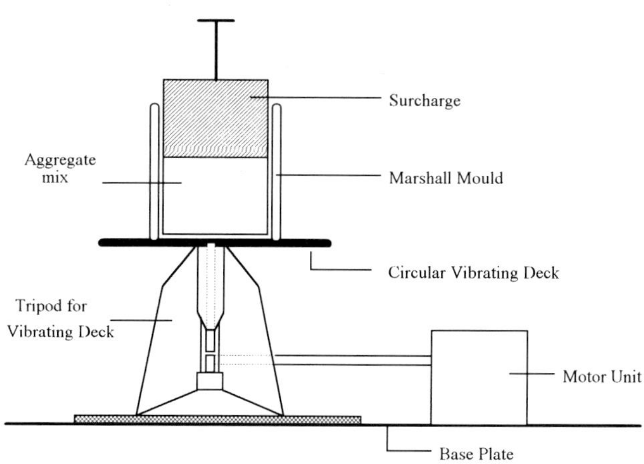

Fig. 2. Schematic diagram of the modified vibrating table set used in this investigation

If softer aggregates are used, fines generated by degradation should be taken into account.

The porosity P was determined from Equation (1).

$$P = 100\left(1 - \frac{D}{D_r}\right) \tag{1}$$

where P = Porosity (%)
D = Compacted density of dry aggregates
D_r = Relative density of mixed aggregate

D_r is obtained from equation (2).

$$Drn = \frac{100}{\sum_{i=1}^{n} \frac{Pwi}{Dri}} \tag{2}$$

where Drn = Relative density of the mixture of n aggregates
Pwi = Percentage of aggregate from bin i
Dri = Relative density of aggregate from bin i

5.3 Determination of target porosity

Three blends each corresponding to the mid-grading, upper envelop and lower envelop of BS 4987 specification limits were compacted and their dry aggregate porosities determined. These values were used to determine the 'target design porosity'.

6 Specimen preparations and further tests

6.1 Experimental design

Marshall specimens were prepared after having determined the appropriate aggregate grading to achieve the target porosity. In accordance to BS 4987, a 4.5% filler is required; 2% of which is hydrated lime. A 4.8% binder content was chosen for all four proposed mixes, including the BS mixes. Aggregates were mixed with binder at 130°C and compacted at 110°C. The modified Marshall impact compactor, designed and fabricated at The University of Leeds, was used. The compactor enables simultaneous compaction of three specimens.

6.2 Permeability

After compaction, specimens were cooled in their moulds. The specimens were then tested for permeability before extrusion to take advantage of the tight bond between the bituminous mix and the mould. A new water permeameter, shown diagramatically in Fig.3, based on the falling head principle was designed and used to quantify the coefficient of permeability. The results were evaluated statistically to determine the standard deviation and coefficient of variation of the results obtained. From the tests carried out to assess the repeatability, a coefficient of variation of 0.87% was obtained.

6.3 Marshall stability

Specimens were extruded and the density obtained by measuring the height and diameter. This was followed by Marshall stability tests at 35°C. The stabilities reported are those resulting from a limiting deformation (flow) of 4 mm. They are as all other results the average of three results.

6.4 Resistance to disintegration

The Cantabrian test of abrasion loss was adopted to quantitatively assess the resistance to disintegration of Marshall specimens in the laboratory. The methodology of this new test on porous asphalt developed in Spain has been reported in detail by Jimenez and Perez [8]. The procedure consists of subjecting a Marshall specimen to impact and abrasion in the Los Angeles drum (without balls) at 18°C. The aim of the test is to determine the Cantabrian loss which is defined as the percentage weight loss after 300 drum revolutions in relation to its initial weight. This test was carried out on the selected proposed mix at varying binder contents and compared it with the BS mixes results.

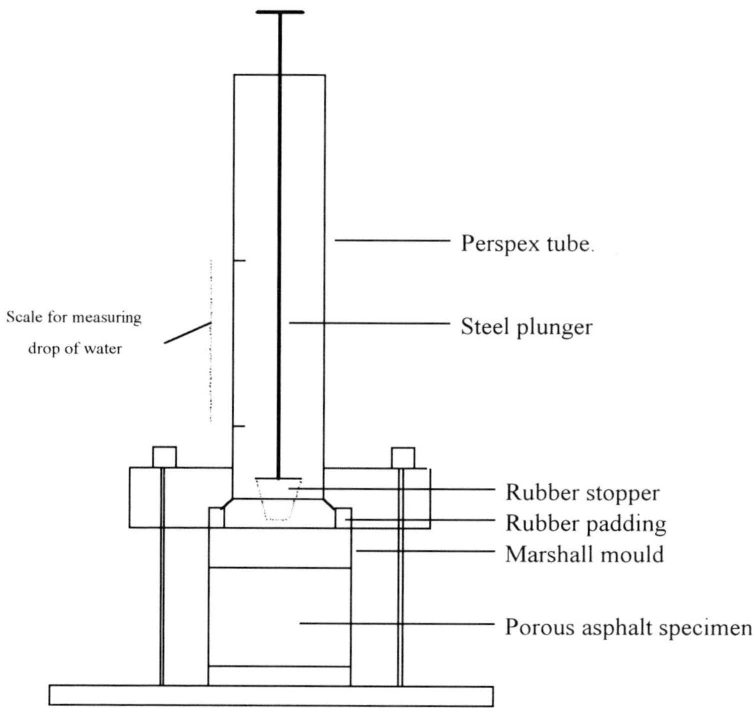

Fig. 3. The water permeameter used to determine the coefficient of permeability of porous asphalt specimens

6.5 Resistance to overcompaction

In service, the performance of porous asphalt is adversely affected by its inability to resist overcompaction, especially during the initial stage of its life. Overcompaction reduces mix porosity and when porosity reduces, all of the benefits associated with an open mix will deteriorate too. Current data on porosity reduction due to the action of traffic depends wholly on field monitoring and measurements. In this study, voids closure due to the kneading action of traffic, hence the resistance to overcompaction, was simulated in the laboratory using the gyratory testing machine .

To assess the resistance due to overcompaction, Marshall specimens were initially prepared. The proposed mixes were prepared at binder contents ranging from 4.5% to 6.0%. Mixes to BS gradation were prepared at binder contents 5.2 $\pm$ 0.5% as recommended in BS 4987. The procedure started with conditioning Marshall specimens of known initial heights and porosities in the oven at 60°C for three hours. The specimens were then overcompacted using the gyratory compactor for up to 110 revolutions at 0.7 MPa. The dial gauge readings every 5 revolutions up to 50 revolutions and every 10 revolutions from thereon were noted. Specimen heights were then measured and porosity reduction with number of gyrations calculated.

7 Results and discussion

7.1 Target porosity

The target or reference porosity as indicated before is the average porosity of blends made in accordance to BS 4987 and equals to 33.8%. This value is the average dry aggregate porosity value of the upper and lower limits of the BS specification.

Table 2. Dry aggregate porosities of the corresponding size range

Bin Designation	Aggregate Size (mm)	Porosity (%)
A	14 - 10	45.14
B	10 - 6.3	44.39
C	6.3 - 2.36	43.87
D	2.36 - 0.60	43.95
E	0.6 - 0.075	42.32

7.2 Porosity of compacted dry aggregate

The porosity values of compacted dry aggregates for the corresponding individual size ranges used in the project are given in Table 2.

The dry aggregate porosity values resulting after mixing and compacting various proportions of aggregates from bins A and B are plotted as curve 1 in Fig. 4. Minimum porosity is achieved when A and B are blended in the ratio of 40:60. Combining bin C aggregate with the optimum of A + B (40:60) produces curve 2 in Fig. 4. The minimum porosity of the three-component system occurs at 45% bin C. This blend gives the most stable coarse aggregate matrix. Therefore adjusting the proportions to express this blend as a combination of A:B:C gives 22%:33%:45%. From Fig. 4, it is obvious that the minimum porosity of a system composed of several components is always smaller than that of a single component.

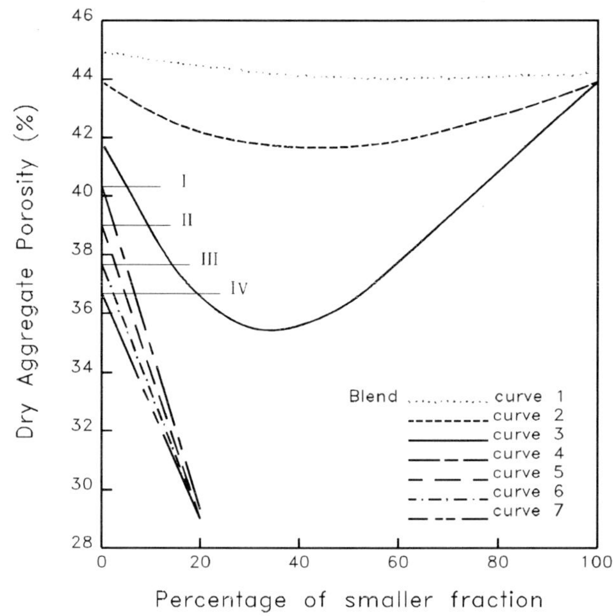

Figure 4. The porosity of the five size aggregates used in this investigation.

The second phase of this process provides data on the influence of the fine aggregate fraction on the dry aggregate porosity. This was done by incrementally adding bin D aggregate starting at the minimum porosity value of curve 2 in Fig. 4 hence constructing curve 3 in Fig. 4. If bin E aggregate is added into the new minimum porosity combination, as is normally done with the design of dense mixes, then the final minimum porosity value of the five aggregate blends would be too low for a pervious mix. Hence bin E aggregates were added beginning from points I, II, III and IV on curve 3 which corresponds to 5%, 10%, 15% and 20% of aggregates from bin D respectively. This results in the final set of curves 4,5,6,7 in Fig. 4 plotted up to porosities of approximately 29%.

The set of curves shown in Fig. 4 can be used to establish any aggregate grading to achieve the target 33.8% dry aggregate porosity. With 4.5% filler content, the cumulative percentage passing the respective sieve is shown in Fig. 5. Mixes prepared based on 5%, 10%, 15% and 20% fine aggregate from bin D were designated mixes G1, G2, G3 and G4 respectively.

7.3 Permeability and stability

The average values of porosity, coefficient of permeability and stability are shown in Table 3.

From Table 3, it can be seen that mixes G1 to G4 and the BS mix are not distinguishable in terms of porosity or stability; however in terms of permeability mix G1 is different than the other G mixes and distinctly different to the BS mix. The effect of fine aggregate gradation on permeability is quite large, progressively decreasing as the ratio of aggregates from bins D to E increases. The lowest value recorded represents a permeability drop of the order of 49%.

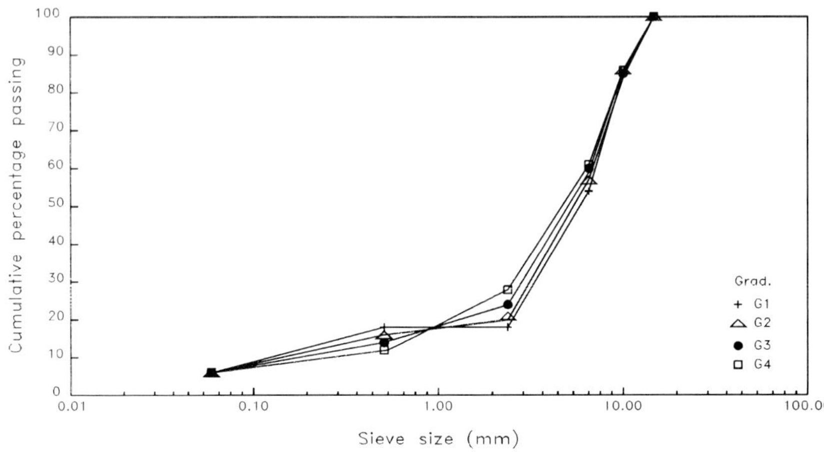

Figure 5. Aggregate gradations investigated.

Table 3 Average values of porosity, permeability and stability of the mixes investigated

Gradation/Mix Designation	Porosity (%)	k (cm/s)	Stability (kN)
BS	17.4 (0.70)*	0.173 (0.008)*	10.7 (0.22)*
G1	17.7 (0.42)*	0.218 (0.020)*	10.9 (0.17)*
G2	17.2 (0.54)*	0.155 (0.006)*	11.0 (0.67)*
G3	17.4 (0.96)*	0.127 (0.028)*	10.7 (0.41)*
G4	16.9 (0.26)*	0.112 (0.013)*	10.6 (0.48)*
Mean of the G gradations	17.3 (0.37)*	0.153 (0.047)*	10.8 (0.18)*

( )* Standard deviation

From these results the porous mix G1 was considered to be an improvement in relation to the BS mix. Therefore it was selected for further evaluation. The difference between the proposed and BS aggregate gradings is graphically shown in Fig. 6.

8 Resistance to disintegration

The experimental results of the Cantabrian Test are shown in Fig. 7. The results confirm the sensitivity of percentage abrasion loss to variations in bitumen content. At 3% bitumen content, both mixes suffered almost complete disintegration. The proposed gradation performed better at bitumen contents exceeding 3.5%. Accepting that 30% abrasion loss is the appropriate limit for acceptance of porous asphalt of adequate performance tested at 18°C [9], the proposed mix exhibits greater resistance to disintegration than the BS mixes. Further tests are being conducted to assess the repeatability of the Cantabrian test and thus to validate the results obtained.

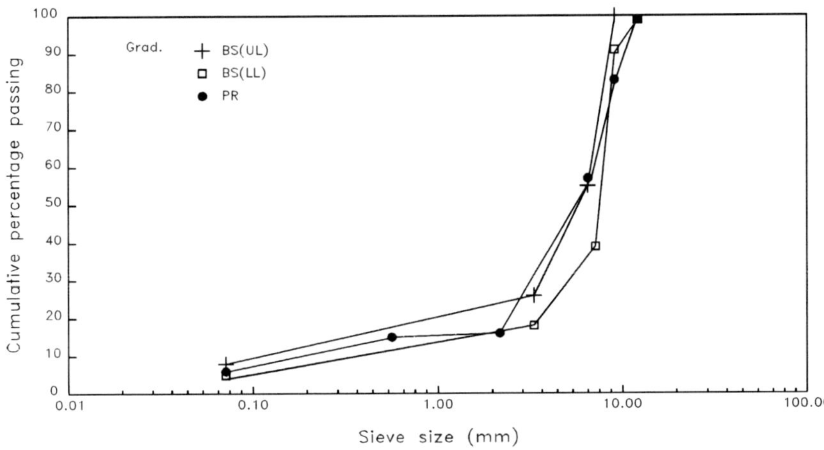

Figure 6. The proposed gradation in relation to BS 4987 gradation limits.

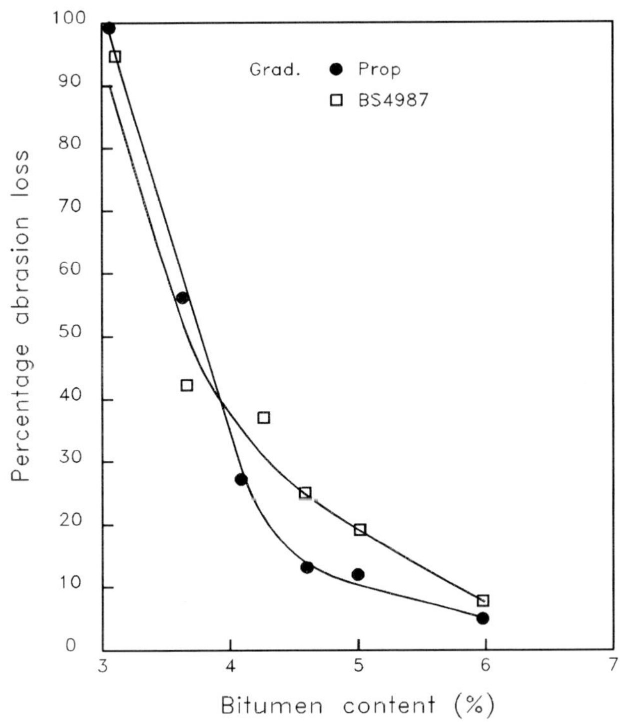

Figure 7. Cantabrian test results of porous asphalt prepared using 100 pen bitumen.

9 Resistance to overcompaction

Fig. 8 shows the porosity reduction with number of revolutions for each bitumen content. Graphs relating relative change in porosity versus number of revolutions are plotted in Fig. 9. Relative change in porosity is defined as the percentage of the difference in porosity at a particular revolution and the initial porosity divided by the initial porosity. Since all curves in Fig. 9 consistently begin at the origin regardless of gradation type, bitumen content or initial porosity values, then the relative merit of each specimen could be assessed. Specimen ranking will be based on this graphical relation at the terminal revolution and the numerical values are summarised in Table 4.

From Fig. 8, it is obvious that porosity of porous asphalt mixes reduces with the application of gyrations. As reaffirmed in Fig. 9, the reduction was greatest during the initial application of revolutions beyond which it over compacts slowly. The initial porosity values on the Y-axis indicate a gradual reduction in porosity with increment in binder content. With the BS mixes, a 1% increase in binder content reduces the porosity by about 2.5%. From Table 1, mixes made using the proposed gradation could better resist overcompaction than the BS 4987 mixes since the proposed mixes yielded lower relative change in porosity values.

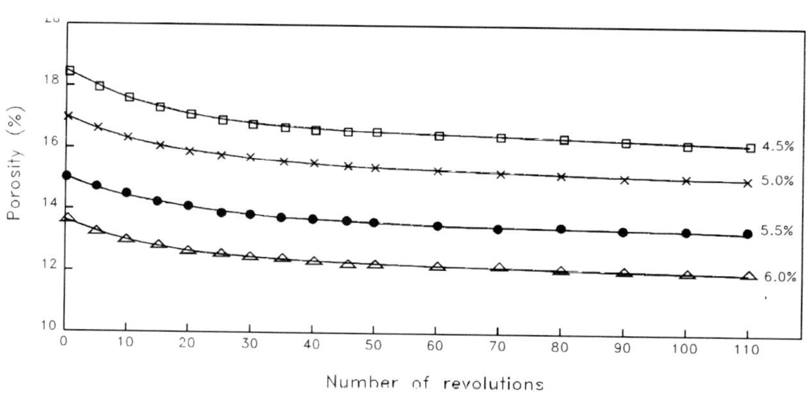

Figure 8(a) Porosity reduction with overcompaction at 0.7 MPa (PR Mixes)

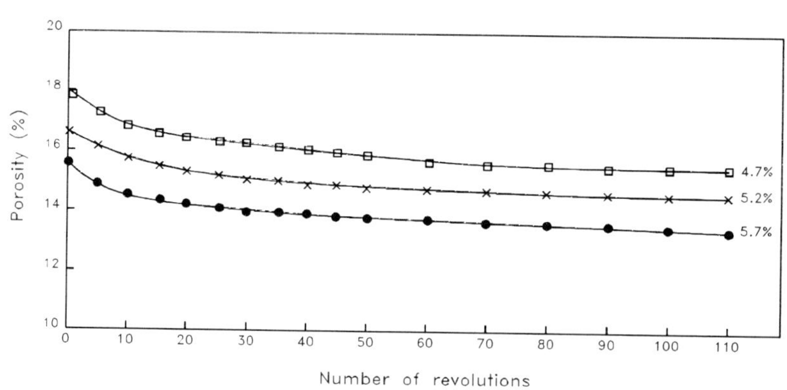

Figure 8(b) Porosity reduction with overcompaction at 0.7 MPa (BS Mixes)

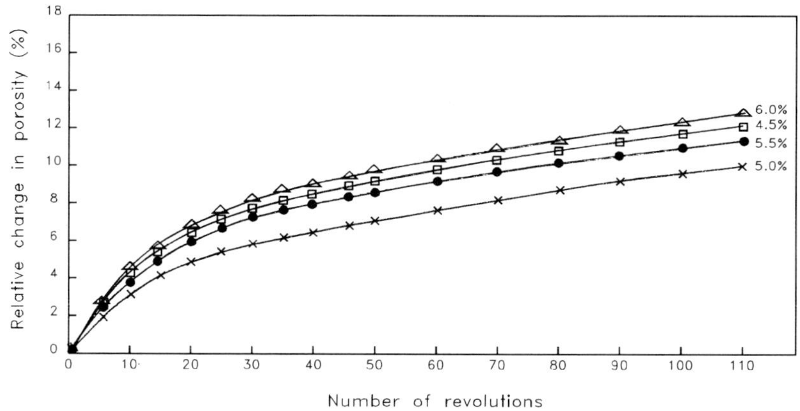

Figure 9(a) Relative change in porosity with overcompaction at 0.7 MPa (PR Mixes)

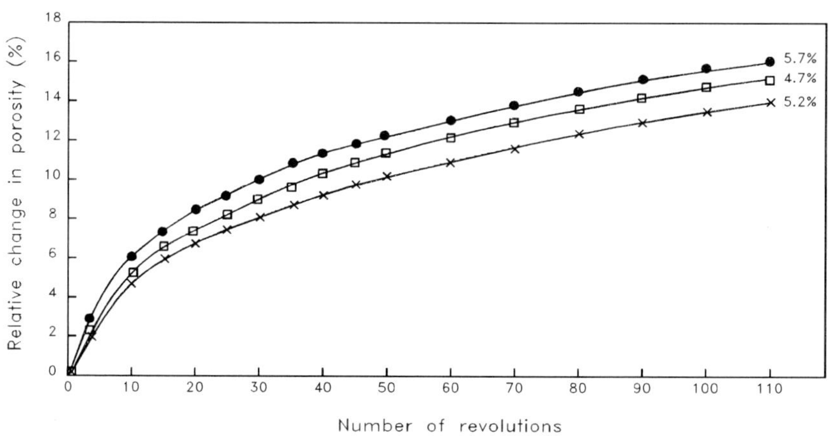

Figure (b) Relative change in porosity with overcompaction at 0.7 MPa (BS Mixes)

Table 4 Relative Porosity Values at the 110th Revolution for Specimens Prepared using the Proposed and BS 4987 Gradation

Type of Mix	Bitumen Content			
	4.5%	5.0%	5.5%	6.0%
Proposed	11.7	9.8	11.0	12.3
	4.7%	5.2%	5.7%	
BS 4987	15.1	14.0	15.9	

9 Conclusions

From the limited study presented in this paper, the following conclusions can be made:

1. A method to design gradings for porous asphalt based on providing a stable coarse matrix of minimum porosity modified by fine aggregate and filler to obtain a target porosity has been developed.
2. A falling head simple permeameter has been designed and tested. Its accuracy has been evaluated and the results obtained have been used to propose a slightly different gradation to the one proposed in the British Standards.
3. The performance of the proposed porous asphalt mix as measured by the Cantabrian tests and resistance to overcompaction is slightly superior to the BS mix.

10 References

1. British Standards Institution (1988), BS 4987:1988 *Coated Macadam for Roads and Other Paved Areas.*
2. Department of Transport (1993), *Porous Asphalt Surface Course*, Volume 7, Section 1, Part 3. HA50/93.
3. Fuller W.B. and Thompson S.E. (1907), *The Laws of Proportioning of Concrete, Transactions* American Society of Civil Engineers, 59, pp. 66-172.
4. Brown S.F., Preston J.N. and Cooper K.E. (1991), *Application of New Concept in Asphalt Mix Design* Proceedings AAPT, Volume 60, pp. 265-286.?
5. Lees G. (1970), *Rational Design of Aggregate Gradings for Dense Asphaltic Compositions*, Proceedings AAPT, Volume 39, pp. 60-90.
6. Furnas C.C. (1931), *Grading Aggregates. Mathematical Relations for Beds of Broken Solids of Maximum Density*, Industrial and Engineering Chemistry, Vol. 23, pp. 1052-1058.
7. British Standards Institution (1980), BS4551:1980 *Methods of Testing Mortars, Screeds and Plasters.*
8. Jimenez F.E.P. and Perez M.A.C (1990), *Analysis and Evaluation of the Performance of Porous Asphalt: The Spanish Experience*, Surface Characteristics of Roadways: International Research and Technologies, ASTM STP 1031, W.E. Meyer and J. Reichert, Eds., Philadelphia, pp. 512-527.
9. Jimenez F.E.P and Gordillo J. (1990), *Optimization of Porous Mixes Through the Use of Special Binders*, Transportation Research Record 1265, Washington D.C. pp. 59-68

3 DURABILITY OF SURFACING AGGREGATE – THE IMPLICATIONS OF CEN TEST METHODS ON CURRENT BRITISH SPECIFICATION REQUIREMENTS

A.R. WOODSIDE and W.D.H. WOODWARD
Department of Civil Engineering and Transport, University of Ulster, Carrickfergus, Northern Ireland

Abstract
This paper considers the methods available to assess the durability of surfacing aggregate. It reports on a comparative investigation of BS 812 and proposed CEN test methods. The work has indicated the limitations of existing specified methods and indicates that certain of the CEN methods may provide better indication of in-service performance.

Keywords: Aggregate, Test Methods, CEN, Micro-deval, Freeze-thaw

1 Introduction

Within the next few years, current British specifications and test methods will be superseded by European specifications and standards. These are currently being prepared by Commite European de Normalisation (CEN) Technical Committees are required to meet the cross-border trading ideals of a single European market where both sides of any border use similar methods to judge specification compliance.

However, few people have actually had the opportunity and practical experience using the methods proposed. This paper outlines an investigation of the testing of surfacing aggregate using both BS 812 methods and those presently proposed as CEN Euro-norms.

The reason why it was initiated was due to the statement in the Specification for Highway Works 7th edition (1991), that once a CEN Euro-norm has been accepted by CEN, then its use will become manadotory and must replace any existing similar British Standard test method. With some CEN methods now being released in final

Performance and Durability of Bituminous Materials. Edited by J.G. Cabrera and J.R. Dixon. © E & FN Spon. Published in 1996 by E & FN Spon, 2–6 Boundary Row, London SE1 8HN. ISBN 0 419 20540 3.

draft form for public comment the implication is that the British highways industry in general has little if any experience with these methods.

This is specially true for British surfacing aggregate specifications which have traditionally been based on Polished Stone Value and Aggregate Abrasion Value. As regards these two methods, the later is likely to be replaced whereas the high levels of PSV aggregate now specified in the British Isles are not used in any of the other European countries.

In an attempt to prepare for this enforced change and period of uncertainty, the Authors have for the past two years been carrying out considerable investigations into the application and implications of these methods given the current practise of surfacing specification based on PSV.

2 Current Specifications for Surfacing Aggregates

Specification limits are placed on the properties of construction materials in an attempt to ensure that they give the required performance both in the long term and during the process of construction. However, with regard to the specification and use of surfacing aggregate within the EEC, the requirements for the British Isles are much different than those of mainland Europe. A basic distinction may be made between that of high PSV and aggregate termed in this paper as "hardstone". These are now discussed.

2.1 The British Isles

In the British Isles, the main guidelines for specification requirements for surfacing aggregate are given in the Specification for Highway Works (1991) and in the Highways, Safety and Traffic Directorate Departmental Standard HD 21/92 (1992). This later document defines the requirements for PSV and AAV.

Depending on traffic density and site conditions, minimum values of PSV and AAV are specified. These have been derived from years of research and in-service measurement of skid-resistance from surfacings such as hot rolled asphalt and surface dressing.

However, a problem has arisen with respect to providing the specified levels of skid-resistance necessary for modern traffic densities which requires the use of aggregate with ever higher PSV to meet this demand. But, aggregate is a natural product where the levels of PSV required are only possible from certain types of aggregate and achieved at the expense of other properties such as strength and soundness.

2.2 Mainland Europe

Within mainland Europe, aggregate requirements are significantly different. This is due to the inter-relation of a range of factors including the distribution of rock-type, local climate, trafficking density and type of use. For example, aggregate used in Scandanavian countries must be resistant to freezing and withstand the abrasive effects of studded tyres. Whereas the main problem in most Mediterrean countries is that the only rock-type available is limestone.

This aspect of availability of aggregate possessing high levels of PSV is a problem with many countries. Traditionally, they have not thought skid-resistance to be as important as other properties such as resistance to freezing and heavy volumes of trafficking.

The result of this is that most European countries have developed specifications which require aggregates resistant to these other properties. As regards a level of skid-resistance, a minimum value may be specified but which is typically significantly less than would be used in the British Isles. However, this problem of poor skid-resistance has been recognised by many European countries, especially as safety is fundamental to the common ideals of the EEC. This has resulted in many of the new developments with surfacing materials such as Porous Asphalt and thin surfacings.

3 European Test Methods - a Brief History

As regards the process of standardisation within Europe, this paper describes the ability of proposed CEN test methods to better assesss aggregate performance and durability. The basic aim of CEN was to produce one test method only for any single specification parameter. Approxiamately 8 years ago each member country was approached to nominate those methods which were felt to offer most potential to meeting this. Where several potential test methods existed it was decided that the one having the best precision would be selected. As regards the British Isles, a general lack of interest at this initial stage has resulted in the PSV and MSSV tests to be the only British Standard methods being considered. All others are likely to be replaced.

The initial ideal of the CEN Technical Committees was that the period of work prior to a European standard being released should have been used as an opportunity to improve on the proposed existing methods, i.e. to change from traditional "recipe" national test methods to those which will assess in terms of "European performance". However, this ideal has often been ignored with the selection process better described as a compromise based on member country strengths and block voting against what may be methods which may go against national philosophy or practise.

The result is that the selection process has gone on much longer than previously anticipated. A stage has now been reached where certain methods are being "forced" through as final drafts for voting without any significant work having being done on the suitability of the method to assess aggregate in the greatly varying conditions of the whole European context.

The practical implication of a basic philosophy of one property / one test / for the whole of the EEC is now causing concern within many of the member states. For example, what if it is shown that high PSV aggregates as used in the British Isles cannot pass the current Micro-deval limits as used in France where a "hardstone" is normally specified.

4 What are the Proposed Methods

Table 1 lists the main methods currently being considered to test surfacing aggregate.

Table 1 Summary of methods being considered by CEN

Property	Proposed CEN method including country of origin	Closest Equivalent BS 812 method
Polishing	Polished Stone Value (British)	PSV
Soundness	Magnesium Sulphate Test (British)	MSSV
Wear	Micro-deval test (French)	AAV
	Studded tyre test (Sweddish)	none present
Fragmentation	Los Angeles (American)	ACV, TFV
Impact	Schlagversuch Impact test (German)	AIV
Freezing / Thawing	Freeze-thaw test (German)	none present
	Freeze-thaw test with 1% NaCL (Icelandic)	none present
Boiling test	Sonnenbrand Test (German)	none present

As the choice of methods is subject to change, this may only be considered as a tentative listing. It can be seen that resistance to polishing and soundness are based on the BS 812 test methods. Whereas, a mix of origins account for the others. For example, the American Los Angeles test, French Micro-deval, Sweddish Studded Tyre test; and the German Impact, Sonnenbrand and Freeze/thaw methods. Also included is the Icelandic Frost Resistance test using 1% NaCl solution which was recently proposed in preference to the German method.

Also included is the closest BS 812 equivalent test method. Whilst some are relatively similar, other methods such as freeze / thaw, do not have an equivalent.

It is also important to recognise that although certain of the methods may be new to the British industry, they may have had a long history. For example, the Magnesium Sulphate test dates from the 1820s, the original Deval test from the end of the nineteenth century and the Los Angeles from about 1910s. Of the more recent methods, the Micro-deval dates from the late 1970s and the Icelandic freeze-thaw with salt from the end of the 1980s.

4.2 Brief description of test methods

As some of these methods are new to most people within the British highways industry, the following gives a summary of each.

4.2.1 Polished Stone Value (PSV)

The proposed method is similar to the BS 812 method.

4.2.2 Magnesium Sulphate Value (MSV)

Again this is based on the BS 812 method except that the result is expressed as percentage passing as opposed to percentage retained in the British method.

4.2.3 Micro-deval test (MDE)

This is French in origin and determines the wear produced by friction between an aggregate sample and an abrasive charge in a rotating drum. Although it may be carried out dry or wet, the wet Micro-Deval test (MDE) is preferred. 500 g of 10 / 14 mm sized aggregate is placed in a steel drum along with 5000 g of 10 mm ball bearings and 2500 ml of water. This is rotated along its horizontal axis for 120 minutes at 100 rpm. The aggregate is then sieved using a 1.6 mm sieve and dried to constant mass. MDE is the percentage mass loss. The method uses a minimum of specialised equipment and a result may be obtained within one day.

4.2.4 Studded Tyre Test

The Studded Tyre test, also known as the Swedish Ball Mill test, is similar to the MDE test in that it assesses wet attrition only the method is designed to assess the "hard" aggregates necessary to resist the effects of studded tyres. 1000 g of 11.2 / 16 mm sized aggregate is placed in a steel drum along with 7000 g of 15 mm ball bearings and 2000 ml of water. The inside of the drum has three ribs. This is rotated for 60 minutes at 90 rpm. The percentage mass loss after sieving over a 2 mm sieve is determined. Again, the method requires little specialised equipment other than the roller and ribbed drum.

4.2.5 Los Angeles test (LA)

This measures resistance to fragmentation using apparatus similar to the America method. 5000 g of 10 / 14 mm sized aggregate is placed in a Los Angeles machine with 11 ball bearings of total mass 4800 g and rotated for 500 revolutions. The LA value is the percentage by mass of material passing a 1.6 mm sieve.

4.2.6 Schlagversuch Impact test (Sz)

This is based on the German method and determines resistance of coarse aggregate to impact. A test sample of 8 / 12.5 mm aggregate consisting of 50 % of the size fraction 8 / 10 mm, 25 % of the size fraction 10 / 11.2 mm and 25 % of the size fraction 11.2 / 12.5 mm is prepared. The mass of test sample is calculated as 0.5 times the aggregates particle density in g / cm^3. This is filled into the mortar of the impact test machine and its surface levelled. A pestle is placed on top which is then subjected to 10 blows of a falling hammer from a height of 370 mm. The crushed sample is sieved using 8, 5, 2, 0.63 and 0.2 mm sized sieves. The Impact Value (Sz) is the average mass of tested sample passing through the 5 specified test sieves. This method may be regarded as an over-specialised AIV where the equipment is very expensive and requires extensive calibration.

4.2.7 Freeze / thaw test

The proposed Freeze / Thaw test is based on the German DIN method but uses the standard test size 10 / 14 mm. 2000g of aggregate is soaked in a test container for 24 hours. Three temperature probes are placed in three of the test samples to control the freeze / thaw cycles. Cold air which is blown around the test samples and causing them to freeze from the outside in. When the probe in the centre of the sample reads - 20° C it holds the temperature for the desired time. Then water is introduced at + 20°C from the bottom of the cabinet to thaw out the frozen aggregates. Following 10 freeze / thaw cycles the test samples are sieved using a 5 mm sieve to determine the percentage mass loss. Similar to the German impact test, the Freeze / thaw test requires the use of a very expensive freezing cabinet with the test taking a number of weeks to perform.

4.2.8 Frost Resistance with 1% NACl

This is Icelandic in origin and assesses frost resistance of aggregate soaked in 1% NaCL solution. 400 g samples of 9.5 / 12.5 mm sized aggregate are subjected to 70 cycles, each lasting 2.4 hours, of a temperature range + 4 / - 4 °C. Following this they are dried and sieved on a 4.75 mm sieve to determine of percentage material remaining. Again the method needs an expensive freezing cabinet.

4.2.9 Sonnenbrand - Boiling test

This German in origin and assesses an aggregates resistance to weathering by mineral alteration and the formation of grey / white spots and radiating cracks typical of Sonnenbrand basalt. It involves boiling crushed aggregate or rock slices for 36 hours in water. Following this the test samples are allowed to dry and then inspected for white spots or signs of cracking.

5 Investigation of proposed CEN methods

The Authors have investigated the use of these test methods using British surfacing aggregate. The majority of work, i.e. PSV, MSV, LA and MDE tests have been carried out at the University of Ulster. Experience with the German Freeze /thaw, Schlagversuch Impact and Sonnenbrand test methods was obtained during a working visit to the German BASt, Cologne, when 14 samples Northern Ireland aggregate were assessed. Experience with the Icelandic Frost Resistance and Swedish Ball Mill tests was gained in co-operatiobn with the Icelandic Building Research Institute.

6 Aggregates Assessed

A wide range of rock-types have been assessed. Although, considering surfacing aggregate, they were not been biased / or restricted by the British pre-occupation with high PSV, but also considered the European requirement for "hardstone". Indeed, in most European countries the high PSV aggregates used in Britain would be considered "too soft" and not meet other minimum specification requirements such as MDE or LA. This important finding will be discussed later in the paper.

Also, the investigations have not been restricted to what are currently perceived as the "higher quality" materials, or even solely to quarry production; but has endeavoured to include low quality and marginal materials as well as samples especially selected to show extremes of performance. By doing so, better inter-relation trends have been possible as well as better insight into the methods ability to assess performance.

7 Testing Details

The investigation attempted to obtain as much information regarding the methods as possible. In general this included the following:

range of values expected from each method
influence of rock-type
factors which affect the test methods
correlation with existing / closest BS 812 method
assessment of results with respect to PSV
assess ability to relate with in-service performance

As well as this list of general assessments, each has been examined in detail. For example, the following factors have been investigated using the MDE method:

influence of test sample factors such as grading, shape, size
relationship between wet / dry Micro-deval
repeatability
effects of sample heterogenity / homogenity
relationship between MDE / AAV
the need for density corrections
relationship between 10 mm MDE / PSV
use of a modified Micro-deval to assess aggregate fines for bituminous mixes

8 Selected Findings

As it is not possible to discuss most of these, a selected few will be detailed further. For example, comparison of post test gradings, MDE / AAV, MSV / Freeze-thaw, and MDE / PSV.

8.1 Comparison of post test gradings

To assess what happens during each method, post test gradings were determined. Figure 1 shows the data for a gritstone aggregate with PSV 61. For the majority of the analysis certain trends occurred. For hard aggregates the AIV, ACV and LA gave similar gradings with LA showing the greatest breakup of the original test sample. As the aggregates become of lower quality, there was a greater spread of gradings, indicating different mechanisms of failure. In all cases the LA broke the test sample up the most.

The gradings obtained for the MDE test were significantly different. In most cases, the grading curve had levelled off at the 8 or 6.3 mm sieve size, indicating failure in a manner different from the impact and fragmentation methods.

As regards the use of post-test gradings, the Authors suggest that they indicate more performance related information than the simple percentage mass passing a certain sieve. For example, in-use performance of chippings for surface dressing or as pre-coats is dependant on the aggregate maintaining its original dimensions not on its ability to remain > 2.36 or > 1.6 mm in size after laboratory testing.

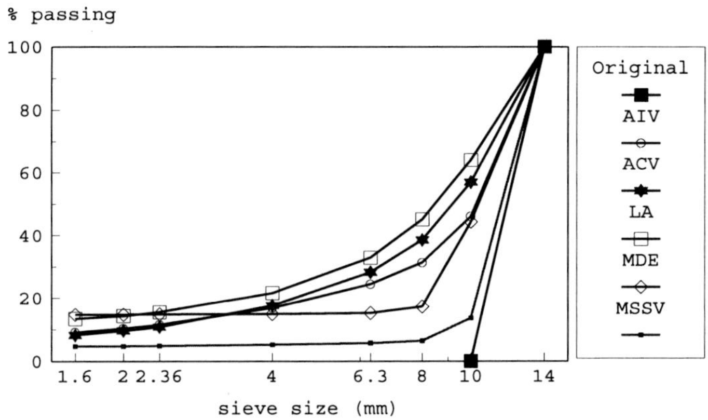

Fig. 1. Post-test gradings for a gritstone aggregate

8.2 MDE / AAV

It is proposed that the French wet MDE test method will replace the AAV test. This is of considerable concern as AAV and PSV have traditionally been used as the basis for British surfacing aggregate specification requirements and is regarded as a tried and tested traditional test method.

Figure 2 shows the relationship between AAV and MDE for a range of rock-types. Despite the differences between the two methods it can be seen that a positive relationship exists. The results also appear to be ranked depending on rock-type with the gradient representing a rock-types susceptability to wear.

As regards meeting specification requirements, most of the aggregates would meet an AAV of < 16. However, if the French MDE for surfacing aggregate of < 16 is used, considerably fewer would meet this value. As regards gritstone, the predominate source of high PSV aggregate, very few aggregates would be assessed as suitable - i.e. evidence of the French requirement for hardstone.

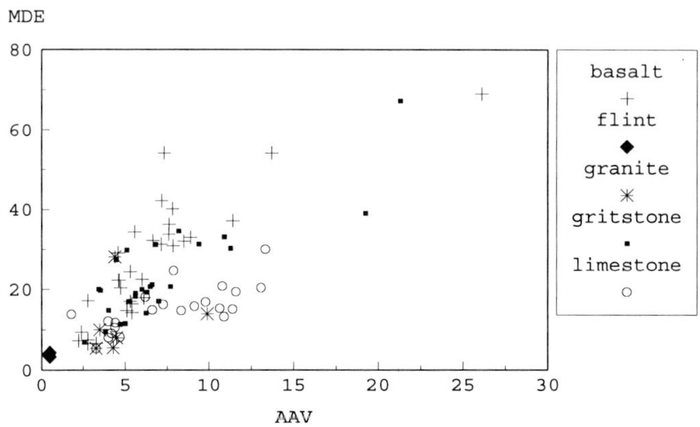

Fig. 2. Relationship betweeen MDE and AAV

Due to this finding, the Authors have investigated the two methods to see which relates closest to in-service conditions and gives the truer indication of quality. To summarise the methods, AAV may be described as a dry abrasion test on specially prepared samples using a standard abrasive; whereas MDE is wet, unconfined and dynamic.

Although they both assess the 10 / 14 mm size, AAV requires removal of flaky aggregate. This typically results in the percentage of material available for testing to be from 30 to as little as 5 %. Clearly this could considerably alter the supposed quality of the agregate if the flaky aggregate was weaker in strength. For example, the shale / siltstone content of a high PSV greywacke. It is also worth noting that a similar case occurs when 10 mm aggregate are deflaked for the PSV test, where for some aggregates, the percentage for testing may be less than 1 %.

Other important differences include the AAV test only assesses 23 chippings whereas the MDE has a 500 g test mass. As regards Health and Safety, the AAV uses a silica sand which comes under the control of COSHH whereas the MDE is a wet ball mill test.

8.3 Freeze-thaw / Frost resistance with 1 % NaCl

A limited investigation was carried out using the German Freeze-thaw method at the BASt, near Cologne, Germany. This involved 8 Tertiary basalts and 4 Silurian greyackes from Northern Ireland. These had been picked to represent the range in quality to be expected from these rock-types. The results are shown in Fig. 3. Also shown are results from a comparitive assessment of the German method and the Icelandic method which uses a 1 % salt solution.

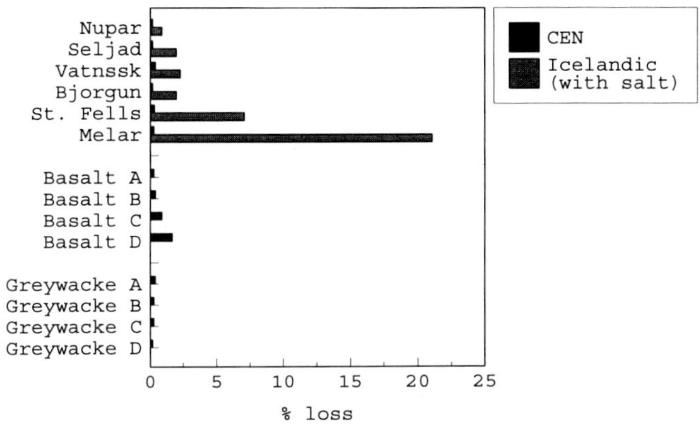

Fig. 3. Freeze-thaw test results

These investigations have raised doubts on the ability of the German method to assess aggregate given the freezing conditions experienced by the Atlantic countries of Europe. It may be argued that the + / 20 ^{0} C of the German method may be more applicable to continental Europe, whereas the greater number of + / - 4 ^{0} C cycles with salt more closely resembles our winter conditions.

8.4 MDE / PSV

The Authors investigated the relationship between PSV and MDE. However, the example given has used a modified MDE test where the aggregate assessed was that which was deflaked for the PSV test. This was done for a number of reasons, i.e. to ensure that there was no difference in quality for the two methods and secondly to see if this method could be specified along with PSV to give both a measure of skid and wear resistance on the same sized aggregate (currently AAV and PSV are done on different sizes). The results for basalt and gritstone aggregates are shown in Fig. 4.

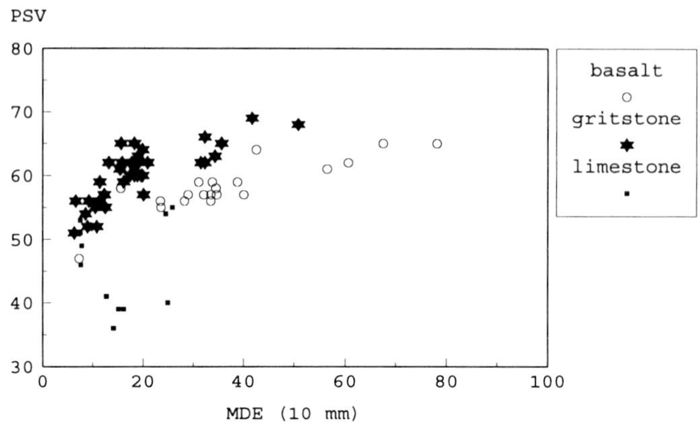

Fig. 4. PSV v. MDE for 10mm deflaked aggregate

It can be seen that a relationship exists between the two methods i.e. increasing skid-resistance is gained at decreasing resistance to wear. It is also apparent that rock-type is important. For example, a PSV of 60 relates to a basalt MDE of approxiamately 45 and 20 for gritstones.

9 Discussion

It is apparent that these methods both assess and use equipment which differ from existing BS 812. For example, the use of wet dynamic tests rather than dry static methods; and the use of freeze / cabinets and ball mills. As regards their ability to relate to in-service conditions, the Authors consider that they relate better than BS 812 methods. This has been illustrated in the examples shown where for example, the MDE has been shown to better assess real-life problems better than AAV. Another reason why the MDE is favoured is that it is a wet test. Once water becomes part of a test unexpected things may happen as shown by the wet Ten Percent fines test where reductions in strength of greater than 50 % may occur. As the typical British road is normally wet then this should be part of the assessment procedure.

What was apparent during this investigation is the important influence of rock-type. This effects all the test methods, generally in different ways. As regards high PSV, all the methods have indicated that this is obtained at the expense of other properties, particularily MDE.

10 Conclusions

The investigation has highlighted certain problems both with the proposed CEN methods themselves and their impact on the British highways industry. The Authors consider that the importance of rock-type must be recognised, that the German Freeze/thaw test may not indicate the types of freezing conditions in the British Isles and that the MDE test is preferable to AAV in predicting the performance of surfacing aggregate.

11 References

Department of Transport. (1991) Specification for highway works. HMSO, London.

Department of Transport. (1992) Highways, Safety and Traffic Directorate Departmental Standard HD 21/92. HMSO, London.

4 THE RELATIONSHIP BETWEEN MINERALOGY, TEXTURE AND POLISHED STONE VALUE FOR GRITSTONE AGGREGATE FROM THE LONGFORD DOWN MASSIF

A.R. WOODSIDE, P. LYLE, M.J. PERRY and W.D.H. WOODWARD
Department of Civil Engineering and Transport, University of Ulster, Carrickfergus, Northern Ireland

Abstract
The recent report by Travers Morgan into High Specification Aggregate has indicated that the predominate source of such aggregates is the Gritstone Trade Group. Greywacke aggregate quarried from the Longford Down Massif forms part of this group. Although there is variation within this group, they predominately consist of medium to fine sand sized particles held together by a matrix of mostly clay and chlorite minerals. The resulting sandpaper type surface texture results in these aggregates having a high level of skid resistance as measured in the laboratory using the Polished Stone Value test method. The results of this method form the basis for all surfacing aggregate specifications in the British Isles. However little is known about the relationship between mineralogy, surface texture and their in-service performance. Indeed doubts are currently being viewed as to the effective use of PSV in relation to skid resistance on the finished surface. This paper considers the relationship between mineralogy, grain size and PSV.
Keywords: Mineralogy, Gritstone, Greywacke, Longford Down Massif, Polished Stone Value.

1 Introduction

It is the aim of the highway engineer to make economic, durable and safe roads. The focus for safety is the tyre/road surface interface. Engineering design is used to prevent skidding at this interface. Such factors affecting skidding at this interface are tyre development, improvement of vehicular braking systems; and the design and construction of the road surface. Tyre development has approached the peak of its development, and with the development of anti-lock braking the emphasis for the improved safety is now with the highway engineer.

Increasing traffic densities and loading, combined with increased performance of modern vehicles require an increasingly

Performance and Durability of Bituminous Materials. Edited by J.G. Cabrera and J.R. Dixon. © E & FN Spon. Published in 1996 by E & FN Spon, 2–6 Boundary Row, London SE1 8HN. ISBN 0 419 20540 3.

demanding specification for road surfaces in order to maintain an adequate resistance to skidding. This has led to considerable effort into locating sources of aggregate capable of meeting modern demands.

This has resulted in the recent report by Travers Morgan (Thompson, et al) for the Department of Transport. In this report, they define such aggregate as High Specification Aggregate (HSA) and consider it to possess the minimum requirements listed in Table 1.

Table 1. Minimum requirements for HSA

Property	Limiting Value
Polished Stone Value (PSV)	$\leq$ 58
Aggregate Abrasion Value (AAV)	$\leq$ 16
Aggregate Impact Value (AIV)	$\leq$ 30
Ten Percent Fines (TFV)	$\geq$ 140 kN
Magnesium Sulphate Value (MSSV)	$\leq$ 75%

The Travers Morgan report stated that resistance to polishing was the single most important characteristic of concluded that the predominant source of such aggregate belonged to the Gritstone Trade Group. Within the Gritstone Group, greywacke and greywacke/siltstone have the highest overall aggregate quality, in terms of measured PSV and AAV.

In the investigation reported in this paper, 38 samples of greywacke aggregate were assessed from 4 quarries located in the Central Belt of Longford Down. Lump samples were selected from individual beds within these quarries, crushed in the laboratory and tested for PSV and petrographic analysis.

2 Geology of the Longford Down Massif

Lower Palaeozoic greywackes, of the Longford Down Massif of northeast Ireland, are quarried as an important source of gritstone aggregate for the construction industry. They are an important resource to the Northern Ireland economy and have been exported to the rest of the United Kingdom and Europe for use as surfacing aggregate.

The peculiar nature of the Longford Down Terrane means that it requires a different stratigraphical description than is normal. In the Southern Uplands three divisions have long been recognized(Peach and Horne, 1899). The Northern Belt consists entirely of Ordovician Rocks; the Central Belt consists of predominantly Silurian rocks (Llandovery age) with inliers of

Ordovician rocks; and the Southern Belt consists purely of Silurian Rocks (Wenlockian age). Recent workers (Anderson & Oliver, Barnes et al.) have traced the lateral extension of the Northern and Central Belts into the Longford Down Massif. It is within this Central Belt that most of the gritstone quarries of northeast Ireland occur.

The distribution and delineation of rock sequences is dominated by strike faults which divide the area into tracts up to about 5km across strike. These may account for much of the variation within the greywackes of Longford Down. The greywackes mostly consist of coarse to medium sand sized particles held together by a matrix of predominantly clay and chlorite minerals. The resulting sandpaper type surface texture results in these aggregates having a high level of skid resistance. Post deposition, the greywackes of the Northern and Central Belts experienced metamorphic conditions of around 350°C and pressures of 2.5-4kbs, which is the equivalent of a burial depth of 9-14 km. This has tended to produce a hard wearing aggregate.

3 Relationship between Aggregate Properties and Petrography

Petrographic examination was carried out with the following determined; percentage quartz (monocrystalline and polycrystalline), feldspar, lithic grains and matrix. Matrix was taken as material < 0.06mm and cryptocrystalline. The size of largest and average fragments was also determined.

From statistical analysis on this limited data set the following relationships with PSV were obtained. These are contrasted with results for greywackes from the rest of the United Kingdom obtained by the TRRL in LR 488.

3.1 Percentage quartz v. PSV

In Fig. 1. it is noticeable that samples from each quarry tend to plot within discrete fields, indicating that the quarries may fall within different tracts with the Longford Down Massif.

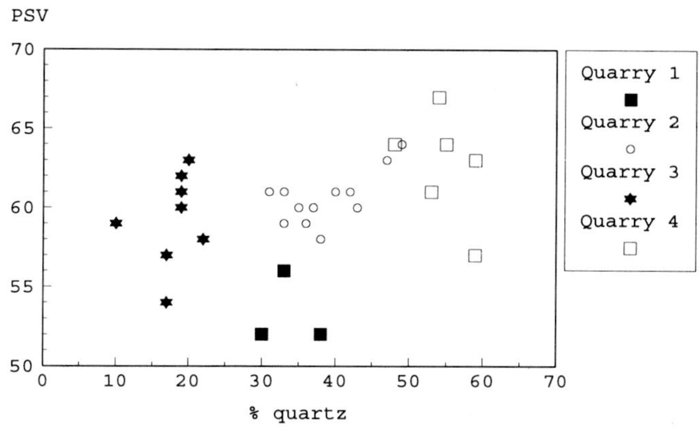

Fig. 1. Percentage quartz v. PSV

From the limited data a weak positive relationship was obtained with an apparent optimum quartz content in relation to PSV at approximately 54%.

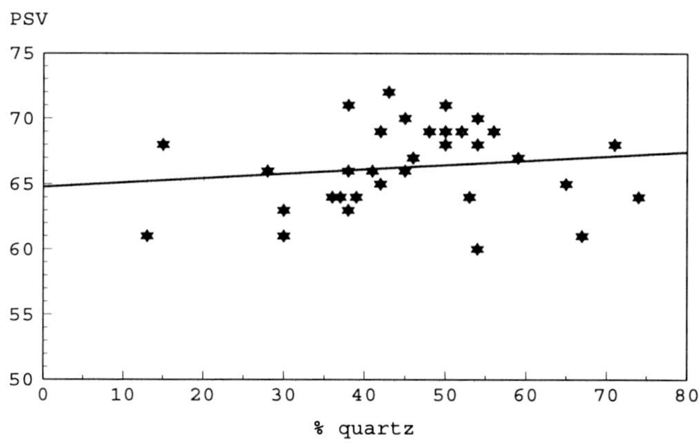

Fig.2. Percentage quartz v. PSV (after TRRL LR 488)

Fig.2. shows the results of greywacke from TRRL LR 488. It can be seen that there is a weak slightly positive correlation.

3.2 Percentage hard grains v. PSV

The relationship between the percentage hard grains and PSV is shown in Fig. 3.

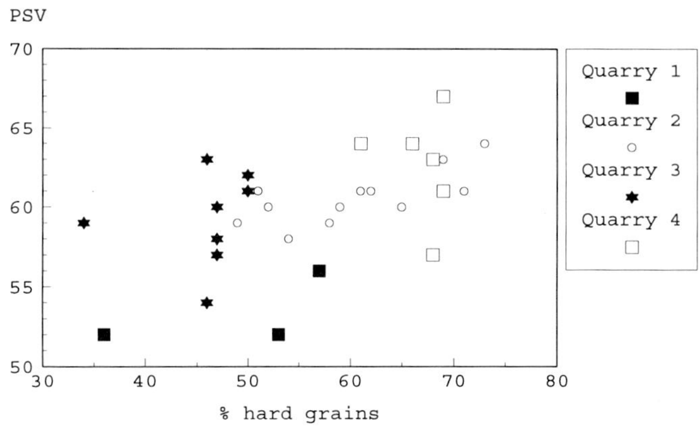

Fig. 3. Percentage hard grains v. PSV

Again a positive relationship exists between the total hard grains i.e. quartz, lithic fragments and feldspar, and PSV. Similar to Fig. 1., the results also tend to plot in fields.

However, as shown in Fig. 4., the TRRL data only indicates weak positive correlation

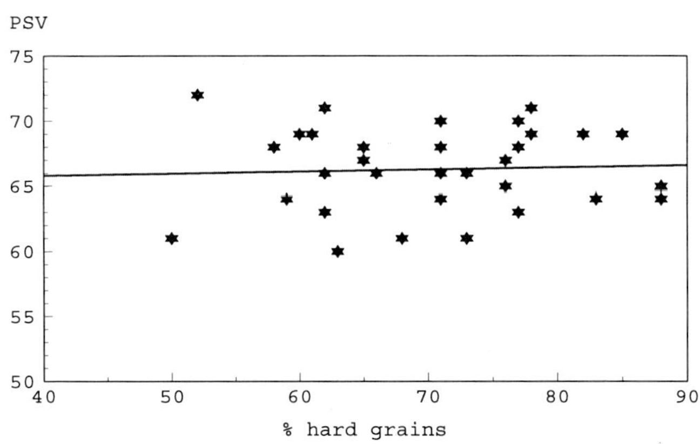

Fig. 4. Percentage hard grains v. PSV (after TRRL LR 488)

3.3 Percentage matrix v. PSV

Fig. 5. shows a negative relationship between the percentage matrix and PSV.

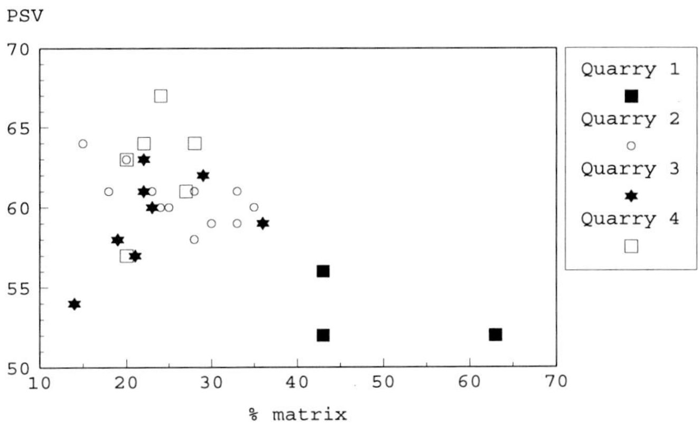

Fig. 5. Percentage matrix v. PSV

However, for the TRRL LR 488 data as shown in Fig. 6., a positive relationship was determined.

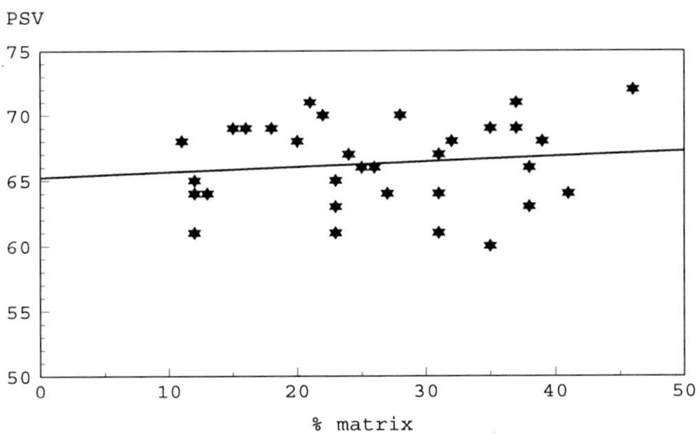

Fig. 6. Percentage matrix v. PSV (after TRRL LR 488)

4 Grain size and its Effect on PSV

The grain size of 13 samples from quarry 1 and 2 were estimated and plotted against PSV. The grain size limits shown in Table 2 were used.

Table 2. Grain size limits

Grade	Predominant grain size (mm)
medium / coarse sandstone	1.0 - 0.6
medium sandstone	0.6 - 0.2
fine sandstone	0.2 - 0.06

As shown in Fig. 7. medium grained greywacke had an average PSV of 62, 3 points higher than that for the coarse grained greywackes. The fine grained greywackes had an average PSV of 57.

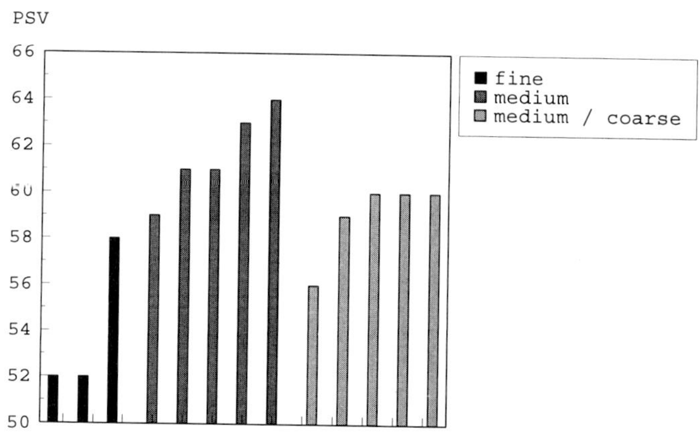

Fig. 7. Variation in PSV with grain size

However in comparison with the data from TRRL LR 488, as shown in Fig. 8., no significant relationship was observed.

5 The effect of post depositional consolidation on PSV

It is interesting to note that the relationships between petrology and PSV of the greywackes investigated in TRRL LR 488 are generally weak. This may indicate that factors other than mineralogy and grain size may influence the resulting PSV. One reason may have been the degree of consolidation that has been experienced by the greywacke since deposition.

TRRL LR 488 used greywackes from different stratigraphical units within the United Kingdom which would have had experienced

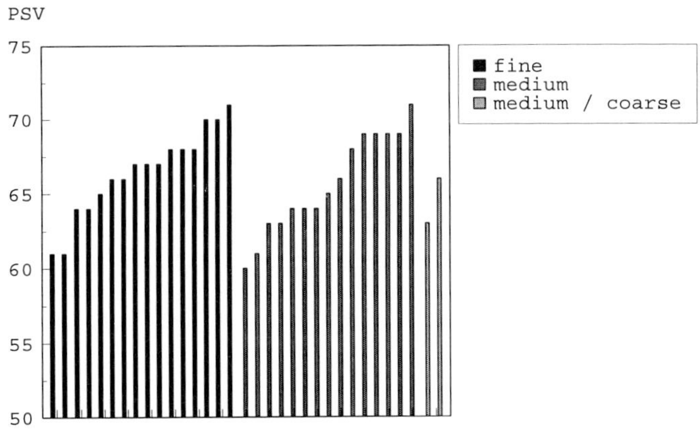

Fig. 8. Variation in PSV with grain size (after TRRL LR 488)

different degrees of consolidation. Whereas the samples taken from the Longford Down Massif would have experienced a similar post depositional history of low grade metamorphism.

6 Relationship between plucking and micro texture

Knill(1960) proposed that variation in PSV was attributed to differential wear of the aggregate. For greywacke, it is thought that hard grains such as quartz, lithics and feldspars, are plucked out of the softer matrix so renewing micro-texture and providing a high PSV.

If an aggregate has undergone a greater degree of consolidation relative to one from another stratigraphical unit, it will have been subjected to greater pressures and temperatures, making bonds between the individual grains stronger. Although this would result in a stronger aggregate, its ability to pluck would be reduced

7 Conclusions

From the work carried out on the four quarries from the Longford Down Massif it may be possible to form a number of conclusions. It appears that the mineralogy of greywacke from different quarries tends to plot within discrete fields. With regard to the percentage of quartz present, PSV increases with increasing content. There may be an optimum at 54% but as yet there is limited data to confirm this.

Between 35%-75% total hard grains there is also a general positive relationship with PSV. Between 15%-65% matrix there is a general negative relationship with PSV.

Medium grained greywackes have higher PSV than coarse and fine grained greywacke.

From available data of greywackes from other parts of the United Kingdom, only weak relationships exist between mineralogy, grain size and PSV. However, factors such as degree of consolidation experienced may account for this.

As the samples from Longford Down have experienced the same post depostional history of low grade metamorphism, it has been shown that variations in mineralogy can account for variation in PSV.

8 References

Anderson, T.B. & Oliver, G.J.H. (1986) The Orlock Bridge Fault: a major late Caledonian sinistral fault in the Southern Uplands terrane, British Isles. Transactions of the Royal Society Edinburgh: Earth Sciences, 77, 203-222.

Barnes, R.P., Anderson, T.B. & McCurry, J.A. (1987) Along strike variation in the stratigraphical and structural profile of the Central Belt in Galloway and Down. Journal of the Geology Society. London, 144, 807-816.

Hawkes, J.R. & Hosking, J.R. (1972) British Arenaceous rocks for skid-resistant road surfaces. Road Research Laboratory, Department of the Environment, Report L.R. 488.

Hosking, J.R. (1968) Factors affecting the results of polished stone value tests. Road Research Laboratory, Ministry of Tranoport, Report L.R. 216.

Hosking, J.R. (1970) Synthetic aggregates of high resistance to polishing part 1 - Gritty aggregates. Road Research Laboratory, Ministry of Transport, Report L.R. 350.

Knill, D.C. (1960) Petrographical aspects of the polishing of natural roadstones. J. appl. Chem. Lond. 10. 28-35.

Peach, B.N. & Horne, J. (1899) The Silurian Rocks of Britain. 1 Scotland, Memoir Geology Survey. U.K.

Thompson,A., Greig, J.R., & Shaw, J. (1993) High Specification Aggregates for Road Surfacing Materials: Technical Report. Department of the Environment, London. Travers Morgan LTD, East Grinstead. by performance.

5 A QUESTION OF FATIGUE?

M.E. NUNN
Transport Research Laboratory, Crowthorne, Berkshire, UK

Abstract
The roadbase is the most important structural layer of the road but any assessment of its structural condition is difficult because of its position in the road; consequently the mechanisms of roadbase deterioration are less clearly understood. Conventional wisdom based on laboratory studies considers fatigue cracking caused by repeated loading from traffic to be the major form of structural deterioration. This paper describes a study in which structural investigations were carried out on selected motorway sites and roadbase material was extracted for subsequent laboratory testing. The measured structural properties of the roadbase from the heavily trafficked near-side wheel path were compared with those of material from the lightly trafficked area, between the wheel tracks in the off-side lane, and with the overall condition of the motorway. The study provided no evidence that fatigue under repetitive traffic loading caused any weakening of the roadbase. The change in the measured residual life of the roadbase was related to material composition, with binder volume and increasing binder hardness with age being the most important variables. This work has important implications for the present criteria for pavement design and maintenance.
Keywords: Fatigue, Structural Properties, Site Investigations, Ageing, Binder Hardening.

1 Introduction

A fully flexible motorway or trunk road consists of bituminous surfacing and roadbase layers laid on a foundation consisting of a granular subbase over a capping layer or subgrade. The roadbase is the most important structural layer of the pavement. A partial or complete failure of the roadbase would require the costly treatment of either a total reconstruction or a thick overlay. On the other hand, a failure of the surfacing usually requires only a thin overlay to restore profile and texture. The strengthening required depends on a correct understanding of the weakening mechanisms and a thorough assessment of the structural condition of each pavement layer. Current nondestructive methods for measuring overall pavement condition are not necessarily suitable for assessing the condition of individual layers.

The assessment of the structural condition of the roadbase is difficult because of its

Performance and Durability of Bituminous Materials. Edited by J.G. Cabrera and J.R. Dixon. Crown Copyright. Published in 1996 by E & FN Spon, 2–6 Boundary Row, London SE1 8HN. ISBN 0 419 20540 3.

position in the road; consequently the mechanisms of roadbase deterioration are less clearly understood. Conventional wisdom based on laboratory studies considers fatigue cracking caused by repeated loading from traffic to be the major form of structural deterioration. Investigation of the roadbase fatigue mechanism in full scale pavements is much more difficult than in the laboratory, and Goddard and Powell (1987) and Thrower (1979) noted that there is little evidence of fatigue cracking in the roadbase of in-service bituminous pavements in the UK. Furthermore, it is well known that bituminous materials become more stiff with time, and this will influence the fatigue resistance of the roadbase material. However, this effect has received little attention in the past in relation to roadbase performance.

Site investigations were carried out on selected motorways and roadbase materials were extracted for subsequent laboratory testing. The aim of this study was to compare the structural properties of samples of roadbase measured in the laboratory with the overall condition of the pavements from which they were extracted. The work, which is reported in more detail by Wu (1992), should improve our understanding of the mechanisms of structural deterioration and will lead to improvements in the design of strengthening.

2 Sites and test methods

Short sections of four motorways (M4, M5, M1, and M62) and two experimental pavements used in TRL's Pavement Testing Facility (PTF1 and PTF6), representing a range of age and traffic loading, were selected for detailed investigation. The selection of the sites was based on previous structural assessment surveys. The sections chosen were representative of the condition of significant lengths of carriageway. Details of the sites are given in Table 1.

Table 1 Pavement sections investigated

Site	Age (years)	Traffic (msa)	Roadbase type	Thickness (mm) Surfacing	Roadbase	Number of fatigue specimens
M4	11	22	HRA	100	130	156
M5	19	66	DBM	100	200	68
M1	23	71	DBM	140*	210	35
M62	21	57	DBM	100	200	59
PTF6	3	3.2	HDM	None	160	64
PTF1	3	3.8	DBM	None	160	96

* 40 mm of HRA wearing course added 9 years prior to this study.

All the motorways and experimental pavements examined had exceeded their nominal design life determined using either the standard in force at the time of construction, Road Note 29 (1970), or the current Departmental design standard, HD 14/87 (1987). At each site test sections of approximately 50 metres long were selected for detailed investigation. The coring pattern for a typical site is shown in fig 1.

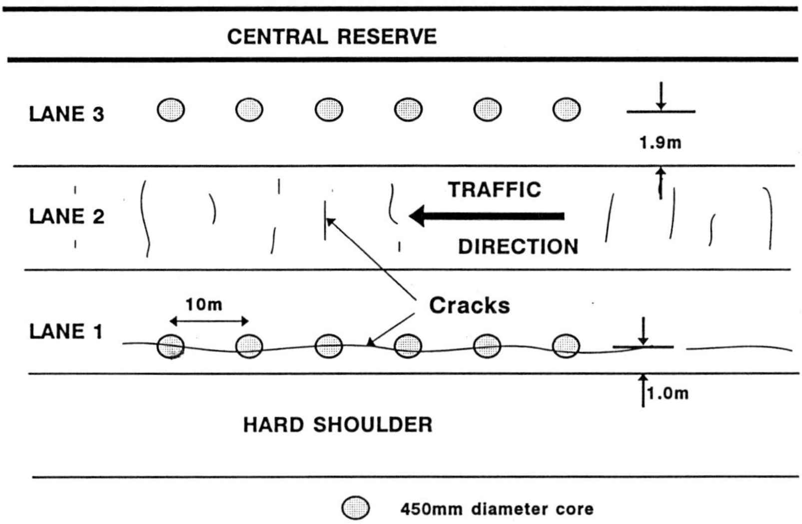

Fig. 1 Pattern of coring on M5

The investigation methods used at each of the sites are summarised below:

Visual examination of surface condition.
Mark positions of cores.
Carry out FWD tests at core locations.
Cut 450 mm diameter cores and carry out FWD tests to measure foundation stiffness through cores before removal.
Remove cores to provide specimens for subsequent laboratory tests.
Measure thicknesses of each bituminous layer and note condition.
Carry out dynamic cone penetrometer tests to a depth of 1.5 to 2.0 metres.
Reinstate core holes.

Cores were cut to enable the structural properties of materials that had been subjected to heavy commercial traffic in lane 1 to be compared to the lightly trafficked material of the same age and composition from lane 3. Fatigue test specimens were cut separately from the upper and lower roadbase layers of the 450 mm diameter cores. The number of cores taken at each site and the number of fatigue test specimens prepared are given in Table 1. These were tested using the TRL laboratory fatigue method described by Goddard, Powell and Applegate (1978).

The composition of the materials and the properties of recovered binder were obtained from representative material samples from each pavement section.

2.1 Condition of pavements

The condition of the pavements examined is summarised in Table 2.

Table 2. Condition of pavements examined

Site	Deflectograph Survey Residual life (years)	Date	Date examined and cored	Visual condition
M4 (W)	5	1988	Nov 88	Good - no surface cracking or rutting
M4 (E)	3	1988	Mar 89	Good - no surface cracking or rutting
M5	0	1989	May 90	Longitudinal crack in wheel path of lane 1 Frequent transverse cracks in lane 2. Lane 3 - good
M1	4	1980	Sep 90	Good - no surface cracking or rutting
M1	17*	1985	Sep 90	*40 mm HRA wearing course added Oct 81
M62	0	1990	Feb 91	Good - no surface cracking or rutting

The visual condition of the sections of M4 and M62 was good, not withstanding the deflectograph method predicting a short or zero residual life.

The two sections of pavement examined in the TRL Pavement Test Facility (PTF) consisted of two roadbase layers with a total thickness of between 140 mm and 180 mm laid on a granular sub-base. The first section (PTF1) was constructed of dense roadbase macadam and the second (PTF6) of heavy duty macadam. The pavements were subjected to over 3 msa of a predominately 80 kN wheel load (equivalent to 160 kN axle load). Cores were cut for testing from trafficked and untrafficked areas of the test pavements. The surface condition of these areas was identical; no signs of cracking, rutting or other distress were evident.

3 Analysis of results

3.1 Fatigue and elastic stiffness

The comparison of the measured residual fatigue life of the bottom layer of roadbase in the heavily trafficked lane 1 and the lightly trafficked lane 3 of the motorways is shown in Table 3.

A comparison of the stiffness and laboratory fatigue life of the lane 1 and lane 3 samples, from the motorway roadbases, showed that the lane 3 samples did not have a longer fatigue life and that any difference was generally not statistically significant. Only the material from lane 3 of the west bound carriageway of the M4 and the M62 had a significantly longer life than material from lane 1, and this difference was mainly due to differences in composition rather than the effects of traffic. Also worth noting is that the presence of the longitudinal crack in the wheel path of lane 1 of the M5 did not appear to have a detrimental effect on the structural properties of the roadbase immediately below it.

Table 3. Comparison of laboratory fatigue life of the roadbase from lanes 1 and 3.

Site	Lane	Number of tests (N)	Elastic stiffness (GPa)	Fatigue life at 100μstrain (x1000 cycles)	Statistically different at 95% level of confidence
M4 (W)	1	29	5.2	25.1	Yes
	3	36	4.6	45.6	
M4 (E)	1	51	5.5	26.9	No
	3	40	5.0	31.3	
M5	1	35	6.1	0.89	No
	3	33	9.1	0.60	
M1	1	19	8.2	0.52	No
	3	11	6.2	0.81	
M62	1	28	4.2	5.2	Yes
	3	31	10.4	1.9	

A separate comparison between the top and bottom roadbase layers showed that, again, there were no significant and consistent differences in fatigue life and stiffness.

These comparisons indicate that the level of traffic loading was not the major factor affecting the residual fatigue life of the roadbase layers. As the two lanes had carried different traffic loads, any effect of traffic loads would be reflected in the differences between the samples. On the other hand, as the widely held understanding that tensile strains induced by traffic loading cause maximum fatigue damage at the underside of the roadbase, it would be expected that the samples from the bottom layer should have been in a worse condition if traffic loading was a major factor significantly damaging the roadbase.

In contrast, the trafficked materials from the experimental pavements constructed in the PTF were in a worse condition in terms of fatigue life and stiffness than the untrafficked materials. The bottom layer of trafficked HDM was too weak to test and the fatigue life of the bottom layer of trafficked DBM was virtually finished. The upper layers were in better condition but their fatigue life was generally only about a quarter of that of the untrafficked material. This was attributed to the fact that the traffic loading on these relatively thin pavements was with wheel loads much higher than those normally encountered on motorways.

3.2 Material composition

Material composition has been found to be a very important determinant of the fatigue life of bituminous paving materials. In this study the fatigue measurements for all four motorway sites were combined and multiple regression techniques were used to relate the fatigue life N_{100} for an initial strain of 100 μstrain to percentage volume of binder V_b, percentage volume of air voids V_v and penetration of the recovered binder Pen. The following equation was obtained:

$$N_{100} = 7.94 \times 10^{-5}\, V_b^{5.77}\, V_v^{-0.079}\, Pen^{1.3} \qquad (1)$$

This equation accounted for a very high proportion, 91 per cent, of the variation in the measured residual fatigue life. The volume of binder accounted for the largest amount of variability followed by the penetration of the recovered binder. The equation illustrates that differences in the fatigue lives between materials from the four motorway sites can be accounted for by variations in material composition rather than by differences in traffic loading.

3.3 Age of pavement

In terms of age, the M4 had been in service for only 11 years compared with 19-23 years for the other pavements. Table 3 gives the measured residual fatigue life of the roadbase material from these pavements. The fatigue behaviour of the M4 is markedly different from the older pavements and while the penetration of the recovered binder from the M4 was about 50, for the other motorways it was close to or less than 30.

While traffic loading was not found to be the major factor for the difference in fatigue life, age will have been an important factor because it is well known that binders age with time resulting in increased binder hardness and a more brittle material.

However, the gradual hardening of the roadbase materials with age, as demonstrated by a shorter residual fatigue life and higher elastic stiffness, may not result in fatigue cracking of the pavement. Provided the surfacing remains intact, it appears that the roadbase is able to perform its structural function for a considerable period of time. Calculations using the relationships developed in this study show that the increase in elastic stiffness with age causes a reduction in the traffic induced tensile strain responsible for fatigue at the underside of the roadbase. This reduction more than compensates for the lower laboratory fatigue life of the aged roadbase. The net effect is that the predicted fatigue life of the road increases as it becomes older. This may be the reason why no positive evidence has been found of fatigue cracking in any of TRL's experimental roads; the available evidence does not support the occurrence of classical fatigue cracking originating in the roadbase.

3.4 Foundation stiffness

The pavements examined were laid on relatively firm foundations, with the equivalent stiffness varying between 90 MPa and 250 MPa. However, no correlation between foundation stiffness and the residual fatigue life or elastic stiffness of the roadbase was found. Neither was the visual condition of the road surface related to the foundation stiffness.

Higher traffic induced tensile strains will occur in roadbases over a weaker foundation. The low sensitivity of roadbase condition to foundation stiffness is, therefore, in line with the observation that traffic loading is not a major factor in the structural deterioration of the motorway roadbases.

3.5 Debonding at interfaces

During pavement coring, layer debonding at the interface between the surfacing and the roadbase, and between the upper roadbase layer and the lower roadbase layer were frequently encountered. However, no debonding was found between wearing course and basecourse. Debonding between the upper and lower roadbase layers had little effect, if any, on the structural condition of the road. No evidence was found to indicate that debonded roadbase layers adversely affected the structural condition of the surfacing or roadbase.

The presence of debonding of surfacing from roadbase observed in this work was associated with the poor condition of surfacing layers. However, it was not possible to determine whether the condition of the surfacing was a consequence of the debonding, or vice versa. It is possible that a failure of a weak bond initially allowed surface cracking to occur, and that this led to further debonding and cracking. With debonding of the surfacing, temperature and traffic induced forces will be less constrained by the roadbase and may give rise to increased damage to the surfacing.

4 Discussion of results

There have been further structural assessments of bituminous roads carried out by TRL that substantiate this work. One recent investigation showed that the roadbase of a 30 year old motorway had stiffened to 19 GPa without any observed structural deterioration of the roadbase.

In another study the consequences of delaying structural maintenance beyond the point when the road was deemed to be in a critical condition were examined. In that investigation, the condition of 5 out of the 7 sites examined was found to be improving with age. Two of these sites were motorways with a fully flexible construction and both carriageways of these motorways were examined separately. The measured deflection on three of the carriageways reduced in the 5 to 9 years period of the study. The deflection remained constant on the other carriageway. At one of the sites, cracking at the surface increased rapidly during the final two years of observation. At that stage a number of cores were taken and they all showed the cracks to be confined to the wearing course. Furthermore, analysis of the recovered binder showed that it had hardened excessively and this was probably the reason for the surface cracking.

Surface cracking did occur in one of the pavements studied and embrittlement of the wearing course through ageing was probably a contributory factor. This form of deterioration from the surface down has received little attention from researchers. The study has demonstrated that even longitudinal cracking in the wheel path is not a good indicator of damage to the main structural layers of the road. Replacement of the wearing course may be all the maintenance treatment required for a road in which surface cracks have recently initiated.

The investigations showed that for fully flexible roads with a large thickness of bituminous material, changes in the structural properties of the roadbase were mainly caused by changes in binder hardness with time rather than by traffic. In the motorways examined there was no evidence that cumulative traffic loading caused a weakening of the roadbase and, consequently, fatigue seems unlikely to be a major cause of roadbase deterioration.

This study indicates that the design of fully flexible roads may be analogous to design of flexible composite roads. In the design of composite roads it is accepted that if roads are designed above a strength threshold they will have a long, indeterminate life in which they are not structurally damaged by traffic. Below this threshold they will deteriorate gradually under the action of traffic. This concept has far reaching consequences in the design and assessment of fully flexible roads and it could substantially reduce the cost of road maintenance. To test and verify the conclusions of this study, it is recommended that further site investigations be carried out on motorways that, according to the current method of assessment, are in a critical condition. If these studies are substantiated a new approach to design and assessment will need to be formulated.

5 Conclusions

1. Structural assessment of pavement sections of four motorways provided no evidence that fatigue under repetitive traffic loading is a major mechanism in roadbase deterioration. This is substantiated by performance data from TRL's experimental roads. The only evidence of roadbase deterioration due to classic fatigue was found with the thin and heavily trafficked experimental pavements in the PTF which had no surfacing layers.

2. The change in measured residual fatigue life of the roadbase was related to composition, with binder volume and increasing binder hardness with age being the most important variables. However, binder hardening increased the stiffness of the roadbase materials and reduced the traffic induced tensile strains.

3. For the motorways investigated in this study, there was no evidence to support the need to remove the roadbase layers in the design of strengthening.

4. Confirmation of the conclusions of the present work should lead to a re-examination of the criteria for pavement design and maintenance.

6 Acknowledgements

The work described in this paper was carried out in the Highways Resource Centre (Resource Centre Manager: Mr P G Jordan) of TRL.

The contribution to the present work of Mr F Wu of the University of Glamorgan is gratefully acknowledged.

7 References

Department of Transport (1987). The structural design of new road pavements. Departmental Standard HD 14/87, South Ruislip.

Goddard, R.T.N., and Powell, W.D.(1987). "Assessing the conditions of bituminous roads," Highways and Transportation, pp26-32, May 1987.

Goddard, R.T.N., Powell, W.D., and Applegate, M.W. (1978). Fatigue resistance of dense bitumen macadam: the effect of mixture variables and temperature, TRRL SR410, 1978.

Road Note 29. (1970). A guide to the structural design of pavements for new roads. Doe, RRL, HMSO, 1970.

Thrower, E.N. (1979). A parametric study of a fatigue prediction model for bituminous road pavements. TRRL Laboratory Report LR892.

Wu, F. (1992). Assessment of residual life of bituminous layers for the design of pavement strengthening, PhD. thesis, The Polytechnic of Wales, 1992.

PART TWO
BINDERS AND MODIFIED BINDERS

6 RHEOLOGICAL PROPERTIES OF CUTBACKS AND THEIR INFLUENCE ON THE PERFORMANCE OF SURFACE DRESSINGS IN THE MINI FRETTING TEST

M.N. FIENKENG and H. KHALID
Department of Civil Engineering, Liverpool University, Liverpool, UK

Abstract
The rheological properties of three cutback binders used in road surface dressing operations have been tested using the Carri-med Controlled Stress Rheometer (CCSR). The properties determined are the complex shear modulus, G* and phase angle, delta. These properties were determined over a temperature range representative of road temperatures during construction and within the first few hours thereafter. A simulative test, called the Mini Fretting Test (MFT), was used to determine the resistance to fretting of the surface dressings systems with the cutback binders over the same conditions. This was to observe how the rheological properties of the binders influence the behaviour of the surface dressing systems and to establish relationships between these two parameters. To this end, G* values were correlated with the MFT results. The binders used in this work include one conventional and two polymer modified binders.
Keywords: Rheological properties, Surface dressing, Carri-med Controlled Stress Rheometer, Complex shear modulus, Phase angle, Resistance to fretting, Mini Fretting Test, Conventional and polymer modified binders.

1 Introduction

It is generally recognised by practising engineers as well as researchers that the performance during the early stages of a surface dressing is of vital importance to its survival. This is due primarily to the fact that most of the reactions that occur leading to the build-up of strength do so in this early stage. It is therefore considered that the knowledge of the changes in engineering and rheological properties of the binders/systems during this critical stage is essential to the understanding of the behaviour of the surface dressings. The rheological properties considered in this work include the stiffness, viscosity and phase angle, while the mechanical property is the ability of the system to resist fretting.

In this paper, two techniques used for measuring the above mentioned properties of binders/systems are presented. The tests used are the Carri-med Control Stress

Performance and Durability of Bituminous Materials. Edited by J.G. Cabrera and J.R. Dixon. © E & FN Spon. Published in 1996 by E & FN Spon, 2–6 Boundary Row, London SE1 8HN. ISBN 0 419 20540 3.

Rheometer (CCSR) and the Mini Fretting Test (MFT). A background to each test is presented along with the results obtained.

2 Carri-med Controlled Stress Rheometer (CCSR)

The stiffness of the binder at any loading time and temperature can be obtained from its viscosity at a shear rate corresponding to the loading time. Measurements of viscosity are normally performed by means of the sliding plate or capillary tube viscometers. These instruments, as highlighted by Taille and Muller (1981), are limited in the range of viscosities that they can measure. This is mainly due to their mode of operation. They all rely on shearing the sample in one form or another which is very difficult at low temperatures when the binder is considerably stiff.

The CCSR is one of a new generation of rheometers which have overcome the limitations of the conventional viscometers. These rheometers apply a different mode of operation to that of conventional viscometers. The CCSR operates in one of three modes, namely creep, flow and oscillation. The creep and flow modes of operation are not directly related to this work and as such, will not be discussed any further. The oscillation mode of operation provides a unique means of measuring the viscosity, stiffness and phase angle of materials at various frequencies, stresses and strains over a wide range of temperatures usually from -100C to +300C. In this mode of operation, a sinusoidal stress or strain wave is generated by the instrument and applied to the sample. The amplitude and phase difference of the resulting stress or strain wave is measured. From this information and that of the input wave, the viscosity, stiffness and phase angle of the sample are calculated. More on the calculations and principles of the oscillatory mode can be found in the books by Walters (1975) and Barnes et al (1989).

The CCSR has a wide variety of measuring systems which can be used. The system selection is based on the magnitude of the applied and anticipated stresses and strains and on the particle sizes in the sample to be tested. In this work, the parallel plate system was preferred to the cone and plate system. The parallel plate system allows for the variation of the gap hence the thickness of the sample used. This is not possible for the cone and plate system where the gap is preset and cannot be altered. Noting that the size of the gap used had an influence on the results, a fixed gap of 1000μm was used throughout the work. The preparation of the samples involved heating the binders to their spraying temperature, pouring a fixed mass on to aluminium plates and spreading the binders to provide a film of even thickness. The samples were then placed in a constant temperature chamber at a set curing temperature for a fixed length of time. At the end of the curing period, a small amount of the samples were recovered and applied to the measuring plate of the CCSR, using a warm spatular.

3 Mini Fretting Test (MFT)

The Mini Fretting Test (MFT) is considered to be one of the tests that best simulate the fretting action of traffic on a surface dressing (Khalid and Walsh 1990, Hoban 1991). The philosophy of the test centres on the planetary motion of a Hobart mixer which is meant to simulate the fretting action of vehicle tyres. This fretting action dislodges the chippings from the surface dressing. The percentage of chippings retained on the test plaque after fretting is considered to be a measure of the binder's ability to resist fretting. On the road, the resistance to fretting is not only provided by the binder (except in the early stages) but also by the interlocking forces due to chipping embedment into the road base. With the MFT, however, the resistance is provided entirely by the binder. This makes the conditions of the test more severe than those on the road. So if a binder were to perform well in the test, it would be unlikely to fail under similar conditions on the road.

Figure 1 shows a schematic presentation of the forces exerted by the test on a single chipping. From the figure, it can be seen that the test does not only exert a vertical load, representing the weight of the vehicle, and a horizontal force, representing the friction at the interface, but also a twisting or rotational force which depends on the point of contact between the tyre and the chipping. This rotational force may result in adjacent chippings, which are not directly affected by the tyre, being dislodged. The ability of the test to simulate these actions or forces makes it ideal for testing surface dressing binders prior to implementation and also for studying the factors that affect the performance of surface dressings in the early stages where little or no embedment has taken place.

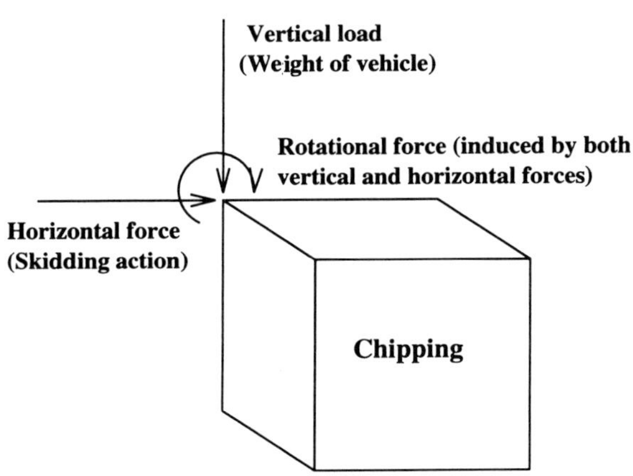

Fig. 1. Forces exerted by the MFT on a single chipping.

The MFT involves the preparation of small diameter samples of dressings which are then subjected to a surface shear by a piece of rubber mounted on a Hobart mixer. The test, originally developed by Tausk et al (1978) is a modification of the well-known Wet Track Abrasion test (ASTM D3910-84) devised by McCoy and Coyne (1964). Further development was carried out at Liverpool University to bring the test to its current stage, which is described in an earlier publication (Khalid and Fienkeng 1993) and in an Institute of Petroleum draft standard, designation BL/92 (1992). For the purpose of this work, the samples were cured and tested at the same temperatures as this was found to be more representative of site conditions in the early stages of the life of the dressings. It should be noted that the percentage of chippings retained on the test plaque after the test, henceforth called the resistance to fretting, is considered to be a measure of the binders' performance under the prevailing test conditions.

4 Results

The results were obtained with the use of 3 cutback binders; one conventional C1 and two polymer modified C2 and C3. C2 contained natural rubber while C3 contained a linear SBS and a linear SIS polymer. The MFT tests were carried out using the standard 3-6 mm graded chippings specified in the Institute of Petroleum's draft standard (1992).

4.1 CCSR results

Figure 2 shows the effect of temperature and curing time on the CCSR results of the 3 cutback binders. The results are presented in terms of the complex shear modulus (G*) and the phase angle (delta). Due to the similarity in the trends of the viscosity and G* results, viscosity was omitted for convenience. The results were obtained at a frequency of 10 Hz because this represents traffic moving at medium speeds and it falls within the range of 0.01 and 20 Hz considered by Denning and Carswell (1981) to represent the range of traffic speeds on the road. With regard to the temperature range of 10 to 40°C, it is one at which surface dressing works are normally carried out in the U.K.

With respect to G*, it can be observed that there is a general decrease as the temperature increases. This is because as the bond strengths within the binder weaken with a rise in temperature, the tendency for the binder to deform under pressure increases. The rate of decrease in G* with temperature is a measure of the temperature susceptibility of the binder and a binder whose properties are least susceptible to temperature changes is considered to be the most suitable. Thus binder C3 will be ranked first in terms of suitability, followed by binders C2 and C1. However, a binder should possess, initially, a high enough G* to enable it to resist the adverse forces that it could be subjected to. On this basis, binder C3 appears to be the least desirable of the three binders with C2 and C1 exhibiting higher G* values.

With regard to the effect of curing time on G*, there is little or no change for all

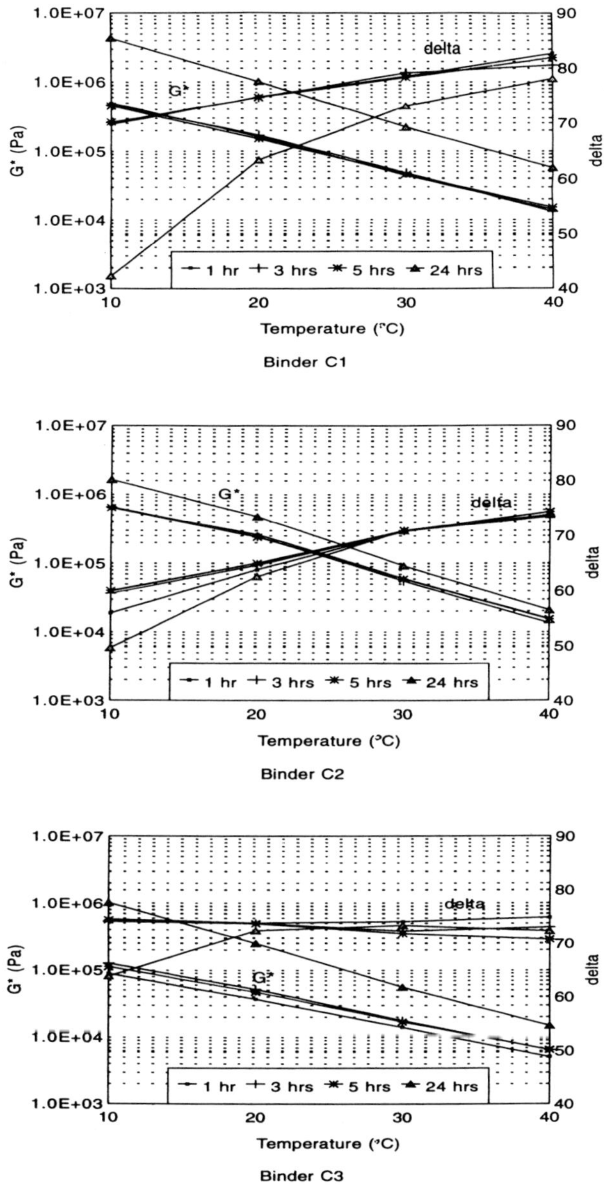

Fig. 2. Influence of temperature and curing time on G* and delta of the 3 cutback binders, obtained from the CCSR at 10 Hz.

three binders within the curing period 1 to 5 hours. After 24 hours curing however, the binders become considerably stiffer and this is reflected in the higher G* values at all the test temperatures. This is because as the curing time increases, further losses in the flux oils within the binders take place together with the ageing as a result of oxidation. The fluxed oils are generally responsible for the low viscosities and stiffnesses of the binders in the early stages.

The variation of delta with temperature can be observed to depend on the type of binder. With binders C1 (conventional) and C2 (polymer modified) delta increases with temperature. With binder C3 however, delta appears to be relatively unaffected within the first 5 hours with changes in the temperature. The rate of change in delta with temperature is considered to be a measure of the temperature susceptibility of the binder and in this case, binder C3 is the least susceptible to temperature changes followed by C2 and C1.

With respect to the effect of curing time on delta, the only apparent change occurs only after 24 hours curing with little or no change observed in the first 5 hours. After 24 hours, there is a general decrease in delta, this being greatest at the low temperatures.

Delta is a measure of the binder's elasticity and is used as an indication of how brittle or viscous the binder is at any temperature and frequency. The trend followed by delta indicates the variation of the binder's elasticity with temperature. For the three binders presented, C1 generally exhibits higher delta values than C2 and C3 especially at high temperatures, indicating a more viscous behaviour and a greater tendency to flow under pressure. Within the polymer modified binders, C3 with higher delta values at low temperatures than C2 is considered best as its delta values are generally unaffected compared to those of C2 with changes in temperature. Thus ranking the binders with respect to G* and delta, binder C3, despite relatively low G* and high delta values in the early stages, will be classified first. This would then be followed by binders C2 and C1. From the effect of curing time on delta, it could be concluded that the binders generally become more elastic, as a result of a decrease in delta after 24 hours, with time.

4.2 MFT results

Figure 3 shows the results obtained from the MFT for the 3 binders as a function of curing and test temperature and curing time. The results show that temperature greatly affects the ability of the dressings to resist fretting. From the figure, it can be observed that there is a general decrease in the resistance to fretting of binders as the temperature increases, which is analogous to the observation made for G*. The rate of decrease is however dependent on the type of binder, with the highest rate observed for binder C1 and the lowest for binder C3.

With respect to the effect of curing time, it can be seen that this varies from one binder to the other and also depends on the curing and test temperature. For all the binders, in the early stages (1 to 5 hours), there is very little improvement in performance with time in the temperature range 10 to 20°C. However, at the high temperatures (30 to 40°C), for binders C1 and C2, there is a definite improvement in performance with time. With binder C3, this increase is very small. After 24 hours

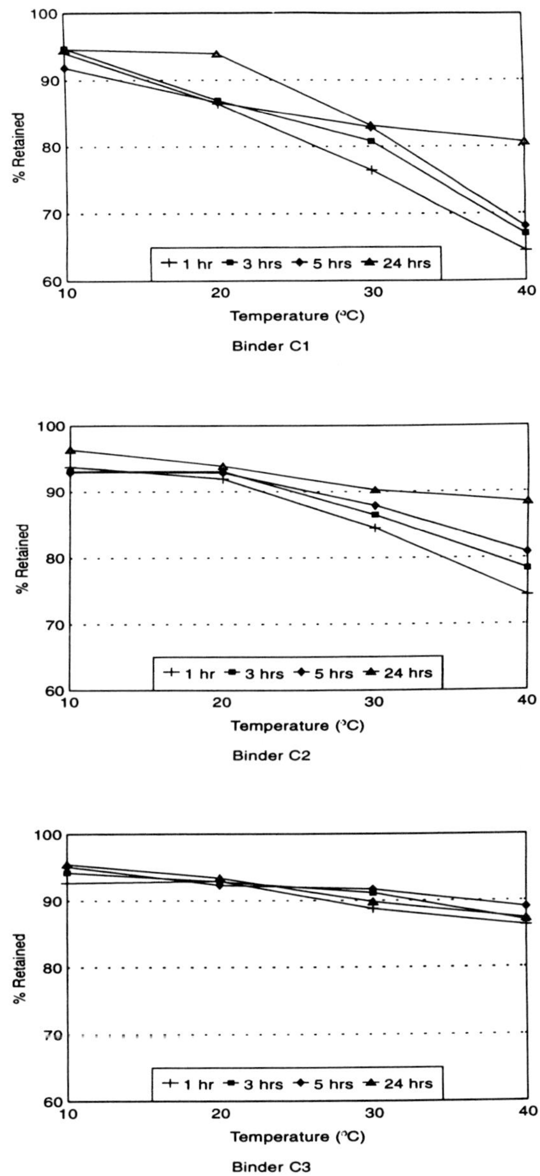

Fig. 3. Influence of temperature and curing time on the resistance to fretting of the 3 surface dressing systems, obtained from the MFT

curing, there is a definite increase in performance for all the binders at all the curing temperatures though the magnitude of this increase is once again dependent on the type of binder, with the highest increase observed for binder C1 and the lowest for binder C3.

In terms of overall performance, binder C3 produces the best performance manifested in the high resistance to fretting, followed by binders C2 and C1. Once again, it can be seen that the polymer modified binders outperform the conventional binder.

The decrease in performance of the binders as temperature rises is due to the effect of temperature on their stiffness, adhesive and cohesive forces. These forces are essential in retaining the chippings and as they decrease with temperature, so too does the performance of the binders. A binder whose properties are less susceptible to temperature changes (C3 in this case) will be able to maintain a uniform performance as the temperature varies.

5 Correlations

One of the aims of laboratory testing is to be able to reproduce, correlate and (thereby) predict conditions and behaviours in the field during implementation and while in service (Lee 1969). Correlations could be used as a means of emphasizing the influence of one property on another. In this case it is the influence of the rheological properties of the binders on the mechanical properties of the system. In this correlation, G* was considered the independent variable and MFT results the dependent variable. This was to emphasize the influence of the rheological properties on the mechanical properties and the point that such correlations may be used to deduce the mechanical properties of the systems from the rheological properties of the binders. Correlations would also provide a means of assessing the reliability of the MFT in determining the resistance to fretting of surface dressing binders.

Figure 4 shows the correlations obtained together with the relevant statistical data. It can be seen that there is a good correlation between G* and the resistance to fretting thereby reinforcing the fact that stiffness has a direct influence on the resistance to fretting of the surface dressings. The correlations also show that the MFT is a reliable means in determining the performance of surface dressing binders in terms of their resistance to fretting.

6 Conclusions

G*, delta and the resistance to fretting have been used to depict the behaviour of binders in their early and quasi-full cure (24 hours) stages. These parameters have indicated the superiority of polymer modified binders over conventional binders. G* has been shown to have a direct influence on the mechanical properties of the surface dressing system as measured by their resistance to fretting in the MFT.

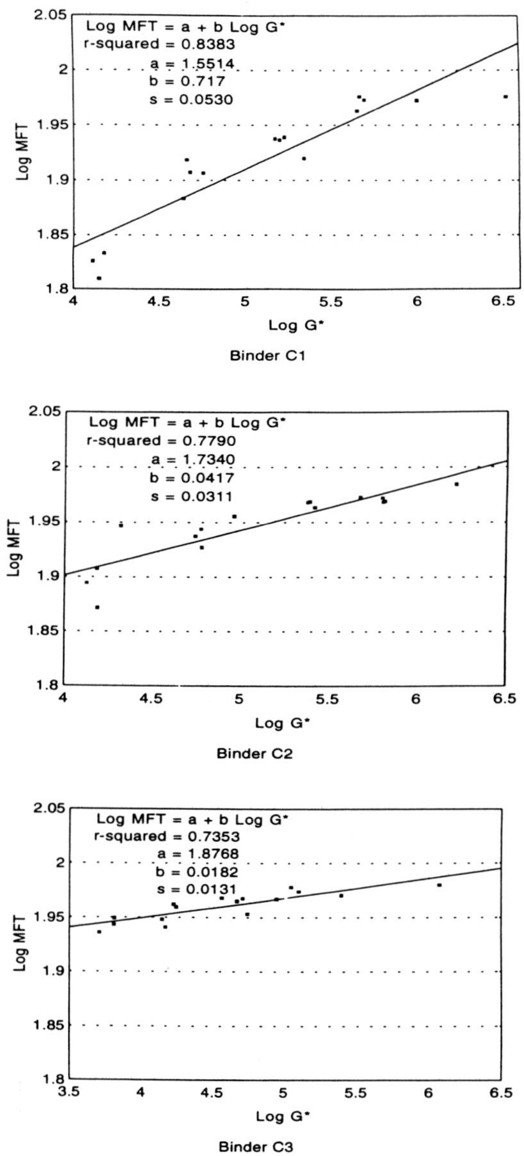

Fig. 4. Correlations between CCSR and MFT results giving statistical data of the regression lines for the 3 cutback binders.

From the limited results presented in this paper, it can be seen that both CCSR and MFT provide useful means of studying the early life behaviour of surface dressing binders. They both provide means of monitoring with time and temperature, the changes in the properties of the binders and the build-up of strength of the systems.

7 Acknowledgement

This work was funded by Shell Bitumen UK, to whom the authors are gratefully indebted.

8 References

ASTM (1984) Effect of heat and air on a moving film of asphalt - Rolling Thin Film Oven Test. **ASTM** D2872-84.

Barnes, H.A., Hutton, J.F. and Walters, K. (1989) **An Introduction to Rheology**. Rheology Series, Vol. 3, Elsevier Science Publishers.

Denning, J.H. and Carswell, J. (1981) Improvements in rolled asphalt surfacing by the addition of organic polymers. **Transport and Road Research Laboratory**, LR 989.

Hoban, T.W.S. (1991) The role of modified bitumen binders for surface dressing. **Highways and Transportation**, Feb. 1991, pp.19-23.

Institute of Petroleum, Proposed method IP BL/92. (1992) Determination of chipping retention properties of surface dressing binders - Mini Fretting method.

Khalid, H. and Fienkeng, M.N. (1993) Factors affecting early life stability of surface dressings with polymer modified binders. **Eurobitume 1993**, Stockholm, Vol. 1a, pp.370-374.

Khalid, H. and Walsh, J.W. (1990) Failure modes of surface dressings with particular reference to the use of polymer modified binders. **1st International Symposium on Highway Surfacing**, Belfast UK, April 1990.

Lee, D.Y. (1969) Evaluation of Marshall Stability and flow values of asphaltic paving mixtures. **Highways Research Record**, No. 272, pp.53-62.

McCoy, P. and Coyne, L.D. (1964) The Wet Track Abrasion Test. Paper presented to the road and paving material session, **ASTM**, Chicago, Illinois, June 1964.

Taille, B. and Muller, J.M. (1981) Rheologie des liants. **Eurobitume 1981**, Cannes, pp.128-132.

Tausk, R.J.M., Scott, J.A.N. and Vonk, W.C. (1978) Setting of cationic bitumen emulsions for surface dressing testing and elucidation of mechanisms. **Eurobitume 1978**, London, pp.176-179.

Walters, K. (1975) **Rheometry**. Chapman and Hall Publishers, London.

7 AN EVALUATION OF THE USE OF A FIBRE-REINFORCED MEMBRANE TO INHIBIT REFLECTIVE CRACKING

C. YEATES
Colas Ltd, Rowfant, Crawley, West Sussex, UK

Abstract
Reflective cracking in bituminous pavements occurs through thermal and traffic induced fatigue, and there are many solutions available to the Maintenance Engineer to prevent or inhibit crack initiation on the road surface.

Much has been written about these causes and solutions, and this review attempts to take into account current thinking, concluding that there is much confusion about where, when and which type of treatment for reflective cracking should be specified.

Through a summary of recent market and technical research, this paper reviews the use of stress absorbing membrane interlayers (SAMI's) in the U.K. and evaluates the use of a fibre-reinforced membrane as an effective and efficient alternative to conventional geotextiles and geogrids.

The fibre-reinforced membrane is manufactured in-situ and spray-applied through specially developed equipment to achieve a continuous, waterproof and flexible mat, capable of inhibiting reflective cracking. The characteristics and technical performance of this membrane are explored.

It is concluded that the use of the membrane as described in this paper, meets an Engineer's need for a practical, economic and speedy solution to inhibit reflective cracking in bituminous pavements.
Keywords: Reflective Cracking, Fibre, Stress Absorbing, Fatigue, Flexible.

1 Introduction

Much has been written about the causes of reflective cracking in pavements and suitable maintenance treatments to inhibit crack initiation. The only one common theme through these studies is that all road surfacings suffer continual structural deterioration from the moment they are opened to traffic. Where the studies differ is in the theories of how and why cracking occurs and the effectiveness of various preventative maintenance treatments. Geotextile fabrics, meshes and geogrids are now well-known and well-marketed as methods of controlling crack propagation.

However, at both the first (1989) and second (1993) World Conferences on Reflective Cracking in Pavements the preface stated:

> "The rehabilitation of cracked roads by overlaying is rarely a durable solution. In fact, the cracks rapidly propagate through the new asphalt layer ... with current financial restrictions, road maintenance authorities have to find solutions with a good cost:benefit ratio. Many solutions have been proposed ... these solutions are supported by numerous studies ... In spite of these efforts, it seems that universal crack repair treatment with good durability is still lacking."

Performance and Durability of Bituminous Materials. Edited by J.G. Cabrera and J.R. Dixon. © E & FN Spon. Published in 1996 by E & FN Spon, 2–6 Boundary Row, London SE1 8HN. ISBN 0 419 20540 3.

2 Reflective Cracking - A Review

Following market research undertaken during 1993, I would suggest that there is much confusion as to when to use a crack repair treatment, where in the pavement structure to install it and there is even less of an understanding as to which type of treatment to specify. Many products are vigorously marketed, but there still appears to be a misappreciation of the nature of the problem to be tackled and a misappreciation of the selection and application of a suitable product to treat reflective cracking with good durability.

2.1 The Problem

Despite, or maybe because of this confusion, reflective cracking continues to be an ongoing problem for Engineers. Cracks on the road surface result in water intruding into the pavement structure, progressively weakening the performance of the foundation layers. This causes soil particles to pump through the crack and results in discomfort for road users and a reduction in the safety of the surface.

2.2 Causes of Reflective Cracking

Current thinking on crack propagation in bituminous pavements states that the origin of cracks is generally related to two mechanisms. Very simply, the first is thermal fatigue whereby regular seasonal temperature variations induce the expansion and contraction of cracks in the roadbase propagating cracks in the overlay. The second mechanism is traffic induced fatigue whereby vertical movements can occur between adjacent lean concrete slabs when wheel loads pass over a crack in the roadbase. This results in shear stresses in the roadbase and wearing course, causing the surface to crack.

2.3 Remedial Techniques

There have been numerous suggestions for remedial techniques to inhibit or even prevent the occurrence of reflective cracking. Generally they can be classified into three broad areas:

(a) Modifications of the overlay to improve its ability to resist stresses and strains induced by the crack movements. Methods here include reinforcing the overlay, increasing the thickness of the overlay or modifying the bitumen used.

(b) Treatment of the origin of the cracks for example stabilizing the slab joints or using modified binder 'plug' to seal the joints.

(c) Placement of a stress absorbing membrane either as an interlayer (SAMI) below the wearing course or as a surface treatment (SAM) on top of the cracked surface. Examples of the first application include geotextiles, synthetic paving felts and geogrids. A SAMI provides a slip plane between the old pavement and new overlay, allowing movement of the old pavement and reducing the potential stress transferred to the new overlay. Examples of the SAM application include modified surface dressings - which will act as a waterproof membrane and stop water penetrating into the pavement layers, but will not have sufficient tensile strength or elasticity to absorb movements in the pavement and inhibit cracking for any length of time. Thus, the use of a surface dressing or slurry seal can only be considered as a temporary solution to the problem.

The fibre-reinforced membrane discussed in this paper falls into the third category. Furthermore, it can be applied as a SAMI or as a SAM and I would suggest is the only effective SAM currently available in the U.K. market.

3 Fibre-Reinforced Membrane - An Alternative

3.1 Description of the Process

The fibre-reinforced membrane is a composite system formed by machine applying two coats of bituminous emulsion binder at a minimum of 0.8 l/m^2 per coat to give a total binder spread rate of 1.6 l/m^2 (although the exact rate of spread will be determined by site conditions).

Chopped glass fibres either 30, 60 or 120 mm in length are sandwiched between the two coats of binder at a rate of 60 - 120 g/m^2 depending on the severity of the cracking.

A dressing of 6 mm or 10 mm aggregate is then applied and rolled into the finished surface. The aggregate allows the SAMI to be trafficked by the construction plant that applies the asphalt wearing course.

The surface is then ready to receive the overlay immediately, or alternatively, the road can be opened to traffic and overlaid at a later date.

3.2 Application

The fibre-reinforced membrane is applied through specially developed and patented machinery.

The applicator comprises of a spray tanker with a 2.4m spraybar on the rear which incorporates two rows of sixteen spray jets for the binder and between the two rows each one of these is matched by a fibreglass feeder/chopper.

The choppers are driven hydraulically and are speed controllable to adjust the fibreglass spread rates. The chopping chamber is charged with a positive air pressure to assist in random lay-down of the fibres. Fibre lengths of 120 mm, 60 mm or 30 mm are achieved by reducing the number of blades on each chopper.

4 Laboratory Research

Two technical studies have been done on the fibre reinforced membrane. The first was conducted in 1987 by Nottingham University and the second, more comprehensive study, was done by Ulster University in 1993. These two studies are now examined.

4.1 Nottingham University 1987

Tests undertaken by Nottingham University in 1987 evaluated glass fibre reinforcement techniques to inhibit reflective cracking in overlays.

The test method employed was a simulation of the situation occurring in practice when a cracked pavement is overlayed.

The sample consisted of a beam of compacted rolled asphalt wearing course on a base of a 20 mm thick laminate. A crack was simulated by a transverse 10 mm gap in the laminate. The fibre-reinforced membrane was applied to the asphalt beam.

A sinusoidally varying vertical load of constant amplitude was applied to the sample. As the test proceeded cracks propagated upwards from the gap in the support. The lengths of the cracks on either side of the sample were measured at intervals to provide the data necessary to determine the relationship between mean crack length and number of load applications. To provide a basis for comparison two 'control' or unreinforced beams were tested.

This is a test used extensively to investigate the potential of various solutions to the problem of reflective cracking and the test conditions have been standardised.

4.1.1 Conclusions

(a) The action of the fibre-reinforced membranes investigated is believed to be that of a stress absorbing membrane interlayer.

It provides strain attenuation in the horizontal direction whilst being capable of transferring loads in the vertical direction without excessive deformation.

(b) With the simulative test used, the fibre-reinforced membrane significantly inhibited the propagation of reflective cracks.

(c) Where permanent lateral movement of the overlay might occur due to the cracks in the underlying material opening, the interface layer appears to reduce the magnitude of the resultant strain in the overlay; spreading it over a greater area. This generally results in a series of unconnected micro cracks which are less damaging that a few wide cracks.

4.2 **Ulster University 1993**
The laboratory research investigated the potential feasibility of using chopped fibre membranes for the reinforcement of bituminous surfacings and the resultant mechanical properties imparted by the matrix - i.e. flexibility and crack control.

Three variables were tested: emulsion type and coverage; length of fibre glass and rate of spread of fibre glass.

Sample slabs of a gap-graded hot rolled asphalt were prepared in accordance with BS 594 (1). These samples were then coated with an emulsion-based binder and chopped glass fibres. Three samples were prepared for each combination of variables to test for repeatability. Due to the random orientation of the fibres and the possibility that this could alter repeatability of performance, all fibre-asphalt slabs were mapped. Fortunately, despite the randomness of orientation, repeatability within the groups was fairly consistent.

Three main test methods were chosen to examine the performance of the SAMI's in delaying reflective cracking. Essentially they consisted of:

* An Instron test rig to determine tensile strength

* A Dartec Dynamic load frame to analyse crack propagation over time

* Trafficking the slabs on a wheel-tracking apparatus.

A review of these test methods and results now follows.

4.2.1 **The Tensile Test Programme**
All samples were tested on the Instron 1114 floor model at ambient temperature, and a crosshead speed of 50mm/minute. Each slab specimen was glued to two steel plates with a gap of 25 mm between them. The plates were then pulled apart and the tensile stress and strain were measured.

Of the variables tested, the combination of K1-70 emulsion with 60 g/m² and 60 mm length of fibre proved the most effective with a tensile strength of 0.77N/mm² compared with 0.57 for the control sample (asphalt slab, no fibres).

The samples where a high polymer content, emulsion binder was used, proved disappointing.

4.2.2 **The Fatigue Test Programme**
All samples were tested on a Dartec testing apparatus (dynamic load frame) using a three point loading system. The samples were flexed sinusoidally using a 20kN load cell with a maximum load set at 0.03kN, an amplitude of 6mm and a frequency of 20 Hz, equivalent to 0.05 seconds loading time. The samples were tested to failure, noting the time for the cracks to propagate the full width of each sample.

Of the variables tested, the combination of K1-70 emulsion with 60 g/m² and 60 mm length of fibres afforded the membrane the longest time to failure. The number of cycles to crack was approximately

3,300Hz and for complete failure, 6,300Hz. The control samples cracked after 2,500Hz and failed at 4,700Hz.

4.2.3 **The Wheel-Tracking Test Programme**
All samples were mounted on supports at a distance of 170 mm apart with weights (approximately 18kg) added to induce cracking, and trafficked at 52 wheel passes/minute, under water at a constant test temperature of 3 degrees centigrade. The number of wheel-passes to failure was measured. Failure of the samples was usually instantaneous once a visible crack had appeared, and the crack started on the top of the sample. This confirms results of research done by the TRL.

The variables using K1-70 emulsion, 30 g/m² and 60 mm lengths of glass fibres appeared to provide the greatest delay to failure and this was very closely followed by K1-70, 60 g/mm² and 60 mm length of fibres. The fibre-reinforced membrane cracked after an average of 15,000 passes compared with 2,500 passes for the control.

The high polymer emulsion binders offered little increased performance when compared with the non-reinforced asphalt samples. Average deformation over time was also measured for each test group. Essentially the polymer binders - i.e. softer binders, deformed at a faster rate.

4.2.4 **Conclusions**
The use of a chopped fibre-reinforced membrane as a SAMI significantly inhibits the propagation of reflective cracking in an asphalt overlay. In each of the three tests conducted, the samples using the fibre-reinforced membrane out-performed the samples where no SAMI was used.

The use of a chopped fibre-reinforced membrane was shown to enhance the performance of an overlay by approximately 30% (fatigue and tensile stress) and wheel-tracking rate by 300%.

The benefits of ease of application of the sprayed in-situ fibre membrane were also noted and compared favourably with the associated difficulties of laying a conventional SAMI. The need for an interfacial binder is also eliminated, thereby reducing the total costs.

5 Conclusion

To date over 150,000 m² of the fibre-reinforced membrane has been used to inhibit reflective cracking in asphalt pavements over the last 3 years. The main conclusions are as follows:

(a) Laboratory research indicates that the use of a fibre-reinforced membrane will significantly inhibit reflective cracking.

(b) The specially designed machine 'manufactures' and applies the SAMI in-situ, making the product quick and easy to apply with no adhesion problems associated with conventional SAMI's. It also presents no problems associated with laying SAMI's around bends or on roads of varying width.

(c) Following an application of aggregates over the membrane, the surface can be trafficked before it is overlayed, making the SAMI both practical and convenient to use.

(d) The fibre-reinforced SAMI membrane is extremely economical. At approximately £1.50 m² all in, it represents a direct cost-saving over other reflective cracking treatments and indirectly through speed of application and through the elimination of the need to raise kerbs and ironwork.

(e) The fibre-reinforced membrane is also very versatile. It can be used as a SAMI below the wearing course or as a SAM on the surface to inhibit cracking starting at the top of the wearing course and propagating down i.e. a waterproof, flexible surface.

In essence, the fibre-reinforced membrane provides the Engineer with an economical, waterproof and flexible surface to inhibit reflective cracking in asphalt pavements.

6 References

Nunn, M. Reflective Cracking in Composite Pavements.

Nunn, M.E. and Potter, J. F. Assessment of Methods to Prevent Reflective Cracking. Paper presented at second RILEM Conference, March 1993.

Rigo, J. M. General Introduction to Second RILEM Conference, March 1993.

Walsh, I. D. An Investigation into Effective Treatment of Reflective Cracking. IHT, 10th National Workshop, Leamington Spa, April 1989.

8 PERFORMANCE CHARACTERISTICS OF CONVENTIONAL AND SBS MODIFIED ROLLED ASPHALT MIXTURES IN VIRGIN AND AGED CONDITIONS

J.N. PRESTON
Shell Bitumen, Chertsey, Surrey, UK

Abstract
The performance characteristics of HRA mixtures measured using the NAT are compared for conventional 50 pen mix compositions and modified mixtures incorporating an SBS binder. The early life curing behaviour is monitored using the same test procedures following laboratory ageing over a twelve month period. Discrepancies between the results from the different material types are explained through rheological assessments of the binders and the potential impact on the development of performance based specifications is discussed.
Keywords: Performance, Stiffness. NAT. Deformation, Curing, Ageing Binder Rheology.

1 Introduction

The move towards performance based specifications for blacktop materials and the requirement for practitioners to have a greater fundamental understanding of bituminous mixtures, has generated an extensive amount of testing in U.K. laboratories in an effort to develop data banks for the generic types of material compositions used in the U.K. Such extensive testing has been facilitated by the availability and universal acceptance of the Nottingham Asphalt Test (NAT) Cooper and Brown (1989), which allows a fundamental assessment of mix samples to be carried out through a suite of easily executable tests.

The Repeated Load Indirect Tensile Test (RLIT), which measures mix stiffness, has become the most popular method of characterising a performance characteristic, because of the ease and speed of the test operation. This parameter has greatest relevance to the roadbase layer where the loadspreading function has a significant influence on the structural capacity and longevity of the pavement. However, there appears to be a trend of applying a similar weighting of importance of the stiffness value to wearing course materials and basing judgements on the competence of rolled asphalts according to such results.

2 Wearing Course Evaluation

An investigation was carried out at the Thornton Laboratory of Shell Research Limited to measure fundamental properties of a hot rolled asphalt wearing course mixture typical of a

Performance and Durability of Bituminous Materials. Edited by J.G. Cabrera and J.R. Dixon. © E & FN Spon. Published in 1996 by E & FN Spon, 2–6 Boundary Row, London SE1 8HN. ISBN 0 419 20540 3.

U.K. surfacing material. A 30% stone composition was selected as the datum material with the binder content determined by the U.K. Marshall method, BSI (1992), in order to be fully representative of the design surfacing mixtures. The binder used in the datum composition was 50 pen bitumen.

It was decided to compare the properties of the conventional mixture with those obtained from a modified 'high performance' wearing course mix incorporating a polymer bitumen. Samples of the modified mixture were manufactured using Cariphalte DM*, a proprietary SBS modified binder supplied by Shell Bitumen. Cariphalte DM has been marketed aggressively since the mid 1980's as a highly flexible binder which greatly increases the fatigue life and deformation resistance of asphaltic mixtures and has been targeted at highly stressed, heavily trafficked sites. Specific details of the performance benefits have been detailed elsewhere, Preston (1991) and Whiteoak (1989), but the modified rheological profile of the SBS bitumen is classically used to demonstrate the improved characteristics imparted to the asphalt, Vonk and Gooswilligen (1989).

3 Preliminary Work

A large population of specimens incorporating 50 pen bitumen was manufactured using the Marshall procedure at the 50 blow compaction level. A total of 67 samples was considered sufficient to encompass the full spread of results which could be expected from the evaluation of a selection of specimens of bituminous material, whilst still maintaining a representative mean value.

Following manufacture, the specimens were allowed to cool to a temperature of 20°C in preparation for RLIT testing in the NAT. This was achieved by storing the specimens in a temperature controlled environment for a 24 hour period prior to testing. The execution of the RLIT tests was carried out in accordance with DD213, BSI (1993), to ensure that procedures should have followed the same format as those of other laboratories. A sister exercise was carried out simultaneously on specimens incorporating Cariphalte DM on a similar population of samples in order to identify whether the "scatter potential" of the test remained unchanged for modified mixtures.

3.1 Initial Results

The overall spread of results obtained from the stiffness testing was disappointing for both sets of material. Figure 1 displays the variation in stiffness for the 50 pen specimens with the total spread of absolute values ranging between 1.5GPa and 3.4GPa. Void contents calculated from a theoretical value of maximum density ranged from 2.2% to 5.0%.

It can be seen that the distribution is approximately normal with both arithmetic mean and modal averages coinciding at slightly over 2GPa. The substantial variation in results, although influenced by the range of void content, cannot be wholly attributable to the volumetric composition of the specimens. Inherent material variation and sensitivity to testing conditions probably accentuate minor differences which are recognised by the very accurate electronic measurements of the test. It is considered that the spread of results does not negate the potential of the Repeated Load Indirect Tensile Test as a candidate method for performance

*Cariphalte DM is a registered trademark.

based specifications. The salient point which arises concerns the minimum population of samples which can be evaluated to give a statistically representative result of a bulk sample of material.

A Similar percentage spread of results was obtained from the mixes incorporating Cariphalte DM, the SBS binder, figure 2 illustrates stiffness values ranging from 0.53GPa to 1.1GPa, with an arithmetic mean of 0.73GPa.

For the mix composition under consideration it can be seen that there is a significant reduction in mix stiffness when Cariphalte DM is used as the binder. For an enhanced performance product these findings may appear paradoxical, so it was considered important to investigate the mechanism responsible for this effect.

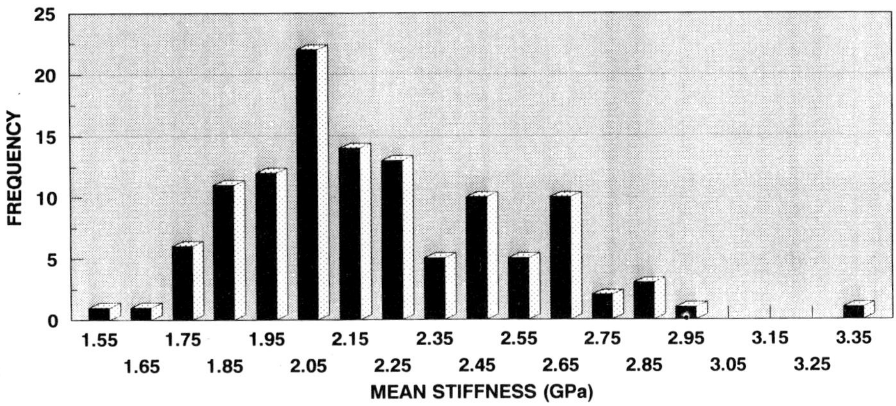

Fig. 1. Range of measured stiffness values, hot rolled asphalt, 50 Pen specimens (all tests after 24 hours).

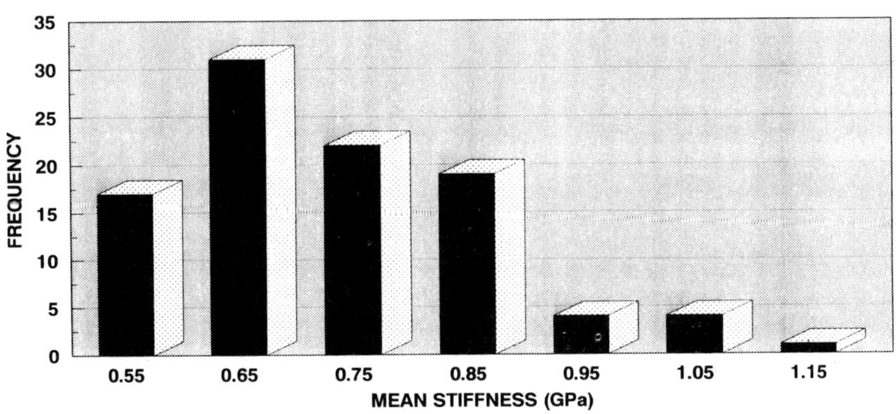

Fig. 2. Range of measured stiffness values, hot rolled asphalt, Cariphalte DM (SBS modified) specimens (tests after 24 hours).

4 Rheological Study of the Binders

Binder rheology was examined using controlled stress rheometry under dynamic loading conditions, with tests carried out within the linear visco-elastic range of the binders. In the tests, sinusoidal strains were imposed through controlled oscillatory shear on samples of binder set within a parallel plate arrangement. Applied stresses were imposed on the samples through the upper plate, with the response of the specimen calculated by measuring the corresponding strain amplitudes.

Due to the visco elastic nature of bitumen, if a sinusoidal stress wave is applied, the resultant stress wave will also be sinusoidal, but will be out of phase by a value delta (δ), Goodrich (1988). The magnitude of delta will vary depending upon the temperature and loading conditions. The relationship between the applied stress and the measured displacement is known as the complex shear modulus, G*, which comprises an elastic component, the storage modulus G' and a viscous component, the loss modulus G". These parameters are related as follows:

$$G^* \ (\mathrm{Pa}) = (G')^2 + (G'')^2 \qquad (1)$$

where

$$G' \ (\mathrm{Pa}) = G^* \cos \delta \qquad (2)$$

and

$$G'' \ (\mathrm{Pa}) = G^* \sin \delta \qquad (3)$$

$$\delta = \arctan \frac{G''}{G'} \qquad (4)$$

It can be seen from figure 3 that over the frequency sweep at 20°C 50 pen bitumen exhibits a predominantly higher complex shear modulus than Cariphalte DM. Only at very long loading times (10^{-4} Hz loading frequency) does the SBS bitumen begin to show a higher value of complex modulus than the 50 pen.

Under the conditions of the RLIT test, the rise time of the load pulse is approximately equivalent to a sinusoidal loading frequency of 2.5 Hz, Nunn and Bowskill (1992),, at which the G* values of 50 pen and DM are in the order of 2.0 Mpa and 0.4 MPa respectively. If the value of poisson's ratio for the systems is in the order of $\mu = 0.5$ then binder stiffness is related to shear modulus by the equation:

$$\text{Dynamic Stiffness} \approx 3G^* \qquad (5)$$

Hence, Cariphalte DM exhibits a stiffness of approximately one fifth of that of 50 pen at the test conditions, explaining the difference in the mean results obtained from the initial evaluation. By increasing the temperature at which the rheological assessment is carried out to 45°C, coinciding with the wheel tracking test temperature, the relationship between the phase angle δ and frequency of loading for the two binders, shows that under conditions of elevated temperature and or long loading times, the SBS binder has a predominantly elastic response to applied loading whereas 50 pen as a viscous response, figure 4.

This rheological data explains how asphalt mixtures incorporating Cariphalte DM exhibit

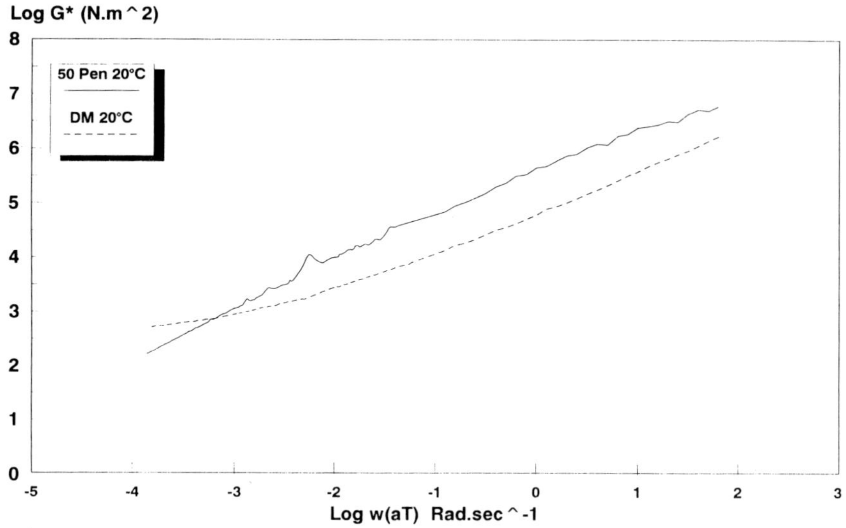

Fig. 3. Rheological analysis: comparison of 50 Pen and Cariphalte DM.

high resistance to deformation at elevated temperatures or long loading times and greater resistance to fatigue at low temperatures. The absolute interpretation of mix stiffness as a performance indicator for SBS modified mixtures may be misleading and must be considered questionable as a technique of assessment.

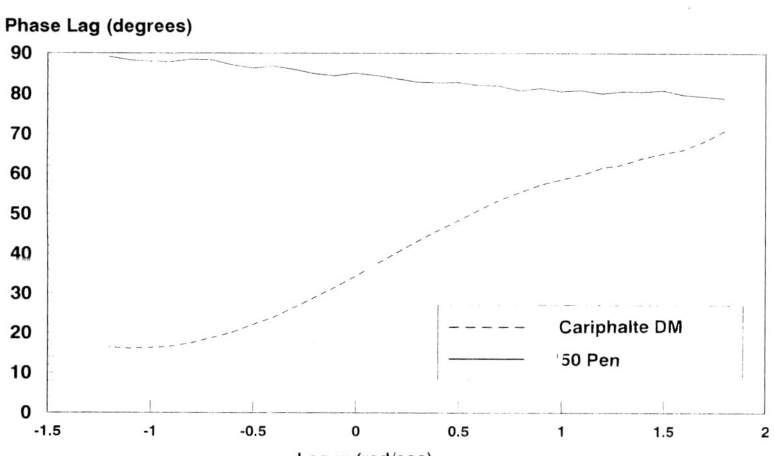

Fig. 4. Phase angle vs frequency, Carri-Med data, Shell 50 Pen and Shell SBS bitumen (Cariphalte DM) specimens, 45°C.

5 The Curing Phenomenon

Short term changes in the properties of bituminous binders, particularly during mixing and hot storage are well recognised, but until recently have been largely ignored in the context of the early life performance of a pavement. Recent findings by the Transport Research Laboratory from investigations carried out on cores of roadbase material suggest that mix stiffness may increase by up to 150% during the first year of service due to curing effects, Nunn and Bowskill (1992). The implications of these findings are considerable when taken in the context of performance specifications and the interaction between material properties and structural capacity.

The ageing phenomenon is generally attributed to binder hardening through oxidation, steric harding and the physico-chemical interaction between the bitumen and the aggregate. Hence, the amount of ageing is likely to be mix specific and dependent upon aggregate type, volumetric composition and the original bitumen penetration.

5.1 A Strategic Evaluation of Early Life Curing

Following the evaluation of the initial properties (24 hour values) of the populations of asphalt specimens, it was decided to conduct repeat testing on the same specimens ate pre-determined time intervals to monitor any change in material performance. Two storage temperatures were selected: 20°C to represent a mean summer temperature, and an elevated value of 35°C to try to accelerate any temperature effects. Any change in material properties was recorded as a percentage change from the original properties of a given specimen and not related to the mean value of the population.

Following stiffness testing, the specimens were then subject to axial load deformation tests in accordance with the procedure set out in DD 185, BSI (1993), in order to assess any early life change in deformation resistance.

Following the mechanical tests on the specimens, binder recoveries were carried out in an effort to correlate changing material performance with bitumen properties.

6 Results

Repeat tests were performed on the specimens at 1 week, 2 weeks, 30 days, 84 days and 164 days. Each value obtained is the mean result from a set of four tests and referenced to the original 24 hour mean of the same set of four specimens. Figure 5 shows the relative change in stiffness averaged for each set of samples of the 50 pen material according to the temperature and duration of storage. The 30 day result appears anomalous, with a marked reduction in stiffness occurring under both storage regimes, suggesting that the validity of these results must be questionable. Overall, the trend of results up to the 168 day value (six months storage) at 20°C does not depict a significant effect of curing on the stiffness of the specimens. At the elevated temperature of 35°C, the 168 day value suggests that material properties are beginning to change with a 17% increase in stiffness from the original value. The research program will continue to monitor the samples up to a 12 month value which should confirm the trends observed to date.

The modified mixtures have exhibited a trend which is similar to the TRL observations with material stiffness showing a steady increase at both storage temperatures over the

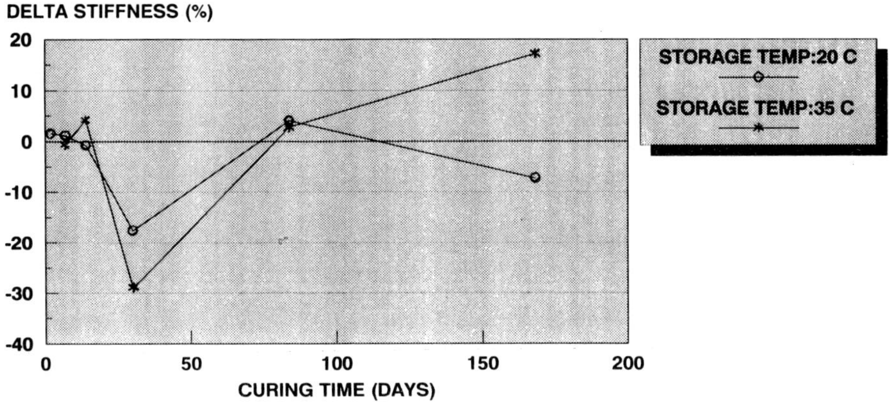

Fig. 5. Change in stiffness with time, hot rolled asphalt, 50 Pen specimens.

duration of storage. Figure 6 illustrates a 30% increase in mix stiffness after 84 days storage (3 months) at 20°C and a 40% increase following 168 days (6 months) storage at 35°C.

As with 50 pen material, final measurements taken after a 12 month storage period will confirm or repudiate the preliminary findings.

6.1 Recovered Binder Properties

An evaluation of the recovered properties of the binders from both sets of test specimens was carried out to attempt to ascertain the contribution of the bitumen to the ageing characteristics. The 50 pen values are shown in figure 7, where it can be seen that a definitive

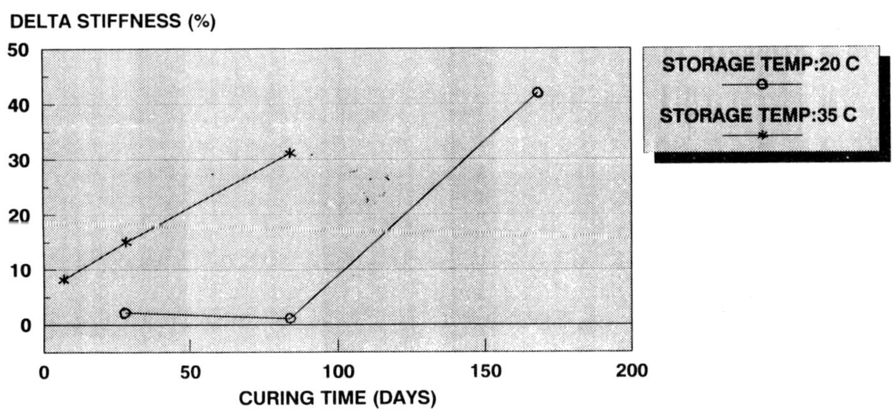

Fig. 6. Change in stiffness with time, hot rolled asphalt, Cariphalte DM (SBS modified) specimens.

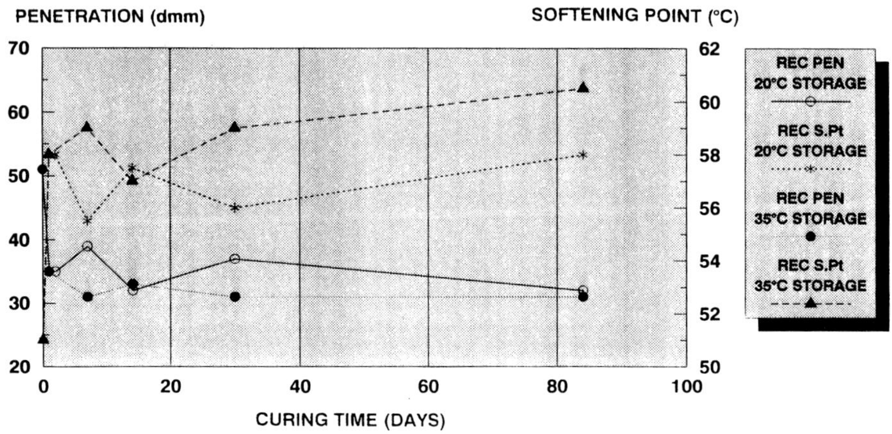

Fig. 7. Recovered binder properties (recovered pens and softening points) for hot rolled asphalt, 50 pen specimens.

trend of bitumen hardening is not apparent at either storage temperature, although an increase in softening point has been established after 84 days notably at the higher storage temperature. This suggests that the significant influences on the curing process for a given mix formulation are dependent upon more parameters than oxidation of the bitumen. The phenomenon of steric hardening, or development of molecular structure which is a reversible process and would not be identified from recovered properties, would appear to contribute to the enhancement of mix stiffness. However, it must be remembered that after six months laboratory curing at 20°C, a trend of increasing stiffness of Marshall briquette sized rolled asphalt specimens was not observed.

The assessment of the recovered binder properties from the modified mixtures highlighted the difficulties which may result from multi-component polymer bitumen systems. The recovered penetrations and softening points from the Cariphalte DM are illustrated in figure 8 where the apparent randomness of results does not form any correlation with the trend of increasing mix stiffness. Because polymer modified binders are complex systems, the breakdown of the original two phase system during the recovery process is unlikely to be re-constituted into the original format subsequently. The irregularities shown in figure 8 confirm this hypothesis where neither penetrations or softening points conform to any pattern at all.

6.2 Resistance to Permanent Deformation

The resistance to permanent deformation of a rolled asphalt mixture could be expected to be influenced through the material curing, as the structural integrity of the mix is highly dependent upon the binder/sand/filler mortar. Figure 9 shows the development of the deformation resistance of the specimens as quantified by the gradient of the final third of the creep curve obtained from the Repeated Load Axial Test. This parameter is referred to as the RLAT rate (analogous to a wheel tracking rate, i.e. deformation divided by time.)

The 50 pen specimens exhibit a general reduction in RLAT rate which appears to become asymptotic at around 0.65 after 28 days storage at both temperatures. The 7 day results show a marked increase in RLAT rate from the 24 hour values, which are consistent at both storage

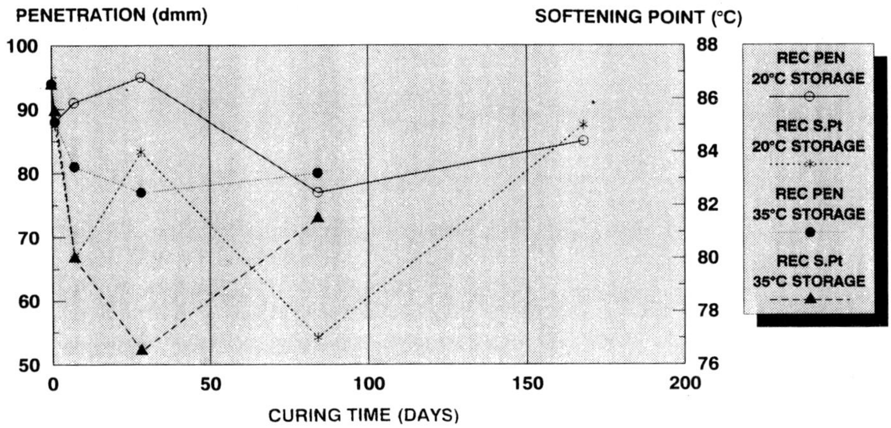

Fig. 8. Recovered binder properties (recovered pens and softening points) for hot rolled asphalt, Cariphalte DM (SBS modified) specimens.

temperatures, but are considered to be anomalous results within the bands of scatter which may occur.

The Cariphalte DM specimens have exhibited a very constant RLAT rate of around 0.35 at both storage temperatures throughout the duration of the investigation. These results are consistent with previously observed performance of asphalts containing SBS polymer modified binders using the wheel tracking tests where deformation resistance is increased by a factor of 2-3 compared to conventional 50 pen materials.

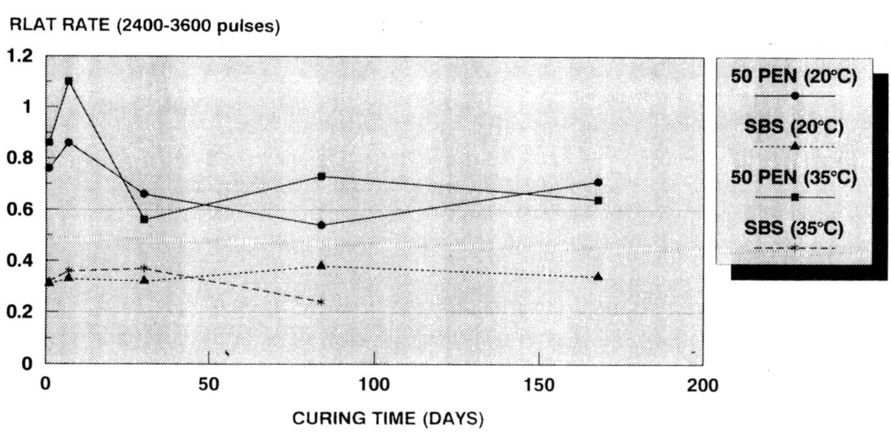

Fig. 9. Curing effect with respect to deformation resistance (RLAT rates), 50 Pen and Cariphalte DM (SBS modified) specimens.

7 Conclusions

From the work carried out on the performance and ageing characteristics of laboratory prepared hot rolled asphalt specimens incorporating conventional and SBS modified binders, the following observations have been made:

1. A large number of specimens is required for testing in order to obtain a representative value of stiffness for a given mix formulation.
2. At a storage temperature of 20°C, specimens of hot rolled asphalt incorporating 50 pen bitumen have not exhibited a trend of increasing stiffness with time.
3. At an elevated storage temperature of 35°C, specimens of hot rolled asphalt incorporating 50 pen bitumen are demonstrating an increasing stiffness after a six month storage period.
4. Mixtures incorporating SBS modified binder have exhibited a trend of increasing mix stiffness with respect to time.
5. Recovered binder properties from polymer modified systems are inappropriate to explain changes in mix characteristics.
6. Deformation resistance of specimens incorporating 50 pen bitumen exhibits a marginal enhancement after 28 days after which a constant value is achieved.
7. No significant change in deformation resistance of the polymer modified specimens has been observed, but they are consistently better than conventional material.

Acknowledgements
The author would like to thank the staff at Thornton Research Centre, in particular Mess J. Ward and D. Webster who carried out the laboratory work and the partners in the Link Research Program at the University of Nottingham.

8 References

British Standards Institution1993. Draft for Development 213. Method for determination of the Indirect Tensile Stiffness Modulus of Bituminous Materials. London.
British Standards Institution BS 598 pt. 107 1992. Sampling and Examination of Bituminous mixtures for Roads and other Paved Areas. Part 107 Method of test for the determination of the composition of design wearing course rolled asphalt. London.
British Standards Institution. Draft for Development DD185 1993. Methods for assessment of resistance to permanent deformation of bitumen aggregate mixtures subject to unconfined iniaxial loading." London.
Cooper, K.E. and Brown, S.F. 1989. Development of a Simple Apparatus for the Measurement of the Mechanical Properties of Asphalt Mixes. Proc. Eurobitume Symposium, Madrid pp 494-498.
Goodrich, J.L. 1988. Asphalt and Polymer Modified Asphalt Properties Related to the Performance of Asphalt Concrete Mixes." Proc. AAPT Vol. 57 pp 116-160.
Nunn, M. and Bowskill, G. 1992. Towards a performance specification for bituminous roadbase." 7th International Conference on Asphalt Pavements Vol. 3 pp 266-279 Nottingham.

Preston, J.N. 1991. Using an SBS bitumen to minimise road repairs. Highways and Transportation Vol. 38 No. 12.

Vonk, W.C. and Gooswilligen, G. 1989. Improvements of paving grade bitumens with SBS polymers." Proc. Eurobitume Symposium, Madrid pp 298-303.

Whiteoak, C.D. 1989. Shell Cariphalte DM. An SBS modified bitumen." Shell Bitumen Review 64 pp 2-5.

9 THE RELATIONSHIP BETWEEN AGGREGATE MINERALOGY AND ADHESION TO AGGREGATE

A.R. WOODSIDE, W.D.H. WOODWARD and
T.E.I. RUSSELL
Department of Civil Engineering and Transport, University of Ulster, Carrickfergus, Northern Ireland
P.R. PEDEN
Tennants Tar Distillers, Northern Ireland

Abstract
This paper considers the relationship between aggregate and adhesion to bitumen. Traditionally, the role of bitumen has been thought to be of more importance. However, work carried out as part of the SHRP Programme in America has shown that failure of the bond is more frequently due to the aggregate. The Net Adsorption test was developed as a simple laboratory method to optimize the match of aggregate and bitumen. The development of this method is first summarized. An investigation of aggregate / binder characteristics in relation to required engineering properties and aggregate mineralogy then is detailed.
Keywords: Aggregate, Bitumen, Net Adsorption Test, Adhesion.

1 Introduction

The detrimental effect of moisture on bituminous materials has long been of concern to highway engineers. Much work has been carried out in an attempt to understand its effect and to develop methods for determining the moisture sensitivity of bitumen / aggregate combinations prior to their use in highway construction.

In 1986, the Strategic Highway Research Programme (SHRP) identified the problem of moisture damage to pavements as one of six major distress areas for investigation. The resulting investigation considered the fundamental processes of both bitumen adsorption to the aggregate and subsequent deterioration of that adsorption in the presence of water. Arising from this has been the development of the Net Adsorption Test (NET) as a relatively fast and simple test method which may be used to quantify the adsorptive nature and water sensitivity of bitumen / aggregate combinations.

Based on fundamental research the test has produced evidence which suggests that rather than attributing adhesion failure to

Performance and Durability of Bituminous Materials. Edited by J.G. Cabrera and J.R. Dixon. © E & FN Spon.
Published in 1996 by E & FN Spon, 2–6 Boundary Row, London SE1 8HN. ISBN 0 419 20540 3.

be solely a "bitumen problem" aggregate characteristics may be more important than previously acknowledged. The Authors consider this to be important as it is contrary to common belief held within the industry which has traditionally tended to blame the bitumen for inservice failure. Rather, the work by SHRP has suggested that factors other than those of the bitumen may be of prime importance and have the major controlling influence on inservice adhesion.

However, the Authors consider that the interpretation of the Net Adsorption data carried out during the SHRP programme was limited in its practical application to the highways industry. SHRP used variables such as bulk aggregate chemistry and total surface area to represent the quality of the aggregates used in the analysis. Whilst aggregate / bitumen adhesion is certainly related to this "chemical approach" the SHRP interpretation failed to consider aggregate mineralogy and the "engineering requirements" necessary for successful in-service performance.

The Authors consider that understanding of these conditions in relation to the information obtained by the Net Adsorption test to be of importance to engineers and the industry in general so allowing the method to be used as a specification requirement. Within the limits of this paper, the Authors shall review its development and discuss the implications of its results in relation to aggregate mineralogy and its engineering properties.

2 The SHRP Investigation

The degree to which a bitumen is adsorbed onto and / or desorbed from the surface of an aggregate is related to its susceptibility to stripping problems. The work carried out by Curtis et al (198) as part of the SHRP research contract attempted to answer a number of basic questions. These included what are the mechanisms of stripping and can one anticipate from simple adsorption and desorption measurements made with aged and un-aged bitumen on different aggregates the likelihood that stripping will occur on a given aggregate and by what mechanism does it occur.

This initial research suggested two mechanisms of stripping. In both cases, water is drawn through the bitumen surrounding the aggregate and unto the bitumen / aggregate interface. This was found to decrease the size of the Gibbs free energy and so reduce the strength of the bond between the bitumen and aggregate.

In the first mechanism, bitumen is pulled free off the aggregate surface by tensile stresses in the bitumen generated by traffic or environmental stresses. The second mechanism is caused by repeated tensile stressing in the bitumen which initiates and then propagates a debonding crack along the bitumen-aggregate interface.

This work also showed specificity in the interactions that occurred between bitumen and aggregate. Further analysis indicated that the amount of a particular bitumen adsorbed or desorbed on an aggregate was dependent upon its bulk and surface chemistry and morphology. It also indicated that the influence of aggregate on adsorption and desorption was much greater than that of the bitumen.

The promising applicability of this initial research was a method which could rank new materials or different combinations of materials. The simple method developed allows the highway engineer to evaluate the affinity of bitumen for aggregate and to determine the water sensitivity of a given bitumen / aggregate pair.

The general test method devised was composed of three parts. First bitumen is adsorbed onto aggregate from a bitumen / toluene solution, then a small amount of water was added to the toluene solution and the adsorbed bitumen that is susceptible to the presence of water is desorbed from the aggregate. Finally the amount of bitumen remaining on the aggregate was determined. This was termed the Net Adsorption and gave a measure of the affinity of bitumen to an aggregate. The difference between the amount of bitumen adsorbed before and after desorption by water gives an indication of the water sensitivity of the bitumen / aggregate pair. The first practical version of the Net Adsorption test was released as SHRP Standard Method #1013 - Measurement of Initial Asphalt Adsorption and Desorption in the presence of Moisture (1990).

However, the original draft version of the Net Adsorption test posed a number of problems as regards the laboratory equipment required ie, the specially adapted chromatography columns and the cost of the peristaltic pumps used to circulate the solution around the system. Therefore a much simpler version of the was devised using routine laboratory equipment.

3 University of Ulster Modified NET method (NETUUJ)

Further developmental work at the University of Ulster, by Woodside et al (1993) has modified the method of expressing the test results. Although the SHRP method of expressing the results is effective in illustrating the moisture sensitivity of the bond it does not give a clear representation between the amount of bitumen initially adsorbed and the quantity desorbed after the addition of water. By re-evaluating the results to express the Ai and An as a percentage of the total bitumen in the solution

(%AiUUJ and %AnUUJ respectively) a more discriminating assessment of affinity and resistance to stripping is possible.

4 Experimental Detail

The investigation detailed in this paper considers the variation in aggregate / binder bond between a wide range of rock-types used for highway construction and a single type of bitumen. This was done to ensure that any variation could be accounted for by the aggregate component.

4.1 Aggregates Used

The aggregates used in the investigation where chosen to represent both a wide range of bulk chemistry composition, mineralogy and engineering properties. Table 1 shows a selection of engineering test data for some of the aggregates assessed in the investigation. It should be stressed that this is a select list.

Table 1. Engineering test data of aggregates

Rock-type	ssdRD	WA	AAV	PSV	MSSV
Carboniferous Sandstone A	2.66	1.15	10.9	72	85.1
Carboniferous Sandstone B	2.68	1.50	3.45	57	80.7
Carboniferous Sandstone C	2.53	3.50	11.30	69	67.1
Silurian Greywacke A	2.68	0.60	5.9	67	98.5
Silurian Greywacke B	2.74	0.49	5.12	64	97.2
Silurian Greywacke C	2.71	0.50	4.04	62	97.0
Silurian Shale	2.79	1.03	12.90	62	85.2
Carboniferous Limestone A	2.66	0.36	4.70	51	98.1
Carboniferous Limestone B	2.74	1.38	13.40	40	77.6
Carboniferous Limestone C	2.72	0.79	7.90	54	93.7
Granite A	2.65	0.55	4.00	54	99.2
Granite B	2.66	0.80	3.50	51	98.1
Granite C	2.62	0.55	4.39	52	97.4
Crushed Flint	2.59	1.54	0.50	38	99.6
Crushed Quartz	2.68	0.58	4.50	51	98.1
Felsite	2.66	0.55	0.64	46	99.3
Basalt A	2.94	1.65	11.4	58	92.9
Basalt B	3.01	1.10	8.9	59	94.7
Basalt C	2.89	5.75	13.7	62	75.0

4.1.1 Discussion of Test Data

The sandstones and greywackes represent what is currently regarded by the industry as high quality surfacing aggregate and are typically characterized by their high degree of skid-resistance as determined by the PSV test. The aggregates assessed include Silurian greywackes and shales, Devonian and Carboniferous sandstones. While all possess a high level of skid-resistance, > 60 PSV, they differ widely in composition and engineering requirements such as strength and soundness.

The basalts assessed were all Tertiary in age and typically vary with regard to their soundness. Although the use of granite as surfacing aggregate is limited due to low PSV values, they may be used as a dilutent with darker coloured surfacing aggregates. Traditionally, their use with bitumen has often resulted in stripping problems.

A wide range of Irish Carboniferous limestones have been assessed. Although they typically have low PSV values, they are an important source of aggregate for bituminous mixed materials and are chemically supposedly the extreme of the silica rich aggregates. A range of very hard aggregates including quartz, flint and felsite have been assessed. Typically these may be categorized as being fine grained, or having a high silica content, a low PSV and a tendency to show stripping related failure.

The aggregate test results indicate considerable variation in quality and acceptability for use as surfacing aggregate. They show that acceptance of an aggregate cannot be made on a single test property, for example PSV, but rather on a combination of results.

4.2 Bitumen Used in the Investigation

Although the test method allows the use of any type of bitumen, this basic investigation of assessing the influence of rock-type used a 200 pen bitumen containing 0.2% adhesion agent. Although not detailed in this paper, other types of bitumen assessed using the NET test ranged from differing sources of penetration grade bitumen to polymer modified bitumens. Work by Walsh et al (1994), in the South of Ireland, is currently developing modified methods for assessing cut-back and emulsion bitumens.

4.3 Solvent

The solvent used in the Net Adsorption Test is HPLC grade Toluene.

4.4 Aggregate Test Sample Grading

The SHRP Net Adsorption test method does not specify a grading requirement for the test sample. Jamieson, of the National Roads Authority, Dublin; and who was involved in the SHRP programme, propsed the use of the No 4 grading as specified in ASTM D 3515 for Sand Asphalt. Table 2 details the required masses of individual sizes required to make up the 50 g test samples.

Table 2. Aggregate test grading used in Net Adsorption Test

BS 410 Sieve Size (mm)	% Retained	Mass of Aggregate (g)
5 - 2.36	8.3	4.15
2.36 - 1.18	25.0	12.50
1.18 - 0.6	16.7	8.35
0.6 - 0.3	23.9	11.95
0.3 - 0.15	13.3	6.65
0.15 - 0.075	6.1	3.05
< 0.075	6.7	3.35

As this is a continuously graded material, a centre specification grading includes all the range of sizes produced during crushing. Although this grading includes material of 4.75 to 9.5 mm in size, this was not part of the grading used for test samples.

4.5 Preparation of NET Test Aggregate Samples

The less than 5 material required for the NET test was obtained by crushing 10 / 14 mm sized chippings in the laboratory. This enabled the NET data to be directly compared to the standard engineering test properties obtained on the 10 / 14 mm size.

4.6 Test Method

A stock solution of approximately 1g bitumen \ 1 litre HPLC toluene was produced for the bitumen to be assessed. This was shaken to ensure that the bitumen was completely dissolved, 4 ml removed, diluted to 25 ml with HPLC toluene and a spectrophotometer reading at 410 nm taken. This was the initial absorbance reading, A1. For each aggregate, two test samples and a control were assessed. The control consists of 140 ml of HPLC toluene (instead of 140 ml stock solution) and 50 g of aggregate. This was included to ensure that the absorbance readings taken during the test were not effected by the toluene or the aggregate. 140 ml of stock solution was then added to each flask containing 50 g of aggregate. These were then placed on a shaker table and shaken for an equilibrium period of six hours at a rate of 200 rotations \ minute. They were then left to settle before 4 ml of solution was sampled. This was again diluted to 25 ml with HPLC toluene before being filtered to remove any fine aggregate particles present. An absorbance reading was then taken (A2) from which the amount of bitumen adsorbed by the aggregate was then calculated.

2 ml of water was added to each flask shaken overnight. After being allowed to settle the same procedure of removing 4 ml of solution was followed and the absorbance reading after the introduction of water to the system (A3) recorded.

The following calculations were used. The subscripts SHRP and UUJ are used to differentiate between the SHRP and University of Ulster methods of expressing the results.

Initial Adsorption (AiSHRP) i.e. the amount of bitumen initially adsorbed onto the aggregate surface:

$$AiSHRP = VC\ (A1 - A2)\ /\ WA1 \quad (1)$$

Net Adsorption (AnSHRP) i.e. the amount of bitumen remaining on the aggregate after water is added

$$AnSHRP = VrC\ (A1 - A3)\ /\ WA1$$

% Net Adsorption (%AnSHRP) i.e. the percentage of bitumen remaining on the aggregate after the test

$$\%\ AnSHRP = (An\ /\ Ai) \times 100$$

Maximum Absorbance Reading (Amax) i.e. theoretical absorbance reading when all bitumen in solution has been absorbed by the aggregate

$$Amax = VC\ (A1 - 0)\ /\ WA1$$

Initial Adsorption UUJ (%AiUUJ) i.e. re-evaluation of SHRP data to determine the percentage of bitumen initially adsorbed onto the aggregate surface

$$\%AiUUJ = (Ai\ /\ Amax) \times 100$$

% Net Adsorption UUJ (%AnUUJ) i.e. re-evaluation of SHRP data to determine the percentage of bitumen remaining on the aggregate after the test

$$\%AnUUJ = (An\ /\ Amax) \times 100$$

Where:
Ai = Initial adsorption, mg / g
V = Volume of solution in the flask at the time Aa is obtained, (normally 140 ml)
W = Aggregate weight, in grams
C = Initial concentration of bitumen in solution, g / l
A1 = Initial absorbance reading
A2 = Absorbance reading after 6 hours
A3 = Absorbance reading after addition of water
An = Net adsorption, mg / g
Vr = Volume of solution in the flask at the time A3 is obtained, (normally 136 ml)

Amax = maximum absorbance reading

5 Discussion of NET Test Data

The SHRP investigation recommended the limits for performance shown in Table 2.

Table 3 SHRP NET Recommendations for Aggregate / Binder Adhesion Performance

Percent Net Adsorption % An (SHRP)	Aggregate / Binder Bond Performance
> 70	Good
55 - 70	Marginal
< 55	Poor

Table 3 shows a summary of the initial, net and percentage net adsorption values, both for the SHRP method and revised University of Ulster method for a select list of aggregates using 200 pen bitumen.
As the only variable to alter during the NET test was the aggregate, the variation in test values may be accounted for by the resulting combination of aggregate and bitumen.

Table 3 NET Test Data

Rock-type	Ai SHRP	An SHRP	%An SHRP	%Ai UUJ	%An UUJ
Carboniferous Sandstone A	2.198	1.834	83.4	74.9	62.4
Carboniferous Sandstone B	1.970	1.567	79.6	69.4	55.4
Carboniferous Sandstone C	1.922	1.444	75.4	67.9	51.2
Silurian Greywacke A	2.087	1.612	77.2	67.8	52.7
Silurian Greywacke B	1.935	1.347	69.7	68.0	47.4
Silurian Greywacke C	2.387	1.515	62.8	81.3	51.1
Silurian Shale	2.140	1.620	75.7	75.2	56.8
Carboniferous Limestone A	1.653	1.344	81.3	58.9	48.6
Carboniferous Limestone B	1.680	1.430	85.1	59.0	50.4
Carboniferous Limestone C	1.722	1.355	77.3	61.2	47.8
Granite A	1.580	1.250	79.4	55.8	44.3
Granite B	1.680	1.370	81.2	59.6	40.4
Granite C	1.470	1.230	83.8	48.3	40.5
Crushed Flint	1.683	1.466	87.1	59.4	51.8
Crushed Quartz	1.308	1.173	89.8	48.2	43.2
Felsite	1.652	1.205	73.0	58.0	42.5
Basalt A	2.166	1.598	73.8	75.2	55.5
Basalt B	1.643	1.393	84.8	52.2	44.2
Basalt C	2.592	2.298	88.6	89.1	78.9

Initial adsorption values ranged from 1.308 for crushed quartz to 2.592 for unsound basalt. Values as high as 2.592 have been obtained for lateritic basalt not shown.

The % Net Adsorption values ranged from 62.8 to 89.8%. Based on the SHRP recommendations this would indicate all of the aggregates shown, except two of the Silurian greywackes, to have good aggregate / binder bond performance. This the Authors believe highlights a problem with the SHRP method of evaluation as it misleadingly shows the granites, quartz, flint and felsite to have the very good values whereas in-service most suffer from adhesion related problems.

However, if the results obtained by the UUJ re-evaluation are used, i.e. %AiUUJ and %AnUUJ, a much different picture is obtained. These results are expressed as a percentage of the maximum adsorption value possible. The results obtained indicate the amount of bitumen adsorbed onto the aggregate from the solution and the amount of desorption due to the introduction of water.

Quite different results may be obtained for the two methods. For example, Granite B gives a %AnSHRP of 81.2% whereas the recalculated UUJ figure shows only 48.4% of the bitumen in solution to be adhered to the aggregate.

Based on the results shown in Table 3 and additional analyses not shown, there would appear to be a ranking based on rock-type i.e., from the fine grained and / or high silica types such as quartz, felsite, flint and granite, the limestones, basalts depending on soundness to the greywackes and sandstones.

If %AiUUJ and %AnUUJ are plotted the resulting data show a linear trend along which rock-type may be grouped in fields. This is shown in Figure 1. Although overlap exists, it is suggested that aggregates to the lower extreme of individual fields ie to the left, tend to be better quality / more sound. As the aggregates move to the right they become more unsound / poorer in overall quality.

Another trend is with PSV values ie they appear to increase to the right for particular groups, the opposite for other aggregate properties i.e. as PSV increases for a particular rock-type, the likelihood for adhesion problems increases.

In terms of achieving adhesion, high PSV aggregates require more bitumen or the use of an aggregate where the degree of desorption is less. For example, the use of a bitumen incorporating the use of adhesion agents or the use of polymer modified bitumen. The NET test can easily determine the effect of this. However, this aspect is outside the scope of this paper.

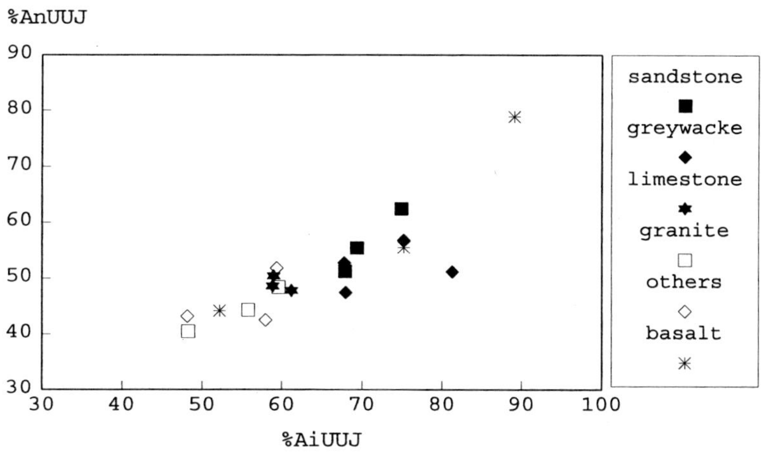

Fig.1. %AiUUJ - %AnUUJ for different rock-types

6 Aggregate Factors Controlling the Aggregate / Binder Bond

Traditionally, adhesion has been attributed to an aggregates chemical composition with high silica and carbonate representing opposites as regards to adhesion. Although this trend generally applies not all aggregates follow it. For example, certain pure quartzites may give good adhesion whereas certain Irish limestones may give stripping problems. Clearly, based on these observations and indeed using the NET data, there would appear to be factors other than simple chemical composition at play.

The NET investigations have shown Irish limestones to have similar adhesion characteristics to acid igneous rocks, thus contradicting existing assumptions. From the results, it is clear that even without knowing the mineralogical composition of the aggregates, there appears to be a ranking of results dependent on geological factors.

The sedimentary rocks can be separated into two groups i.e. the gritstones and limestones. The gritstones include shale, greywacke and sandstone and because of there high skid-resistance properties are used as surfacing aggregate. They also show some of the best adhesion values, with %AiUUJ values ranging from 60 and 80% and %AnUUJ values between 45 and 65%. However, it can also be seen that certain of these aggregates have a tendency to strip in the presence of moisture.

The limestones were shown to have much poorer adhesion than the gritstones tested. The flint and quartz also showed poor adhesion properties.

The igneous rocks tested show the largest range of values with there being a notable difference between the basic and acid types. The values shown in Table 2 relate to basalt which vary considerably in soundness (see Table 1) as indicated by the presence of increasing secondary clay mineral content. These are highly absorbent in nature and may account for the high values obtained for Basalt C, a heavily altered basalt. The inclusion of this very unsound basalt illustrates that the NET test data must be considered in conjunction with other test results.

Whereas acid igneous rocks such as granite and felsite gave low %AiUUJ values of between 40 and 60% they also had relatively small amounts of stripping. However, this could be due to there being less bitumen to strip

7 X-Ray Diffraction Analysis

X-ray Diffraction Analysis was carried out on a range of aggregates. The results are shown in Table 4. Seven had their < 0.075 mm composition compared to that of the whole rock to determine if the finer grained matrix, clays or weaker minerals in the rocks were concentrated in the dust. If this was found to be true the various weight fractions used in the test would have had to be modified; ensuring that the test specimen had the same composition as the whole rock to allow direct correlation.

Table 4. XRD Analysis

Ref.	Aggregate	Sample Type	Mineralogy
1, 2	Lateritic basalt	Whole rock and < 0.075mm	Feldspar, Augite, Olivine, Saponite, Thomsonite, Chabazite
3, 4	Basalt C	Whole rock and < 0.075mm	Feldspar, Chabazite, Haematite, Saponite
5, 6	Basalt A	Whole rock and < 0.075mm	Feldspar, Augite, Olivine, Saponite, Chlorite
7, 8	Sandstone B	Whole rock and < 0.075mm	Feldspar, Muscovite, Quartz, Chlorite
9, 10	Sandstone C	Whole rock and < 0.075mm	Feldspar, Muscovite, Quartz
11, 12	Greywacke C	Whole rock and < 0.075mm	Feldspar, Muscovite, Quartz, Illite
13, 14	Limestone A	Whole rock and < 0.075mm	Calcite, Ferroan dolomite, Quartz
15	Felsite	< 0.075mm	Albite, Anorthite, Quartz

Interpretation of the XRD data indicated that there was not any significant mineralogical differences between the two sizes so allowing the NET results to be directly correlated to the road aggregate with no modification of the test procedure required.

The XRD results indicate that there seems to be a relationship between adhesion and mineralogical composition of the aggregate. The composition of the three basalts was similar despite considerable variation in engineering properties as shown in Table 1. The main difference is the presence of secondary zeolite minerals in the vesicular basalt and hematite and related minerals in the laterite. Saponite, a vein mineral, is found in both the vesicular and sound basalt

Sandstone C was shown to be much purer than Sandstone B. The former was almost entirely composed of quartz with small amounts of muscovite mica, albite and anorthite present. Whereas Sandstone B included all of the these minerals in larger amounts as well as vermiculite \ illite. Greywacke C had similar mineral constituents to Sandstone B, although the actual mineral percentages varied.

Limestone A was found to significantly different to any of the other aggregates tested, primarily due to the presence of calcite and ferroan dolomite. As limestones are usually expected to give better results than those obtained for this particular sample, another factor must be involved. This may be explained by the presence of large amounts of negatively charged impurities which cause the reduction of adhesion. Analysis of this aggregate showed it to have unusually large amounts of quartz present for a limestone. Interaction between this quartz and the calcite may have caused the reduction in adhesion. However, this can not be simply explained by the presence of quartz as Sandstone C gave good adhesion.

The felsite contained quartz, albite and anorthite. Although its composition did not greatly vary from those of some of the other rocks, it also, like the limestone showed poor adhesion qualities. This may be due to the lack of any ferromagnesium minerals present and possibly its crystalline texture.

However, the reasons given to explain the poor adhesion characteristics associated with felsite do not seem to have similar detrimental effects for other rock-types. For example, the sandstones have minimum ferromagnesium minerals and a high quartz and feldspar content, i.e. the factors suggested as reasons for poor adhesion in acid igneous aggregates. They do however have a rough surface texture. Alternatively, the basalt's, although having a low quartz and high ferromagnesium content also have a rough crystalline texture. The NET results for pure quartz showed poor adhesion as expected for an aggregate with a high percentage of quartz. However in comparison, the high quartz Sandstone C gave good adhesion. The difference between the two is that the Sandstone C is composed of individual quartz grains within a matrix, whereas the quartz has a crystalline texture.

These conflicting phenomena for differing aggregates suggests that the aggregate / binder bond exists as a complex inter-relationship between such factors as mineralogy, surface texture characteristics and binder type.

8 Conclusions

The Net Adsorption Test is regarded by the Authors as a means of optimizing the use of aggregate and bitumen in highway construction. The work shown in this paper has indicated its potential as the first simple method to quantify both the initial aggregate / binder bond and how this performs in the presence of moisture, thus optimising the usage of available materials.

The results indicate that for a single type of bitumen large variations are possible. As the bitumen remained the same, then this variation must be attributed to the unique interaction which occurs between bitumen and aggregate in each case. The implication of this is that all types of aggregates cannot be treated in the same way, i.e. that a single national specification limit is not the best way to govern the use of these materials.

9 Acknowledgments

The Authors wish to acknowledge the help and co-operation given by Dr Ian Jamieson, National Roads Authority, Dublin; Dr Margaret O'Mahoney and Geraldine Walsh, Trinity College, Dublin and Queens University Geology Department.

10 References

Curtis, C.W. Lytton, R.L. Brannan, C.J. (1992) Influence of Aggregate Chemistry on the Adsoprtion and Desorption of Asphalt, paper presented at the 71st Annual Meeting of the Transportation Research Board, Washington, D.C.

Curtis, C.W. Stroup-Gardiner, M. Brannan, C.J. Jones, D.R. (1992) Net Adsorption of Asphalt on Aggregate to Evaluate Water Sensitivity, paper presented at the 71st Annual Meeting of the Transportation Research Board, Washington, D.C.

Walsh, G. O'Mahoney, M. and Jamieson, I. (1994) The SHRP Net Adsorption Test: application to Irish surface dressing aggregates and binders, paper presented at the Third International Symposium on Highway Engineering, University of Ulster.

Woodside, A.R. Woodward, W.D.H. Russell, T.E.I. (1993) The Net Adsorption Test, paper presented at Trinity College Dublin.

PART THREE
DESIGN AND PERFORMANCE

10 HOT BITUMINOUS MIXTURES – DESIGN FOR PERFORMANCE

J.G. CABRERA
Civil Engineering Materials Unit, Department of Civil Engineering, University of Leeds, Leeds, UK

Abstract

Modern design of bituminous mixtures involves not only compliance with structural requirements but also compliance with performance requirements during service.

Ensuring satisfactory in service performance and therefore durability is not easy because durability is not a measurable property of a material but rather an attribute which can only be defined as a function which relates the level of performance to the service life. This function is obtained by controlling performance related properties and appropriate environmental variables associated with the exposure conditions for which the design is being prepared.

Researchers in the field of bituminous materials agree that performance can be assessed by one or more of the following properties: permeability, porosity, adhesion of binder to mineral aggregate and long term deformation (creep).

Hot bituminous mixes have to be handled, placed and compacted at high temperature. Some mixes cannot be satisfactorily compacted at lower temperatures and therefore the "in place" resultant mix has low density, high porosity and high permeability.

Assessment of the workability of a bituminous mix during the laboratory design stage should be a requirement of any method of design not only to avoid difficult mixes in terms of handling and compacting but also to allow the optimisation of design by the use of appropriate selected components.

This paper presents a discussion of the Leeds Design Method (LDM) which is applicable to hot bituminous mixtures, i.e. hot rolled asphalt, dense bitumen macadam and asphaltic concrete. It presents results comparing the optimum binder content obtained by the LDM and by the conventional BS method. It also shows the advantages of using a workability parameter during the design stage for optimisation purposes.

1 Introduction

Modern design of bituminous mixtures involves not only compliance with structural requirements but also compliance with performance requirements during service. Ensuring structural adequacy is obtained by specifying minimum values of the known engineering parameters like strength, strain, density and porosity. Ensuring satisfactory in service performance and therefore durability is not easy because durability is not a defined property of a material but rather an attribute which can only be defined as a function which relates the level of performance to the service life. This function is obtained by controlling performance related properties and appropriate environmental variables associated with the exposure conditions for which the design is being prepared.

Performance and Durability of Bituminous Materials. Edited by J.G. Cabrera and J.R. Dixon.
Published in 1996 by E & FN Spon, 2–6 Boundary Row, London SE1 8HN. ISBN 0 419 20540 3.

Researchers in the field of bituminous materials agree that performance can be assessed by one or more of the following properties: permeability, porosity, adhesion of binder to mineral aggregate and long term deformation (creep). This list does not take into account properties which should be measured to characterise separately the component of a bituminous mix before they are accepted as suitable for design.

The production of hot bituminous mixtures is an energy intensive process. The mineral aggregates are heated to at least 150° C, while the binder has to be maintained in insulating tanks at the same temperature. The hot mix is transported, placed and compacted ideally at the same temperature of mixing, but often the mix is cooled rapidly to temperatures below 120° C. Some mixes cannot be satisfactorily compacted at lower temperatures and therefore the "in place" resultant mix has low density, high porosity and high permeability.

The property which allows the production, handling, placing and compaction of a bituminous mix with the application of minimum energy has been defined as "workability"(1). Assessment of workability during the laboratory design stage should be a requirement of any method of design not only to avoid difficult mixes in terms of handling and compacting but also to allow the optimisation of design by the use of appropriate selected components.

Current British specifications for the design of hot rolled asphalt and dense bitumen macadam(2,3) gives as an alternative a design procedure based on the measurement of stability and deformation by the Marshall test. The control of performance is indirectly based on the minimum levels arbitrarily indicated for the different properties. These are the result of long term experience of the British environmental and road loading conditions. Minimum values of stability and maximum values of deformation (flow) together with a narrow range of porosity has given, on average, materials of adequate performance. However, it is recognised that the BS methods should be improved to take into account performance parameters which can be assessed during the design stage. Furthermore the current methods do not take into account any parameter which could given an indication of the workability properties of a bituminous mix.

This paper presents a discussion of the Leeds Design Method (LDM) which is applicable to hot bituminous mixtures, i.e. hot rolled asphalt, dense bitumen macadam and asphaltic concrete. It present results comparing the optimum binder content obtained by the LDM and by the conventional BS method. It also shows the advantages of using a workability parameter during the design stage for optimisation purposes.

2 The Leeds Design Method

The LDM is based on the following parameters and equipment:

1. Compaction of specimens of a selected mineral aggregate combination and variable binder content using the Gyratory testing machine (GTM).
2. Assessment of workability using the same GTM machine.
3. Measurement of air permeability using the Leeds air permeameter (LAP).
4. Measurement of Stability and flow using the Marshall testing machine.
5. Measurement of static stiffness using the "Canik" testing machine.
6. Measurement of vapour diffusion coefficient using a simple "cup method".

The optimum binder content is obtained as an arithmetic mean of the optimum binder contents obtained by the different measured and calculated parameters. The calculated parameters are:

a) mix density;
b) compacted aggregate density; and
c) porosity.

2.1 The GTM compactor

The development and characteristics of the GTM compactor have been fully described in reference (1). The GTM compactor was developed with the aim of simulating more realistically the action of a roller during compaction in the field. The GTM compactor is shown in Figure 1. During compaction the sample is subjected to a shearing action by a gyratory motion of the steel mould containing the bituminous mixture which at the same time is maintained at constant pressure applied via the steel loading plungers whose faces remain parallel to each other during the shearing action. The lower base plate on top of the jack head is fixed, while the sliding upper head rides against roller bearings such that it can slip as the mould is caused to gyrate around the base plate. This combination of static-dynamic energy of compaction can be varied:

a) by adjusting the axial pressure of the hydraulically controlled lower ram;
b) by adjusting the gyration angle with the adjustable roller; and
c) by changing the number of revolutions (gyrations).

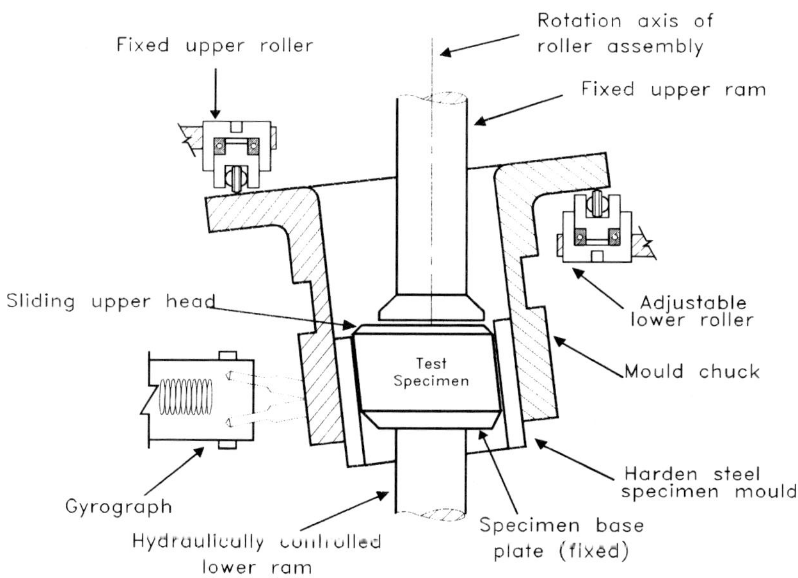

Figure 1 - The GTM compactor

The mould chuck of the GTM compactor is equipped with heaters for the control of temperature of the sample being compacted. It is also instrumented for monitoring the height of the sample while compaction is taking place and it has an automatic counter for controlling the number of revolutions (gyrations of the assembly).

The standard and heavy compaction energy used with the GTM are obtained by a combination of axial pressure, gyration angle and number of revolutions. These were determined after numerous investigations which compared the properties of field compacted cores with laboratory specimens compacted using the GTM compactor (4,5). The compaction parameters are given in Table 1 together with the approximately equivalent energies applied when using the Marshall hammer.

The properties of the specimens compacted by the GTM compactor are substantially different from the properties of similar specimens compacted with the Marshall hammer specially those properties which are performance related. This important aspect of the effects of the mode of compaction will be discussed later.

2.2 Compaction of specimens in the GTM compactor

The procedure to prepare compacted specimens using the GTM is very simple. It consists of the following steps:

a) The mineral aggregates and binder are heated to the appropriate temperature, mixed thoroughly and placed in the GTM mould (diameter 101.6 mm) in sufficient quantity as to produce a compacted specimen of approximately 50 mm height.

Table 1 - Compaction energy parameters for the GTM compactor and for the Marshall hammer

Compaction Level	Axial Pressure, MPa	Gyration angle, degree	Number of revolutions	Marshall hammer No. of blows per side
Normal	0.7	1	30	50
Heavy	1.4	1	30	75

b) The mould is placed in the GTM compactor which has been previously heated to the required temperature. The angle of gyration is adjusted to 1 degree, the dial which controls the height of the specimen is adjusted to zero, and the revolution counter set to 30. Axial energy is applied up to a value of 0.7 MPa or 1.4 MPa and gyration started.

c) After 30 revolutions the mould with the specimen is taken out of the GTM compactor and immersed in cold water until the specimen has reached ambient temperature.

d) The specimen is extruded from the mould, its average height measured and its volume obtained by weighting in air and in water. Compacted aggregate density is also obtained from the knowledge of density, binder content and relative density of the components of the mix.

2.3 Measurement of workability

The data for the measurement of workability of a bituminous mixture are obtained during the compaction of the specimen following the procedure indicated below:

a) After starting the gyration of the specimen as indicated in point b) of the compaction procedure, the specimen height reduction is monitored by reading the height dial gauge at five revolution intervals. These heights are used to calculate the volume of the specimen at 5, 10, 15, 20, 25 and 30 revolutions.

b) The volumes calculated are used to obtain the density and porosity of the specimen at the different levels of gyratory compaction.

c) A graph is plotted relating the porosity at i number of revolutions (Pi) to the log of the number of revolutions (log i). The experimental points should approximate a linear relation of the form:

$$Pi = a - b (\log i) \qquad (1)$$

where:

a,b = constants

The workability of a mixture is expressed by the workability index (WI) which is defined as the inverse of the constant a, i.e. the porosity at 0 revolutions multiplied by 100:

$$WI = 100/a \qquad (2)$$

The workability index can be used effectively to assess the influence of compaction temperature or mix composition, particularly binder content coarse aggregate content, sand morphology and filler type.

Figure 2 shows a typical example of the effect of temperature and binder content on the workability of a standard hot rolled asphalt of 14 mm nominal size suitable for surfacing layers containing 33% coarse aggregate, smooth texture and rounded shape river sand and 10% limestone filler.

Field experience has shown that mixes with a workability index equal or smaller than 6 are difficult to handle and compact.

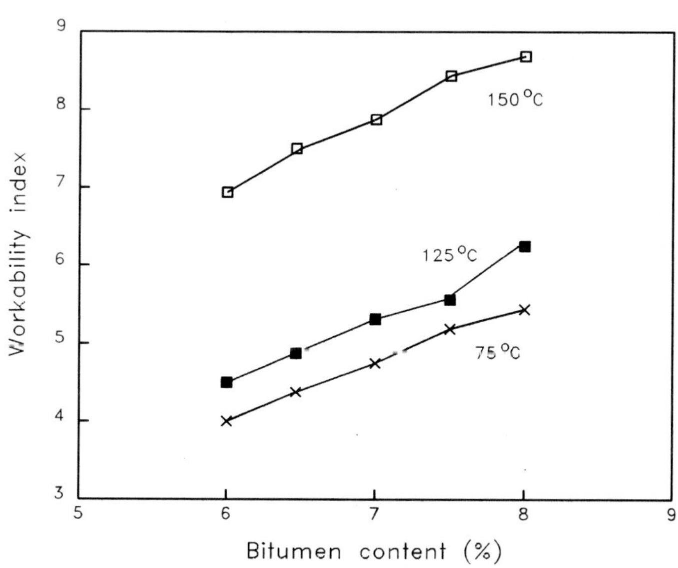

Figure 2 - Influence of binder content and temperature on the workability index of hot rolled asphalt

2.4 Measurement of permeability

The permeability of dense mixtures like hot rolled asphalt and asphaltic concrete is measured using the Leeds Air Permeameter (LAP). This has been fully described in a paper to this conference(6). The test if very effective for detection of variations in binder content and therefore can be used very effectively to control mix composition. The values of permeability are also affected by mode of compaction (see Figure 5 in reference(6). This feature of the test is very useful to assess numerically the differences between laboratory compacted specimens and cores obtained from compaction in the field.

Figure 3 gives an example of the changes in permeability which occur with changes in porosity of a hot rolled asphalt prepared with different fillers (7). Small changes in porosity cause very large changes in permeability. This demonstrates clearly the advantages of using permeability to control porosity and mix composition.

The test is non-destructive and therefore values of permeability are obtained from four identical specimens which are prepared to measurement of other parameters, i.e. stability and creep stiffness.

2.5 Measurement of stability and flow

Values of stability and flow are obtained using the Marshall testing rig which has been fully described in BS 594 (2). The test is part of the BS design alternative. The values of stability and flow give a good indication of the strength properties of a bituminous mixture. It has been suggested that the ratio of stability to flow should be used as an indication of the stiffness of a mix since this ratio correlates quite well with deformation obtained by the wheel tracking test(8). However, the LDM includes direct measurement of creep stiffness and the correlation between the stability/flow ratio with creep stiffness has not been found to be statistically significant.

The current Ministry of Transport specifications (9) regarding minimum values of stability and maximum values of flow are based on conservative appraisal of the performance of hot rolled asphalt in motorways and other paved roads. These limits are given in Table 2. The current values used in USA are also included for comparison (10).

Table 2 - Current Ministry of Transport and Asphalt Institute design criteria

Traffic Level cv/d x 1000	Stability kN	Ministry of Transport Maximum Flow mm	Asphalt Institute Stability kN	Asphalt Institute Flow mm
<2	3	5	2.22	2 - 5
2 - 4	4.5	5	2.22	2 - 4.6
4 - 6	5.5	5	2.22	2 - 4.6
>6	8	4	3.34	2 - 4

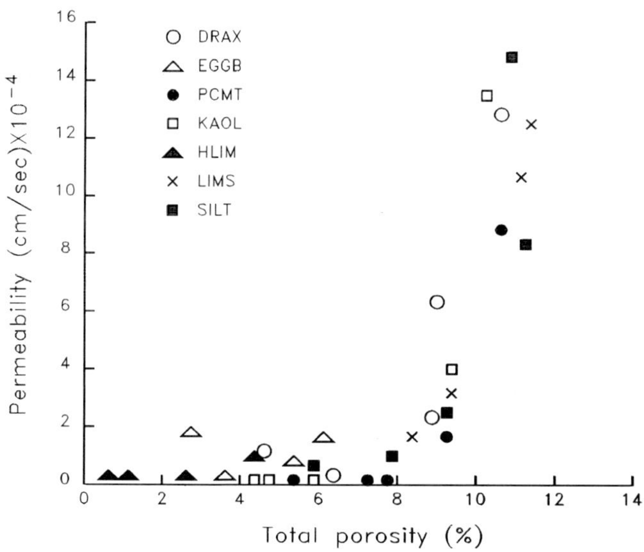

Figure 3. Relationship between air permeability and porosity for a hot rolled asphalt made with different types of filler(7)

2.6 Measurement of static stiffness

The static or creep stiffness is measured using the "Canik" creep testing rig. The Canik machine tests two specimens at the same time. The conditions under which the test is carried out are:

a) Test Temperature 40° C
b) Preloading for 2 minutes at 0.001 MPa
c) Constant stress during test equal to 0.1 MPa
d) Duration of test: 1 hour loading and 1 hour unloading

The rig and its operation has been fully described by Cabrera and Nikolaides (11). The data is obtained manually since the dials are mechanical. A new version is being constructed where the data will be collected via a data logger and processed by computer. The hand operated machine is used for instruction during laboratories and has been found to be very useful as a learning tool. A schematic representation of the Canik machine is shown in Figure 4.

The data obtained from this test is plotted as a relationship between strain and time. This relationship when plotted as a log - log relationship gives a linear equation which can be used to extrapolate values to short times up to 0.1 sec.

The strains measured are used to calculate the creep stiffness values for suitable time intervals and these are plotted as a relationship between creep stiffness of the bituminous mix against the stiffness of the binder at the same time intervals. This equation was suggested by Hills (12) who showed that this relationship is independent of the operating parameters and therefore can be regarded as a genuine characteristic of a mix.

The values of creep stiffness at one hour (Sc1h) are plotted against binder content. If a suitable binder content range has been used the relationship will exhibit a maximum Sc1h value for an optimum binder content. In some instances this optimum value is not apparent. In this situation the relation which is recommended to be used is a plot between the slope of the log creep stiffness of the mix-log stiffness of the bitumen and the binder content from which an optimum binder content can be read.

An example of the logarithmic relationship between creep stiffness of a hot rolled asphalt and log stiffness of the binder used is shown in Figure 5. The Figure also shows the influence of binder content on the relationship.

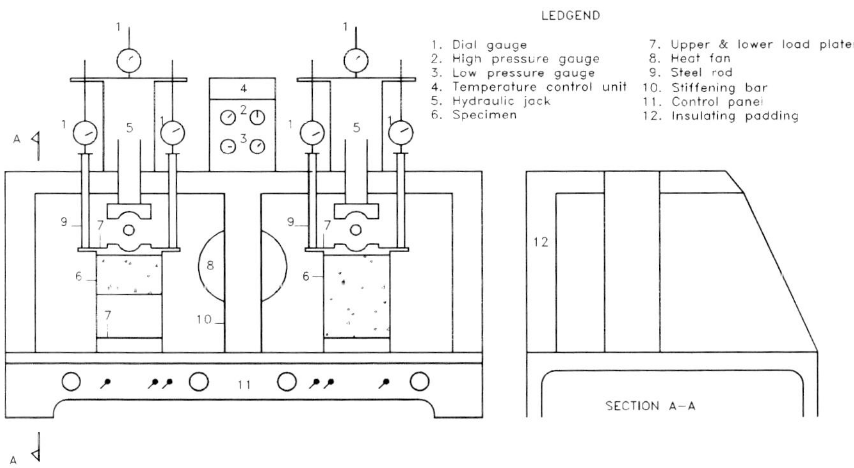

Figure 4. Schematic representation of the "Canik" testing machine(11).

2.7 Measurement of vapour diffusion coefficient

The vapour diffusion test was designed by Cabrera and Al Sayed (13) for the assessment of the resistance of a bituminous mix to water vapour penetration and the consequent bitumen aggregate adhesion loss. Vapour penetration and subsequent damage is a prevalent mode of deterioration in climatic conditions where there are extreme changes of temperature from night to day.

The test is very simple. It is carried out with specimens sliced from the specimens used for the other tests (permeability, stability and stiffness). The dimensions of the specimen for the diffusion tests are: 101.6 mm diameter and 10 mm thickness.

The disk is weighed and placed on top of a suitable plastic container half full of water. The bituminous mix disk is glued to the container by means of silicon rubber glue. The weight of the container, water and disk is recorded and the assembly placed in an environmental chamber maintained at 28 degree C and 5% relative humidity. The container is weighed at suitable intervals up to 8 days.

The water lost plotted against the square root of time gives an linear relationship. An example of the water loss versus square root of time for a hot rolled asphalt containing limestone filler in one case and a filler consisting of limestone plus silt is shown in Figure 6. The slope of this relationship is used to calculate the diffusion coefficient.

It is suggested that diffusion coefficient equal or less than 3 x 10^{-6} cm^2/sec indicate mixtures with adequate resistance to damage due to penetration of vapour and consequent loss of adhesion.

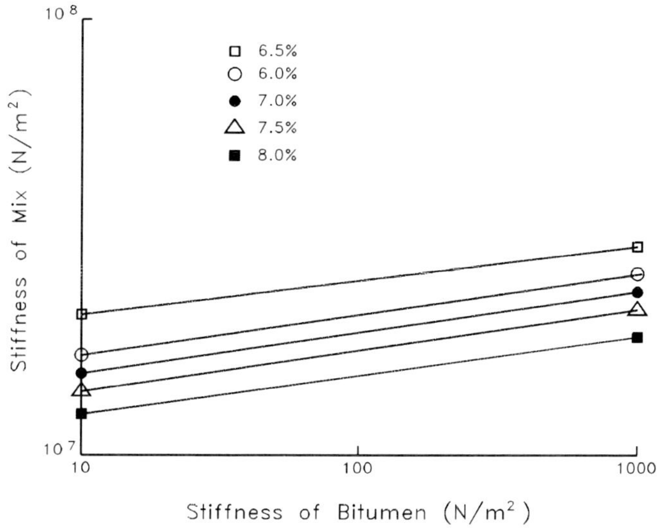

Figure 5. Relationship between Sc of hot rolled asphalt and Sc of its binder. Binder content ranging from 6% to 8%.

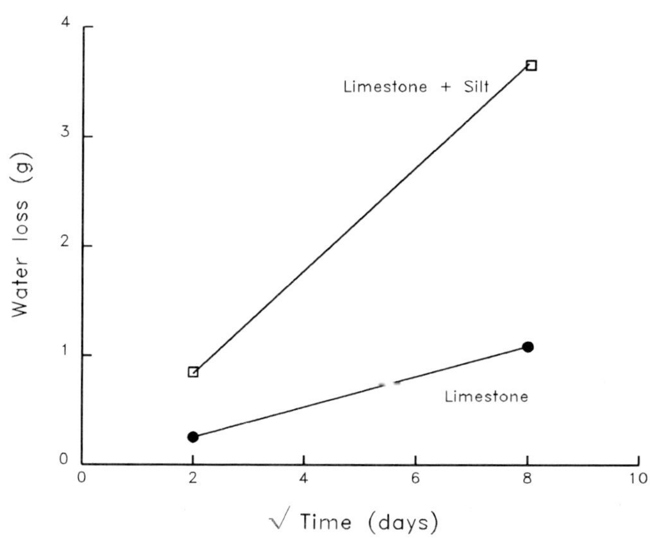

Figure 6. Relationship between water loss by diffusion and square root time for hot rolled asphalt.

3 Determination of the optimum binder content

The optimum binder content is determined as an arithmetic mean of the optimum values obtained from the different tests. An example of this determination is given in Figure 7.

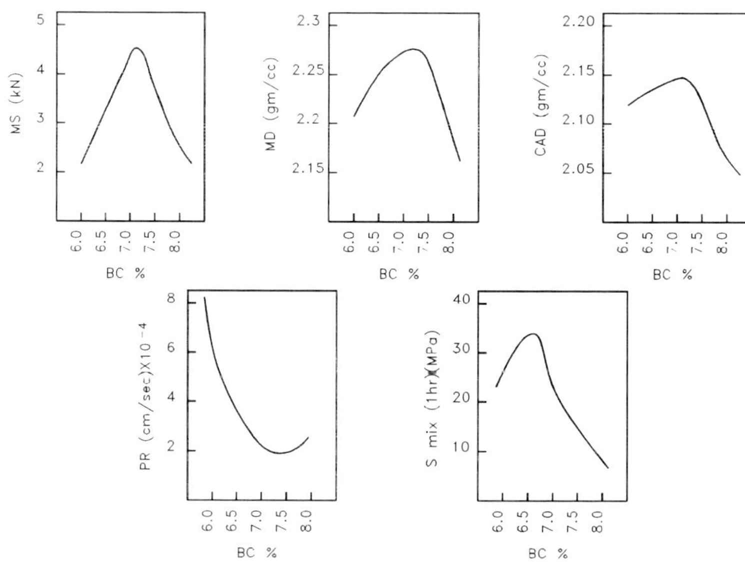

$$OBC\ (UL) = \frac{7.2 + 7.4 + 7.0 + 7.8 + 6.5}{5} = 7.18\ \%$$

CAD : Compacted Aggregate Density. MS : Stability.
MD : Mix Density. PR : Permeability. S mix : Stiffness.

Figure 7. Typical example of optimum binder content calculation

The values obtained by the LDM method are on average 0.5% to 1.0% higher than the values obtained by the BS method. It is the practice of many local authorities in the UK to increase the binder content obtained from a BS design by at least 0.5%; this is based on practical experience which indicates that slightly "richer" mixes than the ones obtained by the BS method give on the whole better performance and satisfactory durability.

4 Conclusions

1. The LDM provides tests which give a numerical indication of the likely performance of bituminous mixtures under service.

2. The tests are simple to carry out, do not require expensive equipment and can be used in central or field laboratories.

3. The assessment of workability reduces the risk of losing valuable materials due to difficulties of handling and compaction and it allows the designer to select the components of the mix to optimise use of thermal energy.

4. Permeability measurements are very sensitive to mix composition and therefore are far more effective in controlling the uniformity and quality of a bituminous mixture.

5. Static or creep stiffness which is measured within the routine design stage can be used effectively to estimate the rut depth of a pavement layer and thus forecast its service life.

5 References

1. Cabrera J G: Assessment of the workability of bituminous mixtures. Highways and Transportation, no 11, pp 17-23, 1991.

2. British Standard Institution. BS 594: Part 2: 1985. Hot rolled asphalt for roads and other paved areas. BSI, London 1985.

3. British Standard Institution. BS 4987: Part 1: 1988. Specifications for coated macadam for roads and other paved areas. BSI, London, 1988.

4. Hussan T Q M: Strength, deformation, permeability and workability of hot rolled asphalt containing pulverised fuel ash. Unpublished PhD dissertation, University of Leeds, UK 1984.

5. American Society for Testing and Materials. Report of Committee D4, Sub-committee DO 4.20, Proc. ASTM, 1975.

6. Cabrera J G and Hassan T Q M Quality control during construction of bituminous mixtures using a simple air permeability test (paper presented to this Conference).

7. Al Sayed M H: The effect of mineral fillers on the performance of hot rolled asphalt mixes. Unpublished PhD dissertation, University of Leeds, UK 1988.

8. Brian D; A design method for gap-graded asphaltic mixes. Roads and Road Construction, No 5, 1972.

9. Ministry of Transport, UK, Technical memo HD/2/89. 1989.

10. Asphalt Institute Mix design method for asphaltic concrete and other hot mix types. Asphalt Institute Manual (MSA-2), March 1979.

11. Cabrera J G and Nikolaides A F: Canik UL a new creep testing machine. The Journal of the Institution of Highways and Transportation, No 11, pp 33-36, 1987.

12. Hills J F: The creep of Asphalt mixes. Journal of the Institute of Petroleum, Vol 59, No 570, 1973.

13. Cabrera J G and Al Sayed M H: Water vapour diffusion characteristics in hot bituminous mixes (in preparation).

11 THE ROLE OF FABRICS IN UPGRADING THE DURABILITY OF BITUMINOUS TREATMENTS

A.R. WOODSIDE and C. ROGAN
Department of Civil Engineering and Transport, University of Ulster, Carrickfergus, Northern Ireland

Abstract
Overlays are normally applied on pavements that exhibit a certain degree of cracking. In addition to enhancing the strength of a pavement, they also correct excessive surface distresses, restore the ride quality and reduce maintenance costs of surface repairs.

In cases where there are limited maintenance budgets, thinner overlays are becoming more and more popular. Without any modification of the overlay system this will result in maintenance strategies that are vulnerable to reflective cracking. This paper examines overlay modification via the addition of four synthetic fabric reinforcements.

To simulate the mechanisms which lead to reflective cracking, three main tests were used. The Wheel Tracking test was used to simulate traffic loading. The INAPOT Tensile test was used to measure tensile strength enhancement and bond strength. The Dartec dynamic load frame was used to assess rate of crack propagation with time.

The paper concludes with a site investigation where one of the fabrics was used to reinforce the asphaltic overlay. As yet few conclusive results are available. Nevertheless the laboratory results suggest that the woven fabric has potential in retarding crack growth rate. What remains to be seen is its effective "crack control" in the field.
Keywords: Fabrics, Overlays, Reflective Cracking, Wheel Tracking, Tensile Strength, Fatigue.

1 Introduction

Overlays are the most commonly used method of rehabilitating deteriorated pavements. However, they often do not perform as satisfactorily as desirable because of existing cracks which propagate through the newly constructed overlay with in a short

Performance and Durability of Bituminous Materials. Edited by J.G. Cabrera and J.R. Dixon. © E & FN Spon. Published in 1996 by E & FN Spon, 2–6 Boundary Row, London SE1 8HN. ISBN 0 419 20540 3.

period of time. This problem of "reflective cracking" is widespread and is considered by some as the most dominant existing pavement problem.

Reflective cracking is caused by one or more cycles of thermal contraction, by repeated traffic loads, or by a combination of these two mechanisms. Existing methods of design do not generally provide crack reflection criteria. However in efforts to minimize or delay occurrence of the problem, alternatives such as increased thickness of overlay, modification of asphalt properties, and placement of stress-relieving interlayers have been attempted. However the degree of success has varied and is often limited.

This paper examines the role of stress-relieving interlayers in the form of continuous woven fabric meshes. A large variety of these fabrics is already marketed for overlay application. This paper will examine four synthetic polymer types currently available on the UK market.

2 Type of Stress Absorbing Interlayers (SAMI'S) used

Table 1 displays the four fabric types which were individually bonded to gap-graded dense wearing course hot rolled asphalt slabs by the heat generated from the warm asphalt mix. The four types were denoted as types A to D.

Table 1. Classification of Woven Fabric Meshes:

Fabric	Structure Description	Mass per Unit Area (g)	Pore Size
A	Woven. Pair of warp monofilaments-on-fibrillated geotextile. Leno weave.	136	Rectangular openings. O_{90} = 790 x 2100
B	Woven. Warp yarn-on-fibrillated geotextile.	97	Rectangular distribution of square openings. O_{90} = 1480 x 2276
C	Woven. Pair of warp monofilaments-on-rolled embossed tape.	123	Rectangular pore openings. O_{55} = 816 x 1952
D	Woven. Monofilament-on-monofilament geotextile, (warp & weft yarns homogenous).	120	Irregular pore openings due to no mechanical interlock between two yarns. O_{15} =3590 x 6622

Key: O_{90} is known as the 90% opening size and corresponds to the size where 90% of the geotextile's openings are the same size or smaller. Similarly, a geotextile with an O_{15} value of 3590 x 6622 denotes 15% of the pore openings as this size or smaller.

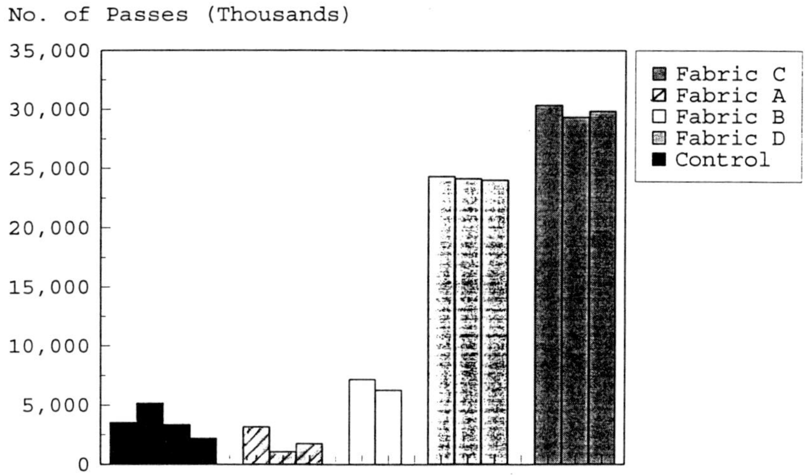

Fig.1. Number of passes to failure for reinforced and unreinforced asphalt

Types A & C displayed the characteristic leno weave (warp). Due to the crimp or inbuilt slack at the intersection of these two individual yarns (warp), the fabric was laid with the weft tapes aligned in the direction of stress.

Both fabrics A and B had fibrillated weft yarns. This is the name for short slits being introduced in to the polypropylene film during its production which in turn enables the matrix to penetrate the fabric. Fabric mesh C however had a rolled embossed tape weft yarn which implies that the surface of the tape was embossed with longitudinal grooves or planes of weakness. This rendered the tape "foldable" and enabled weaving to take place more easily. Fabric mesh D was also a woven geotextile, however due to no mechanical interlock between its two yarns, the pore openings were irregular and easily distorted.

3 Sample Preparation

For the nature of this testing programme, asphalt slabs were made up with similar dimensions to that specified in the wheel tracking test programme. That is dimensions of 305 x 91.5 x 40mm approximately. All samples were compacted with a steel wheel roller and when cool, sliced in half to provide a nominal thickness of 20mm. This in turn reduced the time spent on sample preparation and also helped speed up the process of crack propagation for the respective test methods.

4 Testing Programme

Three main test methods were chosen to examine the performance of the aforementioned SAMI's in delaying reflective cracking. To understand fully the choice of test methods, a reappraisal of crack development now follows. The mechanisms which control reflective crack growth are the horizontal and vertical movements of the original pavement layers. These movements occur as a result of changes in temperature, traffic loading or a combination of both.

At lower temperatures, the pavement surface contracts resulting in movement at the already existing cracks or joints in the original pavement layers. These movements create tensile stresses in the overlay causing a crack to open which will then continue to grow with repeated expansion and contraction of the underlying pavement layers. Another important mechanism that contributes to reflection crack growth occurs due to the influence of traffic loading. Everytime a wheel load passes over a crack in the old pavement, the overlay will be subjected to a shear stress pulse which will be followed by a bending stress pulse and again by a shear stress pulse in the reverse direction.

Hence with a certain degree of simplification, one could state that the traffic loads are mainly acting in the vertical direction, perpendicular on the layers, while the environmental loads are mainly acting in a horizontal direction.

To simulate trafficking, the wheel tracking apparatus was chosen. To determine tensile strength, an Instron test rig was used. Bending and the propagation of cracks with time was determined by the Dartec dynamic load frame.

The following section deals specifically with each test method and the varied performance of the four continuous woven fabric meshes.

4.1 Wheel Tracking Test

All samples were trafficked individually under water at a constant test temperature of three degrees centigrade. Weights of approximately 18kg were added to induce cracking and samples mounted on supports at a distance of 170 mm apart. Time to failure was recorded in simple time increments and later converted to wheel passes (see Fig. 1).

Fabric meshes A and C offered the superior composite when trafficked at a temperature of three degrees centigrade. This was followed by B, the controls (unreinforced asphalt slabs) and finally type D.

Failure of the composite was usually instantaneous once a visible crack appeared in the asphalt mix.

4.2 The Tensile Test

Tensile testing was performed on the Instron 1114 floor model. All slab specimens were bonded to steel clamps by epoxy resin glue and left for an hour prior to testing.

Two main test procedures were performed. The first dealt with the strength analysis of each fabric type and their individual weft and warp yarns. While the second analysed the tensile strength enhancement of the reinforced composites. This was to determine the load carrying ability of each fabric's yarns and secondly the means of strength enhancement to the reinforced composite.

4.2.1 Tensile Strength of the Fabric Meshes

Special fabric grips were used to hold the fabric samples in the Instron machine. Despite the serrated profile of such jaws, slippage or breakage of the fabric near the jaw edge was common. Hence machinery mounting pads and an epoxy resin adhesive were used to eliminate slippage.

In accordance with BS 6906: Part 1 (1), each sample specimen tested had a width of 125 to 130 mm and a gauge length of 100 mm. Although this only provided approximately half of the number of warp yarns and slightly in excess the number of weft yarns when compared to the composite, the results were nevertheless satisfactory. Figure 2 displays a characteristic load diagram for the weft and warp yarns, which were all similar regardless of fabric type tested.

This test programme also highlighted the nature of breakage, particularly for the weft yarns. When a large number of weft tapes are tested together, the weakest failed first, then yielded its load to some others, which soon became overloaded, broke and this led to progressive collapse. Hence this gives rise to a breaking load which is generally lower than that expected on a pro-rata basis, using the failure load of a single tape. This is known as the weakest link theory and occurred for all weft yarns tested, plus the warp yarns of fabric D.

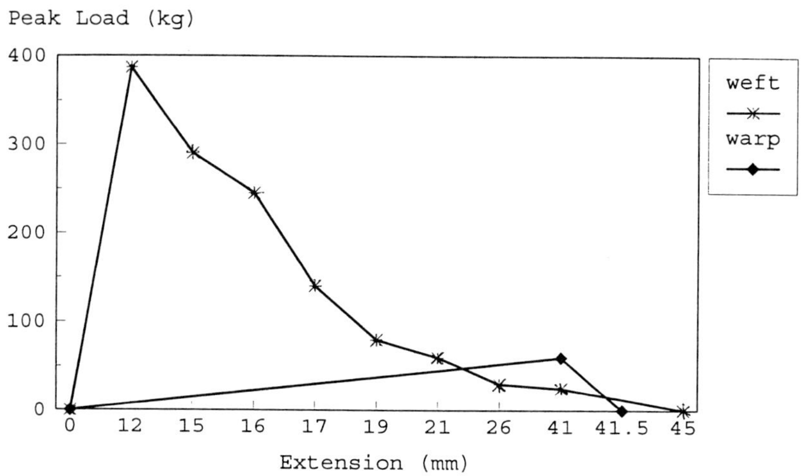

Fig.2. Characteristic load diagram for weft and warp yarns of Fabric C

Extension of the warp yarns was always higher than that of the weft, particularly so for fabric types A, B and C. This can be explained by the inbuilt slack in the leno weaves and folded-up tapes i.e. fibrillated and rolled embossed.

4.2.2 Tensile Strength of the Composite

The application of stress to all samples led to a ductile fracture of the bituminous matrix and consequent reduction in cross-sectional area. Failure of the fabric yarns was however very different and was either typified by complete shearing (debonding) from the matrix or vertical pull-out of the weft yarns. When loaded in tension, the weft yarns began to debond from the matrix, collapsed inwardly and pulled out of the loose warp tapes. Hence the latter, instead of resisting pull-out and clinging to the plugs of the matrix, provided slippage points for the weft tapes and served as an aid to pull-out. The same occurred for fabric type C (rolled embossed weft tape), however the effect was not as pronounced.

Figure 3 displays the tensile strength enhancement of the four sami's over an unreinforced asphalt. Fabric C, A and then B afforded the composite the superior resistance to tensile stress, however the ranked performance to strain varied according to the nature of the fabric weave.

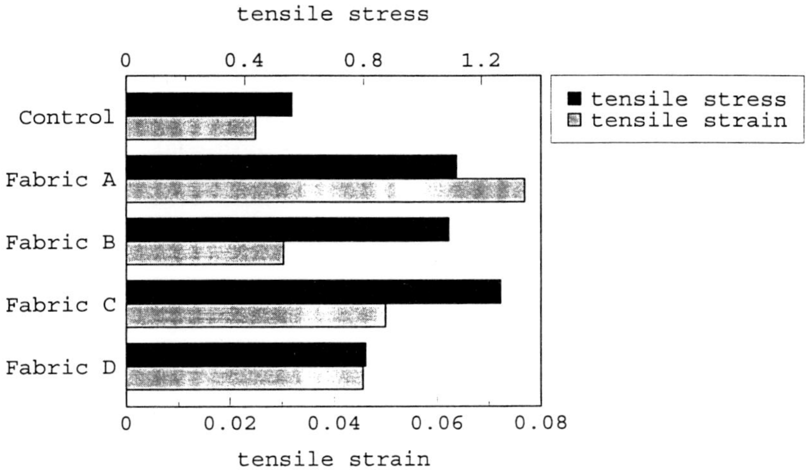

Fig. 3. Tensile strength assessment of specimen samples

The high stress values for fabric types A and C can be explained by the folded weft yarns and their associative loose leno weave warps. Fabric D had no mechanical interlock between its two yarns, hence under applied load, the weft would simply slip out of the woven mesh. Fabric type B produced very little stretch at maximum load (strain), almost similar to the controls and this was attributable to its knitted warp yarn and the tight interlock of its two yarns.

The tensile strength enhancement in the composites did not prove cumulative, i.e. simply knowing the fabric strength and asphalt strength did not produce the total composite strength. Rather the strength of the composite tended to be much lower than this cumulative rating. In addition the very nature in which the composite failed, would suggest that this test was mainly an evaluation of bond strength between the asphalt and fabric mesh.

4.3 The Fatigue Test

Experiments show that a material may fail at a stress considerably lower than its normal tensile strength if this stress is repeated a large number of times. The term fatigue is used for the effects on a material of repeated cycles of stress. Hence fatigue testing here was expressed as the number of repeated load applications to create cracking failure of the material.

All samples were tested on a Dartec dynamic load frame using a three point load system. To aid crack initiation and propagation all samples were mounted on supports 170 mm apart. Using a 20 KN load cell, maximum load was set at 0.3 KN and amplitude at 6 mm. Frequency of load application was set at 20 Hz, equal to 0.05 seconds loading time. Complete failure of the composite was classified as the point where the crack had propagated completely through the asphalt and had reached the fabric mesh. With the aid of two stop watches, the time to crack along the width of the specimen and the time to complete failure was noted.

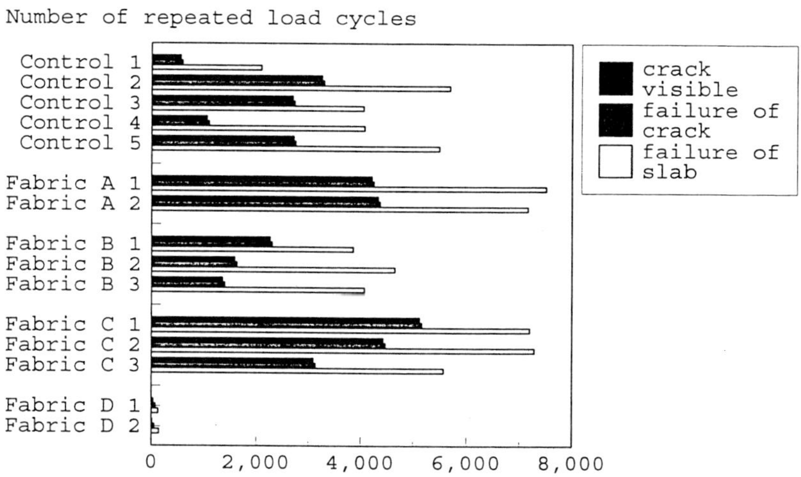

Fig. 4 Stages of fatigue failure under repeated loading

Figure 4 displays the results of this test. Fabric types A and C afforded the bituminous composite the longest time to fatigue failure. In particular fabric A was the best at delaying crack propagation once a crack had been initiated.

4.4 Conclusions

In all the test methods undertaken, fabric type C (the non-fibrillated, rolled embossed geotextile) offered the superior composite. This performance was closely followed by the two fibrillated geotextiles (A and B) and then fabric D. This loss in performance by the fibrillated geotextiles could be attributed to the twisting of the tape prior to weaving. Whereby the twisting prevents the fibrillated tape from opening up fully, this provides surface barriers spiralling longitudinally which hinder penetration by the matrix (via micro-pegging).

In contrast twisting prior to weaving for the rolled embossed mesh was an advantage as it transformed the parallel longitudinal grooves which exist in the parent tape, in to longitudinal helix-like spirals, which protrude above the normal diameter of the parent tape and in turn aid mechanical bonding.

Fabric type D proved to a totally dissatisfactory reinforcing "sami" for bituminous wearing courses. In fact in the majority of tests, the performance of the controls was often superior to the composites reinforced with fabric D.

According to this research, woven fabric meshes are an efficient cost-effective means of reducing or preventing reflective cracking in the surface road network. Indeed, contrary to fibre-reinforced composites, the need for an interfacial binder is eliminated and this in turn reduces the total cost.

5 Site Investigation

One of the fabrics tested in this research was used as a reinforcing "sami" treatment over a lean-mix concrete roadbase on a primary route in Northern Ireland. The flexible surfacing was dug out to reveal three joints in the concrete and a rigid steel grid was applied, followed by blacktop. The road was reinstated to its original level and fabric type A applied as a "sami" within the overlay. As yet the relative effectiveness of this sami geotextile is not yet apparent due to the relative newness of its application. Nevertheless it is hoped that it will delay the onset of reflective cracking and help to increase the overall pavement life.

6 Discussion

This paper has discussed the relative performance of four asphalt overlay fabrics and merited three of them as cost-effective solutions to the repair of cracked road surfaces. A site investigation was also considered briefly since lean concrete is the most widely used roadbase within the United Kingdom. Such roads have excellent load spreading abilities and are resistant to deformation, but regular transverse cracks that originate in the roadbase during curing often appear in the bituminous surfacing.

Until recently it was widely believed that these cracks had propagated upwards from cracks in the lean concrete roadbase, but Burt (2) suggests that in fact the cracks generally initiate at the exposed surface of the bituminous layer and propagate downwards to meet the cracks in the lean concrete. Eitherway the site investigation considered both options and applied a rigid steel grid just over the concrete and a reinforced surface treatment. The effectiveness of these materials in arresting the progression of crack growth will be assessed. Unfortunately due to the small time lapse, the results will not be published here.

It is expected that general conclusions will be obtained within a year or so and conclusions will be refined in subsequent years by monitoring crack initiation and progression in the overlay.

7 References

British Standards Institution. BS 6906: Part 1: 1987. British Standard Methods for Test for Geotextiles. Part 1: Dermination of the Tensile Properties using a Wide Width Strip.

Burt, A.R. (1987). M4 Motorway, A Composite Pavement: Surface Cracking. Highways and Transportation, Vol. 34, No. 12, London.

12 MEASURING THE POTENTIAL COMPACTION PERFORMANCE OF BITUMINOUS MIXTURES

D. FORDYCE, D. MARKHAM, H. IBRAHIM and H. EL-MABRUK
Department of Civil and Offshore Engineering, Heriot-Watt University, Riccarton, Edinburgh, UK

1 BACKGROUND

Flexible pavement layers are formed by an interlocked aggregate framework stabilised by a bitumen, or a modified bitumen binder. The binder partly fills the voids within the aggregate framework. A layer is formed by initially coating the aggregate particles with a film of binder, the mixture is spread, mechanically or by hand, the mixture is then compacted by rolling or impact. Impact is normally restricted to the reinstatement of small, or restricted width areas of material. During compaction the binder is displaced from the surface of the aggregate particles to the void space between the interlocking aggregate framework.

Pavements are designed for service performance. Layers forming a construction have particular functions, the material forming a layer therefore has primary service performance requirements. All materials require to have a stress distribution ability, surfacing layers require to have a degree of impermeability and stability, the wearing course requires to have a degree of durability. All layers require to have a limited potential for 'secondary compaction'. The performance of a material layer will be controlled by the quality of the ingredients used, and their proportions. But, the potential performance of a bituminous material layer can only be achieved if it can be placed and 'fully' compacted. Where a material layer is not 'fully' compacted, secondary compaction will occur in service.

Bituminous materials are specified by recipe formulation, or they are specified by design to meet particular performance criteria. Within the limits of experience, a recipe formulation should ensure that a material can be placed and compacted, and it will have a definable performance life. For situations where required performance exceeds experience materials have to be designed. Commonly, materials are designed for service performance using the criteria of stiffness, and fatigue or creep. Test procedures for these performance criteria vary. Surfacing materials are commonly designed using a partially confined compression test, the stability and flow characteristics of the material being measured. The state of compaction of material is not always a condition of the design procedure. But, even where the compaction condition is a condition of the design procedure, the ability to compact the material is not a design condition. The latter point relates to the design of a material for placement- laying and compaction, as opposed to service.

Currently in the UK the Standards for the specification of bituminous materials for roads and other paved areas is by recipe formulation, except for hot rolled asphalt[1,2]. With hot rolled asphalt the binder content is optimised for a fixed blend of aggregates. There is no requirement to achieve any particular compacted sample density in the material design procedure. The

Performance and Durability of Bituminous Materials. Edited by J.G. Cabrera and J.R. Dixon. © E & FN Spon.
Published in 1996 by E & FN Spon, 2–6 Boundary Row, London SE1 8HN. ISBN 0 419 20540 3.

standards for the laying and compaction of hot-mix bituminous materials for roads and other paved areas provide guidance only on laying conditions and on compaction plant, rolling temperatures and rolling procedures[3,4]; there is no guidance on the degree of compaction to be achieved with a material layer. One exception to this is with dense roadbase macadam and dense basecourse macadam. For these materials there is an end result specification for the degree of compaction achieved with a layer[4]. The end result specification uses the ratio of material density achieved on site to the maximum density achieved with a sample core re-heated and subject to a method of secondary compaction in the laboratory. The end result specification is material specific.

With the reinstatement of openings in pavements the Standard for the specification of hot-mix material follows that for roads and other paved areas. Guidance on the compaction of hot-mix material is a method specification[5]. Permanent cold-lay material can be specified as a surfacing with reinstatements. Permanent cold-lay material is specified by design to meet a service performance specification. The specification defines the number passes of compaction plant over an area. No guidance is given on the compaction of cold-lay material as the required performance is measured at the end of a two year service period.

In summary, the state of site compaction of bituminous pavement material is not a design consideration in the UK, except for roadbase and basecourse macadams. No laboratory material design procedure in the UK includes a compacted voidage or compacted density requirement. The ability to compact a mixture will be controlled by specific environmental conditions, but guidance only is provided on compaction in terms of providing general limits to material and environmental conditions for laying and compaction.

The consequence of a lack of knowledge about the compaction performance of bituminous materials is that the process remains an art only, skilled operators using their feel for performance during material laying and compaction. The difficulty is with new materials: materials including modifiers in the bitumen, or particularly today, the increasing use of cold-lay materials. The sensitivity to environmental conditions is important to define with a material: the temperature sensitivity of a mixture, or the water sensitivity of a cold-lay mixture. There is a need to understand the action of the ingredients during the process of aggregate interlock and binder displacement at one level, and there is a need for a procedure to define potential performance at another level.

2 MEASUREMENT OF COMPACTIBILITY

Material compaction includes three processes, it is: the interlocking of the aggregate framework; the displacement of the binder from around the aggregate particles to the voidage space within the interlocking aggregate framework; and, the consequent reduction in air voids within the material. Compaction is not solely the reduction in air voids within a material. The movement of the bitumen as it displaces to the void space requires work to be done. The interlocking of the aggregate requires work to be done. The work comes from compactive effort. Mixtures where the viscosity of the binder is low, through temperature or pen grade, will require less work to be done. Aggregate gradings with limited finer fractions require less work to be done. High pen grade, open graded materials are consequently easier to compact compared with lower pen grade dense graded materials.

Compaction effort comes from the applied surface stress, which is roller weight and drum diameter dependent, and dependent on the number of roller passes. Increasing compactive effort comes from increasing the roller weight, decreasing the drum diameter and increasing the number of roller passes. The effectiveness of the compactive effort depends on the stiffness of the supporting structure to the layer being compacted, the layer thickness and the ability of the material to provide self confinement under the roll. Thin layers being compacted which are supported by a stiff substrate will ensure maximum compactive effort from a roller. As the resistance to internal movement reduces, and therefore less work is required to cause internal movement, the self confinement ability of a material may reduce. Here compactive effort will result in material displacement and not compaction.

Where compactive effort exceeds resistance to internal movement compaction will occur, assuming self confinement is possible. Compaction will cease when compactive effort equals resistance to internal movement. The degree of aggregate interlock at the equilibrium point will define the potential of the material to exhibit 'secondary compaction' under traffic loading. Where the aggregate is not fully interlocked after compaction, but the residual voidage is low, the material may lack stability after trafficking. Here, if additional interlock of the aggregate does occur, this will further reduce the residual voidage; the residual voidage may be reduced to a value which results in material displacement under traffic loading. This effect is more likely with fine dense mixtures, such as wearing course hot rolled asphalt.

Compaction is the processes of aggregate interlock, binder displacement and voidage reduction. Compactibility is the rate at which these processes occur. Where compactive effort greatly exceeds resistance to internal movement, when the effort is applied the material will reduce volume quickly. With rolling compaction, one pass of the roller with a highly compactible material will result in a high percentage of the potential closing up of the aggregate framework and binder displacement to occur. A highly compactible material will require fewer roller passes to achieve an equilibrium condition where no further closing up of the aggregate framework and binder displacement will occur. It would be expected that with each pass of the roller the degree of additional closing up of the aggregate framework will reduce as the difference between the compactive effort and resistance to internal movement reduces. Measured in material density terms, the increase in material density per roller pass will reduce. The density increase profile with roller passes will be upwardly curvilinear, becoming asymptotic to a limiting value. The nature of the curvilinear profile will be indicative of the compactibility of a material.

With a sample of material in the laboratory, compaction is applied with the sample fully confined in a steel mould. Compaction may be direct drive, through a single acting piston, or double acting piston arrangement, such as with the Duriez compaction procedure[6], by impact, such as with the Marshall compaction hammer, or by a kneading action, as with the gyratory testing machine compactor. The kneading action of the gyratory testing machine is believed to provide a more realistic action to the system during compaction, one which simulates the action of a roller. With any of the sample compaction systems, the rate at which aggregate interlock and binder displacement occurs will be more rapid with more compactible materials. In terms of a density plot, the density increase profile will be upwardly curvilinear in the same manner as that for roller compaction. Although material compactibility may be detected from a similar behaviour pattern with laboratory compacted samples as with layer rolling, a laboratory compacted sample would not necessarily indicate a limitation with material self confinement. In other words, no indication would be given with sample compaction as to whether a material layer could sustain the weight of a roller without material displacement.

2.1 Previous laboratory investigations

There have been a relatively large number of investigations into the compaction performance of bituminous pavement mixtures. Powell, in a Road Research Laboratory publication in 1971[6], reviewed the literature available at that time on the compaction of bituminous materials. Work was divided into laboratory investigations, field studies and pilot-scale laboratory experiments. Powell described the work of Lefebvre[7] who recorded the density increase with material subject to an increasing number of blows with the Marshall compaction hammer. The density profiles of two materials is reproduced in Figure 1. Lefebvre drew a straight line through the points and used the gradient of the line as an index of compactibility. The plot is semi-logarithmic, but for the natural sand and gravel is clearly curvilinear. The data was later converted to a

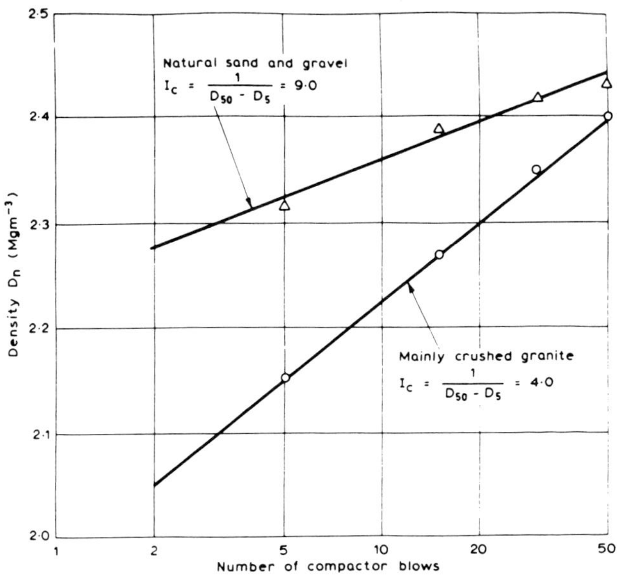

Figure 1: Variation of Density with Compactive Effort (After Lefebvre[7])

'compaction resistance index'[8,9] . The compaction resistance index is the gradient of the data points plotted as a straight line when the density is expressed as the ratio of density achieved to design density. This manner of plot avoids the difficulty of comparing densities with different material which contain aggregates of different specific gravity. McLeod and McLean reproduced this work with hot dense graded asphalt concrete samples[10]. Both Lefebvre and McLeod and McLean defined compactible mixtures as those having a steeper profile of mix density against logarithm of blows applied to the sample.

Renkens[11] also used the Marshall compaction apparatus to investigate hot mix asphalt. But, Renkens found an exponential relationship between number of hammer blows and sample density, when plotted to a natural scale. This would account for the curvilinear profile of points on the semi-logaritmic plot of Lefebvre with natural sands and gravels. More importantly, Renkens compacted mixtures over a range of temperatures, from 75°C to 155°C. He found that around 115°C compaction became more difficult. This characteristics was reproduced by Markham with the equipment described in this paper[12]. Measuring compactibility over a range of temperatures is important to the provision of a pavement layer the properties of which require to be maximised.

The compactibility of emulsion-based material has also been investigated. The Hveem compaction procedure is used to provide a 'light' kneading action, this is followed by a double plunger action static load. With the static load, a 100 mm diameter sample is loaded to maximum load value at a plunger drive rate of 1.3 mm/min, the load is then held constant for one minute. Puzinauskas and Jester found an approximately linear relationship between dry sample density and level of static load[13].

Cabrera has used the gyratory testing machine to define the workability of hot rolled asphalt samples[14]. He recorded the thickness of samples over an increasing number of revolutions of the equipment and plotted porosity against the logarithm of number of revolutions. This resulted in a straight line plot. The nature of the plot is no different from that using the Marshall compaction hammer. The number of revolutions reflects the compaction energy applied to a sample. The gradient of the line would directly relate to the definition of compactibility used by other researchers, such as McLeod and McLean. But, Cabrera calculated an index which is the inverse of the porosity of the sample at zero revolutions; he called the index a Workability Index. Workability is a measure of the potential for compaction.

3 A MEASURE OF MATERIAL COMPACTIBILITY USING THE DURIEZ PROCEDURE

A method to assess the compactibility of bituminous materials has been developed at Heriot-Watt University based on the Duriez compaction procedure. The loading frame is shown schematically in Figure 2 and the mould is shown schematically in Figure 3. The frame consists of an overslung beam with a load cell attached. The beam is attached to a base unit which houses an electrical motor and mechanical gearbox. The gearbox drives a shaft attached to a horizontal bed-plate. The bed-plate is driven upward at a constant rate of drive. The drive rate can be varied. The mould is a split cylinder mould machined internally to 100.2 mm. The mould is 150 mm in height. A close fitting machined piston, 100 mm in diameter, fits into the top of the mould. The walls of the piston are vertically fluted. The piston has an internal shaft which is used to adjust the total length. There are two base-plates with the mould, a shallow base-plate and a deep base-plate. One kilogram of material is placed in the mould fitted with the shallow base-plate. The piston is fitted and the height of the internal shaft of the piston adjusted until the top of the shaft is located with the load cell. The output on the load cell indicates when this position is reached. A computer is used to activate the motor and therefore start the drive of the bed-plate. The drive can be cut-out at a pre-set value of load measured by

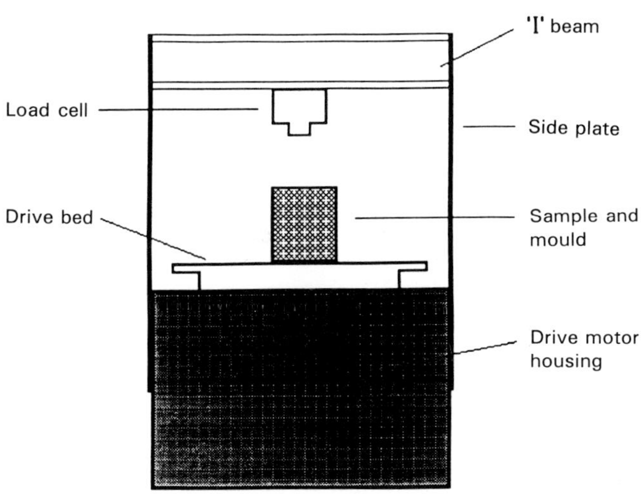

Figure 2: Schematic Diagram of the Direct Drive Compactor

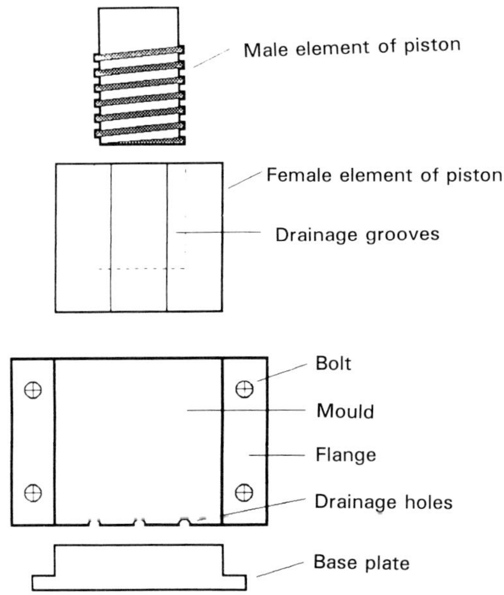

Figure 3: Schematic Diagram of the Compactor piston, Mould and Base Plate

the load-cell. An initialising load of 0.2 kN is used, so that all samples have the same pre-conditioning. The load is then removed by driving the bed-plate downward and the shaft length re-adjusted until the load reading is again zero. A pre-compaction load of 10 kN is then set and the material loaded. A relaxation period is pre set into the computer so that it records the load drop-off after the motor drive is stopped. The pre-set relaxation period is commonly 40 seconds. This record measures what residual movement exists within a sample. On removing the load the deep base-plate is used with the following loading-relaxation cycles. Forty second relaxation is used with a load of 30 kN. The load-relaxation cycle is repeated until there is no measurable relaxation with a sample. Figure 4 shows schematically the arrangement of the parts of the total equipment. Figure 5 shows a load-relaxation plot using a block of steel in the equipment, the load being 30 kN. Figure 6 shows a typical plot with a cold-lay material with an additional 20 kN load-relaxation cycle.

3.1 Data with hot-mix material

Wearing course hot rolled asphalt and dense macadam mixtures have been assessed with the compaction procedure described. Mixtures were compacted over a range of temperatures, from 155°C down to 75°C.

Figure 7 shows two plots with hot rolled asphalt: for material at 155°C, and for material at 95°C. The nature of the plots is distinctly different. The time to load cut-out at the 10 kN load for the material at 95°C is twice that for the same material at 155°C. The first load-relaxation cycle at 30 kN is distinctly different, there is significantly more relaxation with the material

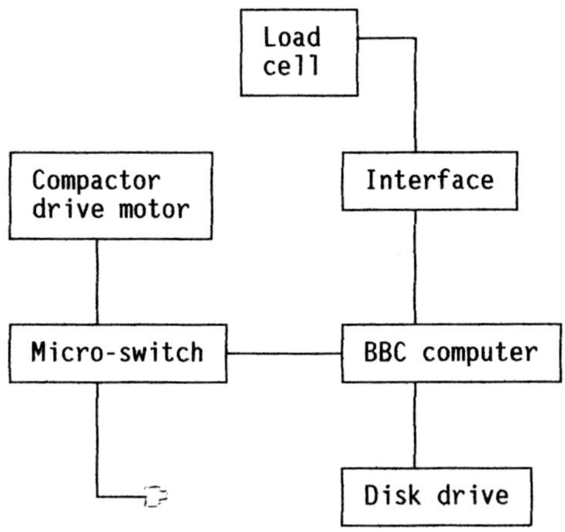

Figure 4: Schematic Diagram of Direct Drive Compactor System Elements

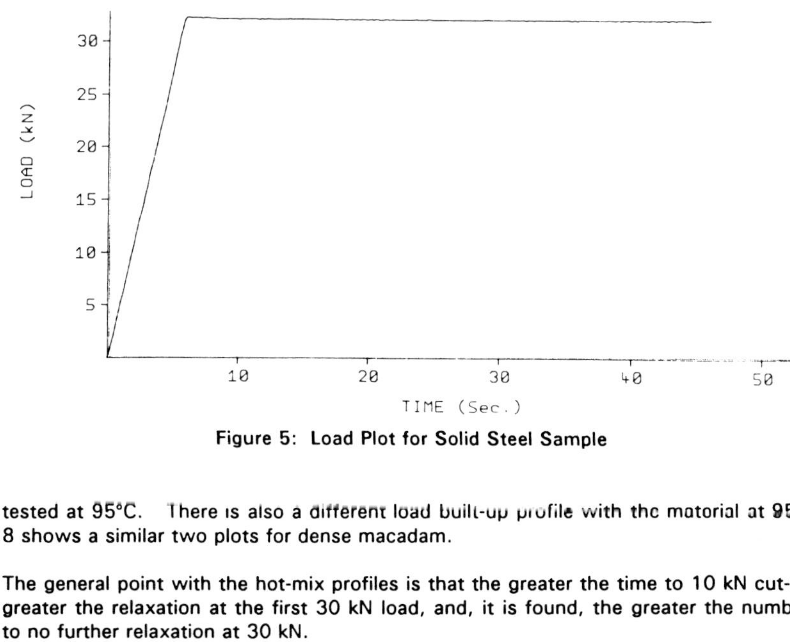

Figure 5: Load Plot for Solid Steel Sample

tested at 95°C. There is also a different load built-up profile with the material at 95°C. Figure 8 shows a similar two plots for dense macadam.

The general point with the hot-mix profiles is that the greater the time to 10 kN cut-out load the greater the relaxation at the first 30 kN load, and, it is found, the greater the number of cycles to no further relaxation at 30 kN.

More specifically, with the hot rolled asphalt mixtures the time to 10 kN cut-out is greater than with macadam mixtures at the same temperature. During the time to the 10 kN cut-out the aggregate framework is being formed and the binder is being displaced. Asphaltic mixtures contain a greater aggregate surface area compared with macadam mixtures and more binder is being displaced. This is reflected in the movement measured with asphaltic material.

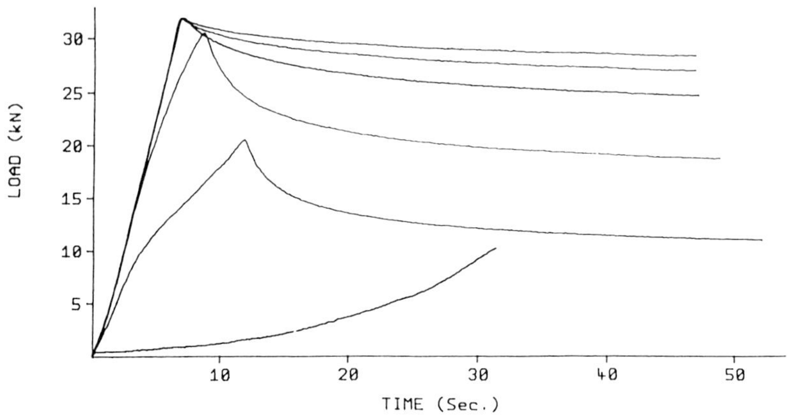

Figure 6: Multiple Load-Relaxation Cycles with a Cold-Lay Mix

At cooler temperatures the viscosity of the binder is greater. With both materials, the longer time to the 10 kN cut-out load at 95°C compared with 155°C reflects the amount of binder being displaced. At the higher temperature the binder displaces more ready from around the aggregate particle surfaces. Much of the displacement takes place during the initialising stage when the 0.2 kN is applied. The aggregate framework will be initialised also during this early loading phase. The shorter time to the 10 kN cut-out load with material at higher temperatures therefore represents the movement of some of the residual binder into the void space during the initialising loading phase.

With the first 30 kN loading cycle, the profile initially is steep and apparently linear, it then inflects to a curvilinear profile. The degree of curve differs, in particular, for the asphaltic material between the two temperatures. With the lower of the two temperatures the aggregate framework is still being formed, the curvilinear profile above the 10 kN load reflects the internal movement occurring within the material: aggregate interlock is occurring, along with binder displacement. The reduced profile above the 10 kN load with the higher temperature material reflects the degree of aggregate interlock and binder displacement which has occurred. The degree of internal system movement is reflected in the relaxation after the 30 kN cut-out load.

With the macadam mixtures the cooler material has a longer time to the 10 kN cut-out load, but the profiles for the two mixtures on the first cycle are essentially identical. This reflects the lower volume of binder being displaced, compared with hot rolled asphalt. The aggregate framework apparently forms more readily, or it locks in a particular pattern. To check whether the framework is locked in a semi-compacted state, or not, the voidage within a dry compacted aggregate blend is made and used to compare with the calculated voidage within the compacted aggregate framework of the mixture. A vibration hammer procedure is used at Heriot-Watt University for compacting dry aggregates. It has been found that with both the asphaltic and macadam mixtures maximum compaction is achieved with both materials within the cyclic loading system.

It is important to note here that the comparison of state of compaction with this work is with a dry compacted aggregate framework, not a second laboratory compaction system- typically 50 blows of the Marshall hammer. The reason for this is that with hot rolled asphalt materials Marshall compaction is found not to have a definable relationship with rolling compaction in

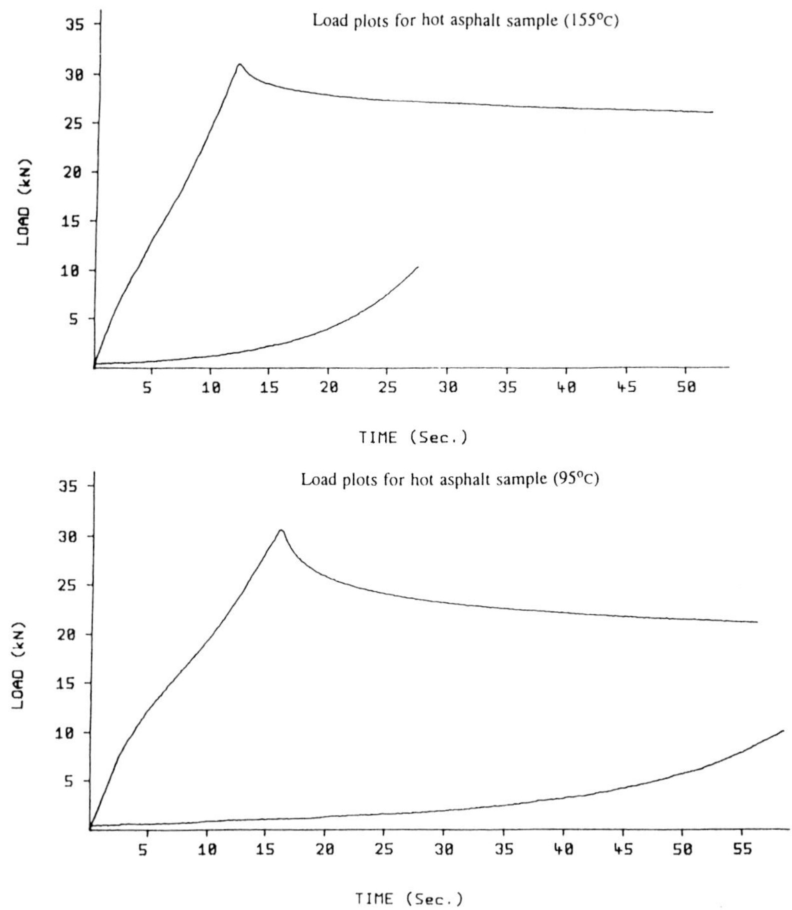

Figure 7: Load Plots for Hot Rolled Asphalt Samples

terms of the measured residual voidage within material. From extensive measurements of different compaction methods the dry compaction system used has been shown to create maximum interlock with all aggregate gradings examined.

3.2 Second analysis of compaction data

The profiles of relaxation can give information on how the internal structure of a material is changing during a method of compaction. The profiles themselves do not provide a useful single measure of compactibility. As there appeared to be a relationship between the time to 10 kN cut-out and relaxation and number of cycles to maximum compaction, this time was used as a single measure of compactibility. The time to 10 kN cut-out is an identical measure to the 'compaction resistance index' defined by Lefebvre and Robertson[9]. The time to 10 kN cut-out has been plotted against material temperature, as shown in Figure 9. The results show an

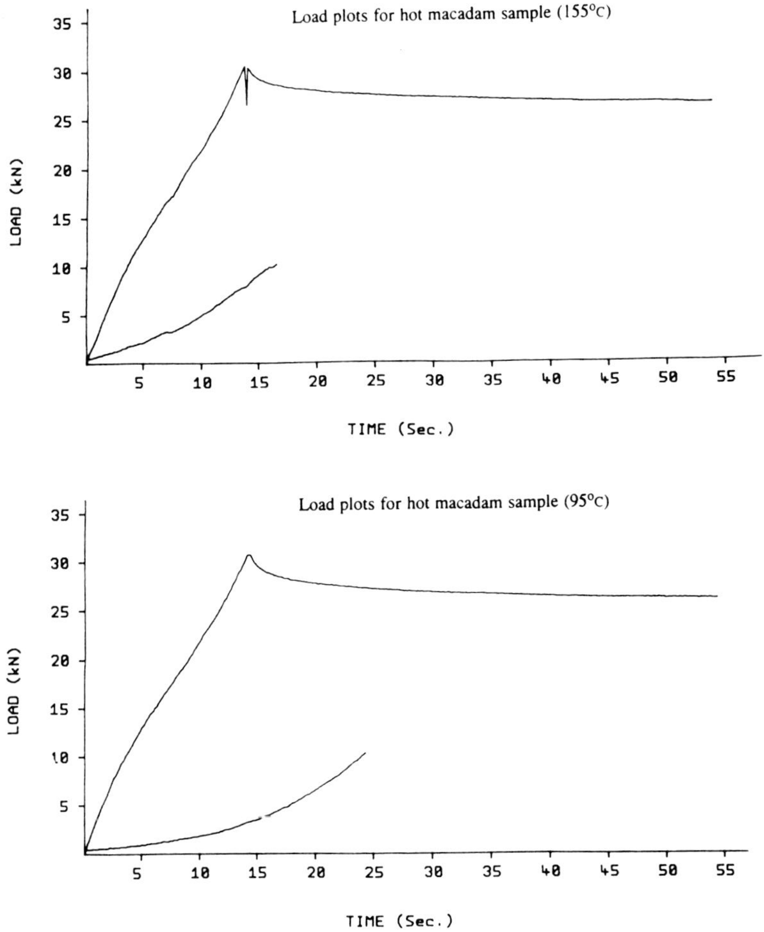

Figure 8: Load Plots for Hot Macadam Samples

interesting relationship. With the hot rolled asphalt, from 155°C down to 115°C there is little difference in the time to 10 kN cut-out load. There is a significant shift in the time to 10 kN cut-out between 115°C down to 95°C. Below 95°C there again is little difference in the time to 10 kN cut out load. This data is consistent with that found by McLeod and McLean: there is a significant increase in resistance to forming an interlocked aggregate structure at a temperature below 120°C. But, the interesting point from this data is that there is a step shift in the resistance to compaction over a c20°C temperature drop.

The explanation for the negligible difference in time to 10 kN cut-out between 155°C and 115°C results from the self compacting nature of the material, or certainly low resistance to compaction during the initialising 0.2 kN load cycle. As the system is contained the self-confinement of the material is not being assessed. There will be an upper compaction temperature for a material at which the self-confinement of the material reaches it limit and a

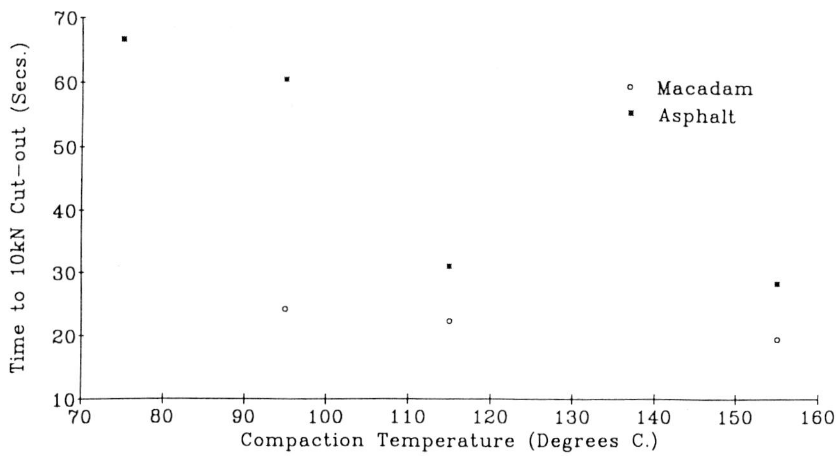

Figure 9: Plots of Time to 10 kN Cut-Out Load with Temperature for Hot Mixes

mat no longer can be compacted effectively. Measurements using a high-temperature triaxial cell developed at Heriot-Watt University[15] have found for the material tested in the compaction system that the upper compaction temperature lies around 135°C. Sample depths can be measured during the compaction process. Knowing the final depth of a sample allows a simple calculation of sample density, or contained residual voidage. Although not known with the material data shown in Figure 9 the density of compacted samples at temperatures above 135°C may be higher than those below 135°C. It was found from the high-temperature triaxial work that the upper compaction temperature was a consequence of the expansion of the binder volume with temperature. If consistent with the compaction system then sample density should increase at high temperatures.

Below 95°C apparently little significant increase in resistance to compaction occurs. It is known that material can be compacted at low temperatures, but the number of roller passes has to increase significantly. What is not known yet is the implication of the shift in time to 10 kN cut-out load in terms of increased number of roller passes.

With the macadam mixes there is no significant change in the time to 10 kN cut-out load. The macadam mixes includes 100 pen grade bitumen. As such, 95°C is unlikely to be near the transition region in terms of increase in resistance to compaction.

3.3 Cold-lay mix compaction data

Figure 6 shows data with a cold-lay macadam to compare with the hot-mix data shown in Figure 8. Figure 10 is a second cold lay material of similar aggregate grading to that of Figure 8. The material shown in Figure 10 has a 40 second relaxation period applied at the 10 kN cut-out load. The importance with the profiles in Figures 6 and 10 is the amount of relaxation with a macadam mixture. This reflects the difference in the rheology of the binder, even although the base bitumen pen grade of the emulsion is the same as that with the hot mix macadam. Without the 40 seconds relaxation at the 10 kN cut-out load, the time to the first 30 kN cut out would be significantly higher than appears. With cold-lay material the time to the first 30 kN cut-out load may be a more useful index to allow hot and cold-lay material to be compared.

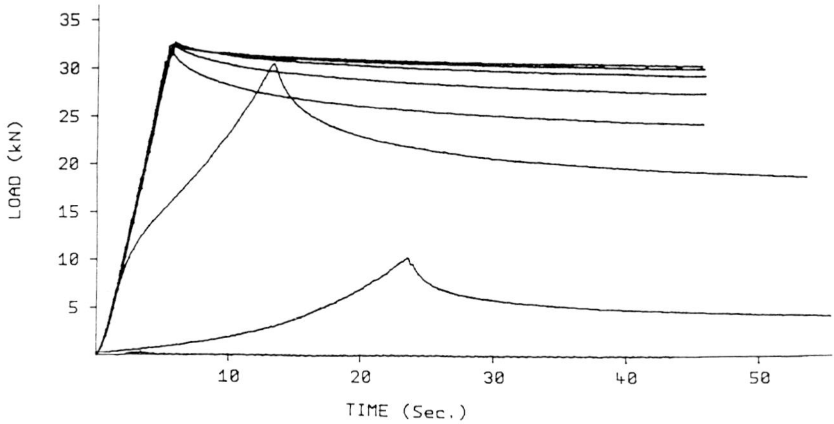

Figure 10: Load Plot for Cold-Lay Macadam

Figure 11 shows an equivalent plot of data relating time to 10 kN cut-out load against water to emulsion ratio with a cold-lay material research project. In the project emulsion mixes were mixed in the laboratory using the same aggregate grading but varying the emulsion and water content of the mixes. Mixes were left to break prior to compaction. The compaction procedure is able to differentiate between mixes quite clearly. To interpret the data, increasing the water content of a mix has the same effect as increasing the temperature with a hot mix: the time to 10 kN cut-out load reduces, because of the self compacting nature of the material. Increasing the total liquid content of the mixes increases the time to 10 kN cut-out load, because a greater binder film thickness has to be displaced.

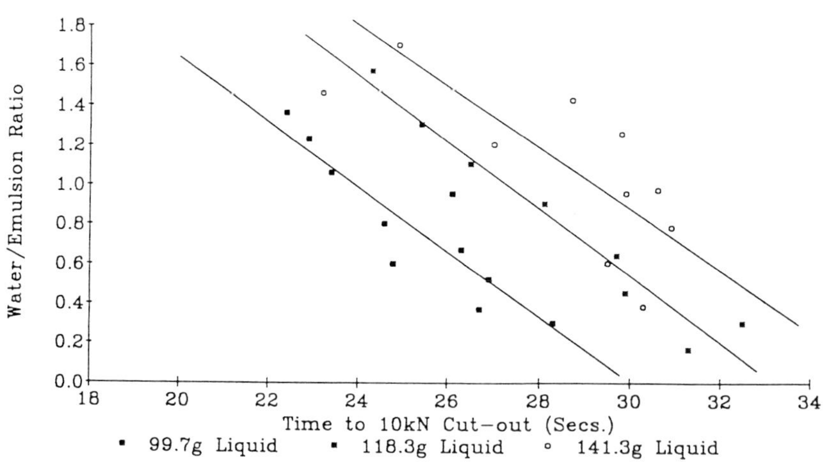

Figure 11: Time to 10 kN cut-out Load with Cold-Lay Macadams

The interpretation of the cold-lay data is as with the hot mix data, the greater the time to the cut-out load the more compaction effort is required, either through a heavier roller, or a greater number of roller passes. In addition, what the system can also detect is pore water pressure build-up within emulsion mixes during compaction. This is reflected in the load increase rate, particularly with the first 30 kN load cycle. Such a situation may require a smaller load during rolling, but the roller to be used a greater number of times.

4 CONCLUSIONS

The compaction performance of bituminous material has been studied by a number of researchers over the past 30 or more years. The common approach taken with the definition of compactibility of material is through the profile of densification of material with increasing compaction effort. Compaction effort is applied to samples most commonly through the Marshall compaction hammer. The gyratory testing machine has also been used. Neither pieces of comapction equipment can easily detect what is happening during the process of aggregate interlock and binder displacement. The duriez compaction procedure, with a load cell and personal computer, employs a simple procedure which can provide significant detail about aggregate interlock and binder displacement during material densification. It can also detect pore pressure build-up with emulsion mixtures during material densification.

Using a single measure of compactibility, time to 10 kN cut out load, which is similar to the compaction resistance index, the sensitivity of hot mixes to temperature change can be defined. The data produced with hot mixes using the duriez compaction procedure replicates work using the Marshall compaction hammer, but with a much simpler procedure.

From the hot mix data the temperature range over which a significant increase in resistance to compaction occurs is important to site operations to ensure material layers are effectively compacted. Cold-lay mix data suggests that these materials have a different sensitivity profile to key performance characteristics: added, or contained water at compaction, or emulsion content. Cold mixes have an increasing resistance to compaction with decreasing total water content and increasing total liquid content through increased emulsion content. There appears to be no change in compaction resistance as with hot mixes. This is significant to the compaction performance of cold-lay materials.

The profiles of load build up during the compaction process are important to the interpretation of potential material compaction performance. Pore pressure build up can be detected with cold-lay material. Additional relaxation with cold-lay material reflects the need for the binder to be given time to displace, compared with hot mix material. This can be translated to the manner in which material should be compacted: fewer passes of a heavier roller or more passes of a light weight roller.

5 REFERENCES

1. British Standards Institution, BS 594 : Part 1 : 1992, "Hot Rolled Asphalt for Roads and Other Paved Areas, Part 1. Specification for Constituent Materials, and Asphalt Mixtures", London, 1992.
2. British Standards Institution, BS 4987 : Part 1 : 1988, "Coated Macadam for Roads and Other Paved Areas, Part 1. Specification for Constituent Materials and for Mixtures", London, 1988.
3. British Standards Institution, BS 594 : Part 2 : 1992, "Hot Rolled Asphalt for Roads and Other Paved Areas, Part 2. Specification for the Transport, Laying and Compaction of Rolled Asphalt", London, 1992.

4. British Standards Institution, BS 4987 : Part 2 : 1988, "Coated Macadam for Roads and Other Paved Areas, Part 2. Specification for Transport, Laying and Compaction", London, 1988.
5. HAUC SWP 163, 1991, "Specification for the Excavation and Reinstatement of Openings in Highways", 1991.
6. Brennan, M. J. and Sweeney, A., 1986, "The Duriez Compaction Procedure", The Journal of the Institution of Asphalt Technology, No. 38, pp 12-15.
7. Lefebvre, J. A. 1965, "Effect of Compaction on the Density and Stability of Asphalt Paving Mixtures", Proceedings of the 10th Annual Conference of the Canadian Technical Asphalt Association, Vol. 10, pp 23-109.
8. Ruiz, C. L. and Dorfman, B., 1968, "Sobre 1a Medida de la Compactacion y de la Comapctibilidad de las Mezclas Asfalticas del Tipo Superior", Comision Parmanente del Asphalto, Buenos Aires Argentina Decimoquinta Reunion del Asfalto, Mar del Plata, pp 189-208.
9. Lefebrve, J. A. and Robertson, W. D., 1969, "Effect of Mineral Aggregate Characteristics on the Compactibility of Asphalt Paving Mixtures", Proceedings of the 14th Annual Conference of the Canadian Technical Asphalt Association, Vol. 14.
10. McLeod, N. W. and McLean, J. A., 1974, "A Laboratory Investigation of the Compaction of Dense Graded Asphalt Concrete", Proceedings of the 19th Annual Conference of the Canadian Technical Asphalt Association, Vol. 19, pp 377-450.
11. Renkens, P., 1986, "The Effect of the Compaction Temperature on the Compactibility of Rolled Asphalt Mixtures", Die Asphaltstrasse, pp 210-225.
12. Markham, D. E., 1993, "The Development of Apparatus for the Investigation of the Properties of Emulsified Bituminous Materials", Unpublished M.Phil. Thesis, Heriot-Watt University, Edinburgh.
13. Cabrera, J. G., 1991, "Assessment of the Workability of Bituminous mixtures", Highways and Transportation, pp 17-23, Nov. 1991.
14. Khweir, K., 1991, "The Influence of Material Ingredients on Asphalt Workability", Unpublished Ph.D. Thesis, Heriot-Watt University, Edinburgh.

13 PERFORMANCE ASSESSMENT OF SPANISH AND BRITISH POROUS ASPHALTS

H. KHALID
Department of Civil Engineering, Liverpool University, Liverpool, UK
F.K. PÉREZ JIMÉNEZ
Department of Highway Engineering, Polytechnic University of Cataluña, Barcelona, Spain

Abstract
Porous Asphalt is currently gaining in prominence in the UK after recent legislation that will help in enhancing its use to a level comparable to that in other European countries. Apart from its obvious advantages of improving safety on major roads, porous asphalt has also been found to aid in the reduction of noise levels and of vehicle operating costs. Its disadvantages have been attributed to inadequate structural support, inadequate durability and premature loss of porosity due to clogging of the voids.

In Spain, porous asphalt has been used extensively and has a well established performance record. The material is predominantly designed with the aid of the Cantabro Test, developed originally at the University of Cantabria and is based on the abrasion loss undergone by specimens after being rattled in a Los Angeles machine without the steel balls.

In the UK, apart from the Binder Drainage Test propounded by the Transport Research Laboratory and which is conducted on uncompacted mix specimens, there is no laboratory tool which can be used to aid in porous asphalt mix design. This paper attempts to arrive at a mix design method which takes into account the parameters referred to above which address the limitations of porous asphalt, namely strength, porosity and durability. Non destructive tests for the determination of the elastic modulus and permeability have been proposed to make maximum use of the prepared samples before introducing them to a Cantabro-like test for abrasion loss determination.

Porous asphalt grading specifications from BS4987 have been adopted and designed according to the proposed method. The performance of these mixes was then compared to that of mixes designed according to the Spanish specifications and inferences made thereof. Recommendations for further refinement of the method have also been made.
Keywords: Porous Asphalt, Mix Design, Field Performance, Laboratory Tests.

1 Introduction

Porous asphalts are bituminous mixes with specially selected aggregate

Performance and Durability of Bituminous Materials. Edited by J.G. Cabrera and J.R. Dixon. © E & FN Spon. Published in 1996 by E & FN Spon, 2–6 Boundary Row, London SE1 8HN. ISBN 0 419 20540 3.

gradings so that, when compacted, have about 20% interconnected air voids. These voids allow rain water to percolate through the layer which is normally constructed on an impermeable basecourse layer, and drains out through the road edges into appropriately placed drainage pipes. Porous asphalt descended from the porous friction course material which had been originally developed by the then Air Ministry and RRL for use on airfield runway pavements to avoid aquaplaning and skidding in wet weather (Fabb 1992)). Unfortunately, it has not been used adequately in the UK due to some limitations which will be outlined herein, despite the fact that these limitations may be outweighed by its important advantages.

Porous asphalt has been used extensively in Europe. Figure 1(Fabb 1992) shows the 1990 level of use by many European countries together with the percentage of total production. It is interesting to note from this figure that porous asphalt is used in widely differing climates; hot and dry, e.g. Spain to cold and wet, e.g. Austria and Germany. In the UK, only a limited number of trials have been reported (Daines 1992, SERC 1993) where sections of the A38 and the M1 were laid with porous asphalt using conventional binders and binders containing polymers or other additives. From this limited use of the material, it is hoped that sufficient experience will be built up to enhance the use of porous asphalt to levels corresponding to those in other European countries.

In this article, the function, advantages and disadvantages of porous asphalt are outlined, and the existing Spanish experience in its design, use and performance is highlighted. A preliminary exercise has been undertaken with the objective of utilising some laboratory tests as design tools in a suggested mix design procedure.

2 The Case for Porous Asphalt

A typical porous asphalt would be composed of about 75% of coarse aggregate (retained on 2.36mm sieve), 4% binder and the rest of fine aggregates and filler. It is so designed to contain about 20% air voids which makes it very porous. When used as a wearing course, its interconnected voids allow the water to pass through and drain away in the manner shown in Figure 2 (Fabb 1992), thereby preventing aquaplaning.

2.1 Advantages of Porous Asphalt

Most of the following advantages have been précised from a PIARC Report (Lefebvre 1993) on porous asphalt produced by a Working Group formed by the Technical Committees on Flexible Roads and Surface Characteristics. The Working Group included workers in the field from Belgium, Germany, Italy, Switzerland, Holland, France and the Spanish co-author of this article.

2.1.1 Safety aspects

(a) Aquaplaning

One of the major hazards when driving in the rain is aquaplaning. A layer of water builds up between the tyre and the road, breaking the

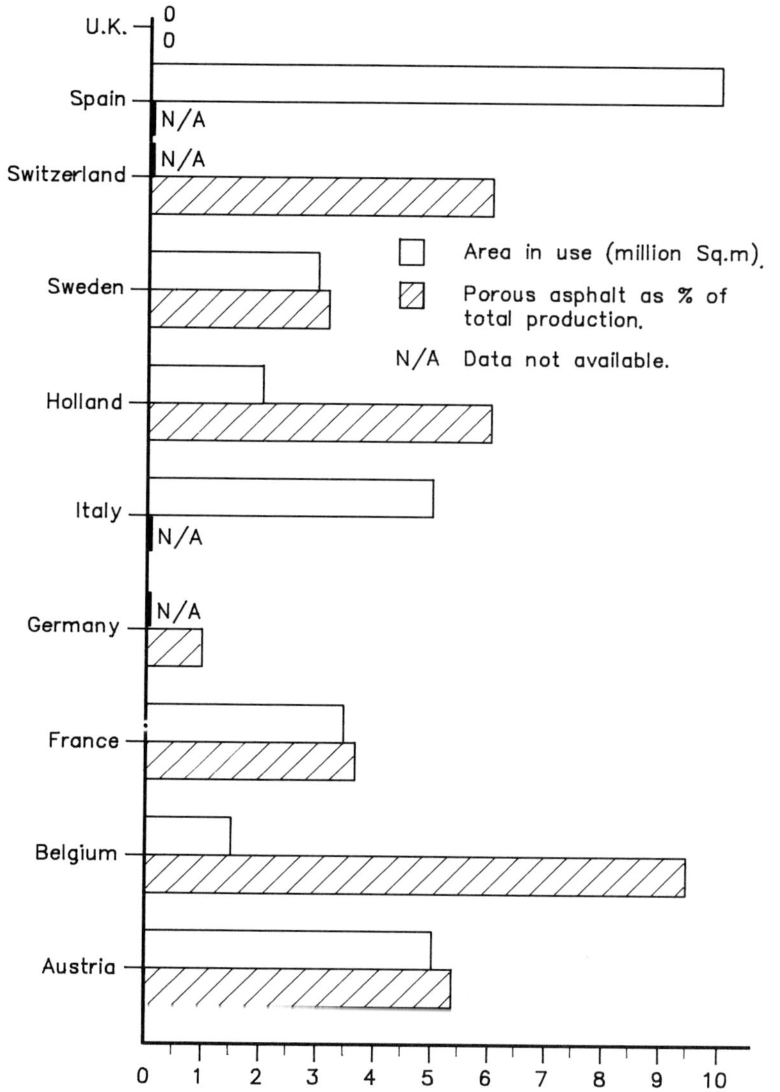

Figure 1. The use of porous asphalt in some European countries in 1990 (after Fabb 1992).

contact between them. The tyre (and thus the car) will "float" on the water, and consequently it is quite difficult to steer and to brake. Reference is made to Figure 3 which shows the pressure P in the hydraulic wedge which lifts the tyre from the surface. Aquaplaning depends primarily on the amount of rain and the drainage capacity of the road, the driving speed and to a certain extent on the tyre profile. On porous asphalt, even when the passages are cluttered with debris and dirt, aquaplaning does not occur at normal speeds. For this reason, porous asphalt was first introduced on airfield runways.

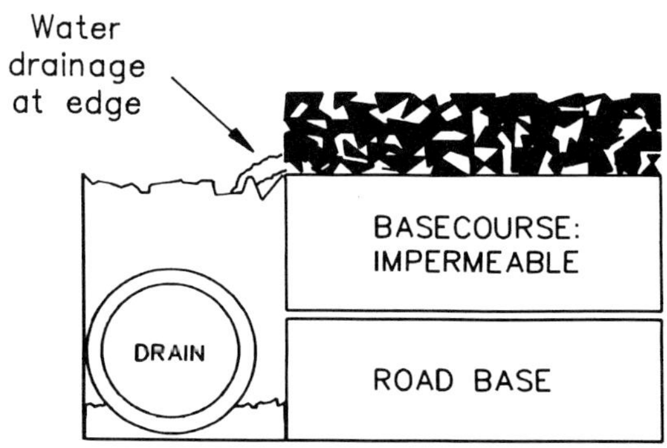

Figure 2. Section through a typical porous asphalt pavement showing its function.

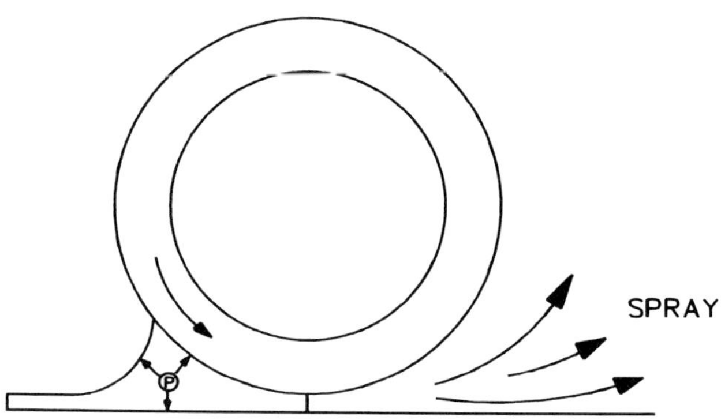

Figure 3. Tyre / water / pavement interface showing aquaplaning 'lift–off'.

(b) Splash and spray

Rolling wheels throw up water from pools on the surface "splash" and they also mist the surface water "spray". Physically, the two phenomena are quite different, but in practice they can be addressed at the same time as they usually occur together. The water droplets reduce the visibility in the atmosphere in the same way as rain and/or fog. The reduction in visibility is usually more severe than in real fog because the droplets in splashing and spraying are larger than those in fog, and have higher densities. A real fog of this density would precipitate as drizzle, and would disappear rapidly. Measurements of the visual range in splash and spray "clouds" are not available, but it will often be below 5 to 10m. Such a short visual range is not experienced in real fog. Poor visibility is restricted to the rear and the sides of vehicles and particularly of trucks. The effect of splash and spray may therefore be quite different from the effect of fog. A further effect of splash and spray is the misting and dirtying of windscreens, as the water is usually contaminated. Often, windscreen washers and wipers cannot cope, particularly when salt has been spread on a snowed road.

(c) Glare reduction

Users of motor vehicles have to observe the road ahead. At a normal driving speed, the road must be viewed from a considerable distance, e.g. 50 to 100m ahead of the vehicle. As the eye height in cars in the UK is about 1.05m above the road surface, the road is observed under a glancing angle of 1 degree or less. When viewed under such angles, most surfaces reflect the incident light very strongly because when the surface is smooth, it usually looks like a mirror. The water from a small rain shower on a traditional asphalt is enough to produce this aspect. Porous asphalt, on the other hand, shows, even when wet, a predominantly diffuse reflection, and even when observed under a glancing angle. The diffuse reflection is important both in daylight and in darkness. In daytime, a horizontal surface with mirror reflection reflects the sky. Even the overcast sky is relatively bright, and consequently the road surface is almost as bright as the sky. Furthermore, all surface characteristics of the road disappear under the reflected light, particularly road markings which tend to be invisible. The effect is similar at night under overhead lighting.

On unlit roads, where the night-time visibility has to be ensured by applying retroreflecting devices, the effect of water is yet more severe. The water layer prevents the light from car headlamps from penetrating in the retroreflectors, thereby rendering them invisible. Porous asphalt reduces the amount of water on the road, and the situation can be improved even further by applying profiled road markings (corrugated lines, or retroreflecting road studs). Finally, a diffuse reflecting road surface enhances the economy of road lighting installations by ensuring a higher road surface luminosity and a better uniformity.

On unlit roads, porous asphalt road surfaces prevent in dry, and even more so in wet conditions, the mirror reflections that are caused by headlights of oncoming vehicles. In this respect, porous asphalt may contribute considerably to the reduction of glare caused by such headlights.

(d) Skid resistance
Rain may reduce the skid resistance of road surfaces considerably even when no aquaplaning takes place. Porous asphalt may counteract this effect, even when the surface stays humid. The skid resistance of a wet porous asphalt at high speeds will be higher than that of a wet traditional asphalt, but not equal to that of a dry road. At low speed, the skid resistance of a wet porous asphalt is not higher than that of a wet traditional asphalt.

2.1.2 Economic aspects

(a) Fuel consumption
It has been reported (Fabb 1992) that there can be a 2% saving in fuel consumption for vehicles running on porous asphalt. This is attributed to the vibrations in tyres which generate the noise and increase the rolling resistance, thereby increasing fuel consumption. The savings figure indicated above depends on the alternative surfacing type with which porous asphalt is compared. The rougher the alternative - as is likely in the UK in view of the exigence on texture depth - the greater the savings with porous asphalt.

(b) Tyre wear
It has been suggested (Fabb 1992) that due to the enhanced macrotexture of porous asphalt, it stresses the tyres much less than conventional asphalt, thereby reducing the rate of tyre wear. It may be difficult to study this aspect in isolation, but general indications point to the fact that reducing the stresses on tyres aids in reducing tyre wear.

2.1.3 Environmental aspects

(a) Noise reduction
On average, the noise level resulting from dense traffic on porous asphalt is about 3 dB(A) lower than on traditional asphalt. There are reasons to believe that the noise reduction goes down as the surface gets older, probably due to clogging of the voids.

On most concrete roads it is yet 5 to 7 dB(A) higher than on asphalt. Particularly when residential or recreational areas are located near motorways or trunk roads, the noise reduction properties of porous asphalt are important. In the Netherlands and in Belgium, these noise reducing properties were, much more than the road safety aspects, the incentive to apply porous asphalt on a large scale in recent years. The same considerations led to a reduction of the application of cement concrete roads, irrespective of their many "strong points" particularly as regards maintenance and life. Research on the construction of concrete road surfaces with drainage and noise reducing properties is presently under way in the Netherlands.

Noise reduction comprises two elements. The noise caused by the rolling tyres is reduced. The second is that most of the downward noise from the car itself is absorbed to a certain extent, contrary to traditional road surfaces where it is reflected back.

(b) Noise barriers
The major reason of the popularity of porous asphalt in the Netherlands is the legislation that forces the road authority to install noise barriers along the roads where noise levels at the locations of the dwellings is higher than a legally permitted maximum. Noise barriers can be dangerous and are very expensive. The noise reducing properties of porous asphalt (particularly in new condition) allowed the road authorities in a number of cases to abstain from erecting noise barriers. In this respect porous asphalt has a clear positive cost/benefit ratio in direct economic terms, and all the other benefits were considered as being secondary. Noise reduction, in general, is the main "selling point" of porous asphalt. This automatically limits its use to high speed rural roads as the noise reduction is highly dependent upon speed.

(c) Driver comfort
The level of noise is reduced inside the vehicle as well as outside with the use of porous asphalt. In Germany, porous asphalt is sometimes referred to as "Whisper Asphalt". This factor aids in making driving less stressful and more pleasant.

(d) Use of waste rubber
Waste rubber from old car tyres is a major environmental problem in many countries. Using granulated waste rubber in asphalt roads is a way to dispose of it. Experience showed, however, that this application was often more than waste disposal, where many characteristics of asphalt roads improved by using granulated rubber. The most important improvements are the noise reduction and the increase in life. When using granulated waste rubber in porous asphalt, the increase in life was sometimes dramatic. Increases in life from some 9 to 12 years have been reported (Lefebvre 1993).

2.2 Disadvantages of Porous Asphalt

2.2.1 Strength
Traditionally porous asphalt has not been considered as a contributing layer to the overall structural integrity of the pavement. Nevertheless, a range of structural equivalency factors has been quoted in relation to conventional dense asphalts by several workers, from about 80% in Holland (Van der Zwan et al 1990) down to 50% in France (Sainton 1990). It should be noted that, given a void volume of about 20%, direct comparison of porous asphalt with, say a Hot Rolled Asphalt may not be justified. To this end, on a ton to ton basis, it is the opinion of the authors that there is very little difference between the two types of material.

2.2.2 Durability
The service life of porous asphalt has been generally considered to be shorter than that of a traditional dense asphalt. This is mainly attributed to premature clogging of the voids which leads to ineffective drainage of surface water. To tackle this problem, encouraging trials have been carried out in Belgium (Van Heystraeten

and Moraux 1990) to de-clogg porous asphalt surfacings using a suction sweeper with a water jet. If these trials were to prove successful, it means that the "useful life" of porous asphalt can be extended considerably. This notwithstanding, it must not be forgotten that invariably dense asphalts require maintenance/replacement within a few years of their construction. This, coupled with the fact that rutting is not common in porous asphalt, contrary to traditional asphalt, the relative service lives of the two materials are clearly changing in favour of porous asphalt.

2.2.3 Construction costs

It is difficult to give a general indication on the differences in construction costs between porous asphalt and traditional asphalt. It is generally assumed that porous asphalt is more expensive to build than traditional asphalt. The main aspect seems to be the need for aggregate with a higher resistance to polishing, and a more precise processing during construction as regards the temperature of the mix and the air temperature. However, as road authorities, designers and contractors gain more experience in applying porous asphalt, this extra cost factor may decrease. Furthermore, the seemingly high cost of aggregates may vary from one country to another. On the one hand, porous asphalt requires more aggregate with high PSV than a surface treatment, as it is used throughout the mixture and not only in the rolled-in top layer, but the mean value of PSV might be lower. Further, for reasons that are equally obvious, the comparison depends on the reference surface. Porous asphalt may require similar binder to a close graded asphalt, but less than asphaltic concrete or Hot Rolled Asphalt.

Another factor that may influence the construction costs of porous asphalt is the need to adapt the road markings. Since part of the road marking material is drained away, more material is needed. On the other hand, as a result of the greater macrotexture, road markings may have a longer practical life, even if the top of the markings is worn off rapidly. There is a suggestion that ordinary painted road marking on porous asphalt is almost as good as thermoplastic markings on traditional asphalt (Lefebvre 1993).

Still another factor is the need to have a water-tight layer underneath the porous asphalt top layer. In some cases, this requirement may increase construction costs.

2.2.4 Winter maintenance

On porous asphalt more salt is needed in winter which increases the cost of winter maintenance. It has been noticed in Switzerland that snow and icing rain can form quicker on porous asphalt than on dense asphalt because deicing salts do not remain on the surface (Isenring et al 1990). Moreover, it has been suggested by Dutch workers (Van der Zwan et al 1990) that by increasing the frequency of liquid salting operations on porous asphalts can help overcome many of the difficulties encountered. In Belgium (Van Heystraeten and Moraux 1990) it has been reported that there has been very little difference in behaviour in snowy weather between porous and dense asphalts when the former is spread intensively with liquid deicing salts.

3 The Spanish Experience

The first application of porous asphalt if Spain was in 1980 on four experimental road sections on one of the northern highways located in a region of frequent rainfall (Ruiz et al 1990). Initially, the objective was to use these mixtures in rainy areas in order to improve traffic safety and comfort on wet surfaces.

The favourable results obtained from these mixtures have promoted the construction of new experimental pavements and small projects to be carried out in the following years. But in 1986, this material started to be used extensively, after initial doubts about its durability were eliminated. Now, the purpose for using this material has changed. It is not only used to improve driving conditions in the rain, but also to provide a durable surface, with a smooth, safe, and quiet ride in any type of weather.

Currently Spain has 30 million m^2 of porous asphalt roads. Porous asphalt is being used for all types of traffic conditions and for any type of roads and highways. The most notable projects are th 44km (about 500,000m^2) on Highway N-VI, between Las Rozas and Villalba, with some 20,000 vehicles per day per carriageway, 2,000 of which are trucks (13 tonne axle load), the 70km (about 800,000m^2) on the toll road between Bilbao and Behobia, with about 9,000 vehicles per carriageway, of which 1,200 are trucks, and the 33km (400,000m^2) in private toll roads, with traffic varying between 800 and 1,800 trucks per day.

The most common practice at present is to use 4cm layers with 10 or 12.5mm aggregate gradings, with very little sand, and 4.5% of pure or modified bitumens which results in a voids content of about 20%.

3.1 Material Requirements

3.1.1 Aggregates

Considering that the material is for a thin, open, top layer, coarse aggregates which show great resistance to fragmentation, good and stable microtexture, and adequate interlock are called for.

Fragmentation of aggregates can lead to particle losses, ravelling, and the closing up of the surface texture by the separate fines. An abrasion loss value (Los Angeles machine) of 20% is considered as a maximum. For the same reason, a flakiness index below 25% is required.

Frictional characteristics of the surface make a nonpolishing aggregate necessary for maintaining a good, durable microtexture. Spanish specifications set PSV's above 0.45 for traffic volumes of more than 800 trucks per day per lane, and 0.40 for other traffic volumes.

Limestone is frequently used as fine aggregate because of its adhesion to the binder. Mineral filler is always added (commercial limestone dust or cement). To avoid the presence of detrimental fine dust, a sand equivalent value above 50% is required.

3.1.2 Aggregate grading

The selected aggregate grading primarily influences the water drainage capacity, resistance to particle losses, resistance to plastic deformation and macrotexture of the mix. For porous asphalt two grading envelopes have been defined, namely P12 and PA12 and are given in Table 1. The P grading envelope has a discontinuity in the 2.5mm size, whereas the PA envelope has a discontinuity in the 5mm size. These gradings usually need three commercial aggregates; 2.5mm to dust, 5 to 2.5mm and 10 or 12 to 5mm. Using them, mixes with about 18 to 23% voids can be obtained.

Table 1. Grading envelopes of the Spanish porous asphalt mix specifications

	% Passing						
GRADING	20mm	12.5mm	10mm	5mm	2.5mm	0.63mm	0.08mm
P - 12	100	75-100	60-90	32-50	10-18	6-12	3-6
PA - 12	100	70-100	50-80	15-30	10-22	6-13	3-6

The maximum particle size has been set at 10 or 12.5mm for both gradings, although the 10mm top size is generally used. With this size, sand patch depths are also related to the thickness of the layer being used (4cm).

3.1.3 Binder type

In porous mixes, because of the open texture, a thick film of binder coating is sought in an attempt to offset early aging. From this viewpoint, binders with high viscosity would be preferred. On the other hand, hard bitumens would take less time to reach a critical hardness of the binder. For this reason, an equilibrium is necessary.

In selecting the binder, other factors to consider are weather and traffic volume. Soft bitumens tend to bleed under high temperatures and can lead to plastic deformations in the mix, particularly under heavy traffic volumes. In cold climates, hard bitumens can produce brittle mixes.

Taking all this into account, in Spain the grades of binder specified are 60/70 and 80/100. The former is recommended in areas of mild and hot climates for heavy traffic. But the binders more commonly used are polymer (EVA and SBS) modified bitumens or bitumens with fibre addition. The main purpose for using such binders is to improve the resistance against particle losses with very open mixtures through a higher cohesion, and get longer durability through thicker films of binder because of the higher viscosity. A reduction in the thermal susceptibility of the mix is also sought in an attempt to get higher consistencies with high temperatures and more flexibility with low temperatures. Currently, 75% of porous mixes existing in Spain have either a polymer modified binder or a conventional binder with fibres.

3.2 Mix Design Approach

The design of porous asphalt is based on:

(1) A minimum binder content to ensure resistance against particle losses resulting from traffic and a thick film of binder on the aggregates, and
(2) A maximum binder content to avoid binder runoff and have a good drainability in the mix.

The resistance to particle losses is analyzed through the Cantabro test seen in Figure 4 an abrasion and impact test conducted in the Los Angeles rattler, without balls and at controlled temperature, on Marshall samples compacted with 50 blows on each side. The results are given as the weight loss, in percentage, after 300 drum revolutions. The maximum abrasion loss value admitted is 25% at 25°C. With this test, a minimum amount of binder is determined.

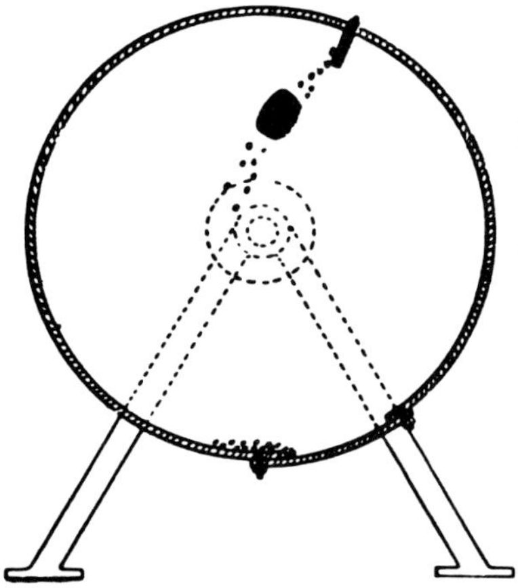

$$P = \frac{P_1 - P_2}{P_1} \times 100$$

Where: –

P = Abrasion loss (%)

P_1 = Initial weight of the sample

P_2 = Final weight of the sample

Figure 4. The Cantabro test for abrasion loss determination.

The calculation of voids is made on the same Marshall samples, considering the volume which is geometrically determined. For a specific grading, a minimum amount of voids is set (20%) to define a maximum content of binder. Also there is a maximum binder content to prevent drainage of the bitumen from the aggregates, although this is not yet under specification, but for which there exists a test (Pérez Jiménez and Gordillo 1990) similar to the binder drainage test used by the TRL in the UK (Daines 1992).

In the design of porous mixes, the Cantabro test after immersion and some laboratory permeability tests have been used, but they are not under a standard yet either. Sometimes indirect tension and wheel tracking tests have also been used. This method usually gives binder contents of about 4.5% for normal specific gravity aggregates. With these in practice, no major problems of particle losses or binder runoff have been encountered.

3.3 Performance of Existing Surfaces

In the first application of porous asphalt, a conservative approach was taken primarily using mixes with a moderate content of voids (15 to 18%). The good durability of mixes with voids contents of more than 20% in the experimental road sections and the closing up observed in the mixes with a low voids content has meant that since 1986, the more open mixes have been preferred.

The measurement of in-situ permeability was conducted by means of the LCS (Laboratorio de Caminos de Santander) Drainometer developed at the University of Santander in 1981. It is a variable charge static outflow meter, shown in Plate 1, used to measure the time necessary to drain 1.735 litres of water through a pavement surface of $7cm^2$ area. The voids content is related to the time of water drainage by means of a power equation (MOPU 1988) to enable the evaluation of voids volume and the extent of their closing up at any point in time.

Plate 1: The LCS Drainometer.

For the mixes laid in 1980 in Santander with voids content less than 20% the initial drainage times (LCS Drainometer) on the highway varied between 30 and 75 sec. The texture depth measured with the sand patch varied between 1 and 1.5mm. The Sideways Force Coefficient (SFC) measured with the SCRIM at 50km/hr gave values of between 0.50 and 0.70. Values of the Skid Resistance Tester Coeficiente (SRC) between 0.45 and 0.70 were found.

The traffic level on this road is 5000 vehicles per day per lane of which 700 are trucks. It is a rainy area used mainly for agricultural traffic. Some of the laid sections maintained a certain drainage capacity after 7 to 9 years. The LCS Drainometer values gave a wide scatter from 300 sec. to more than 600 sec., which is an indication of the closing up of the surface voids. This can be caused by various types of debris material, the silting up of internal voids due to the movement of fines or densification due to the action of tyres. The SFC and SRC values were 0.50 to 0.60 and 0.50 to 0.70, respectively, after the same period. The sand patch depth was 1.2 to 1.5mm.

Despite the decrease in drainage capacity, all the sections, including those that were closed up, remained dry in light rains or immediately after heavy rains, with a marked difference in this aspect compared to conventional dense graded mixes.

For the mixes laid in 1986 on the same road, but with voids content more than 20%, the initial drainage time varied between 15 and 25 sec. and the sand patch depth between 1.3 and 2.2mm. The SFC and SRC values were 0.60 to 0.80 and 0.54 to 0.80 respectively.

The mixes tested in the experimental pavements of Santander had, after 7 to 9 years of service, drainage times of between 150 and 300 sec.

In the Las Rozas-Villalba road (44km), with 2,000 trucks per day per lane, the mix initially had a voids content of 22% and registered initial drainage times of about 20 sec. After 2 years, the interval of drainage time values was between 20 and 50 sec.

4 Experimental Work

Crushed granite aggregates from the Croft quarry in Leicestershire have been used to manufacture the porous asphalt mixes used in this study. The PSV of the 10mm aggregates was reported by the supplier to be 0.47. For the British mix, the grading suggested in BS 4987 : Part 1 : 1988, Table 39 was used with two straight run 100 and 200 pen binders with Softening Points of 44 and 40°C respectively. For the Spanish mix, the P12 grading with 100 pen binder was used, which corresponds to the British 10mm size pervious wearing course stipulated in BS 4987. These two gradings required three commercially available aggregate sizes, namely 10mm, 6mm and 3mm to dust. The two grading curves are shown in Figure 5.

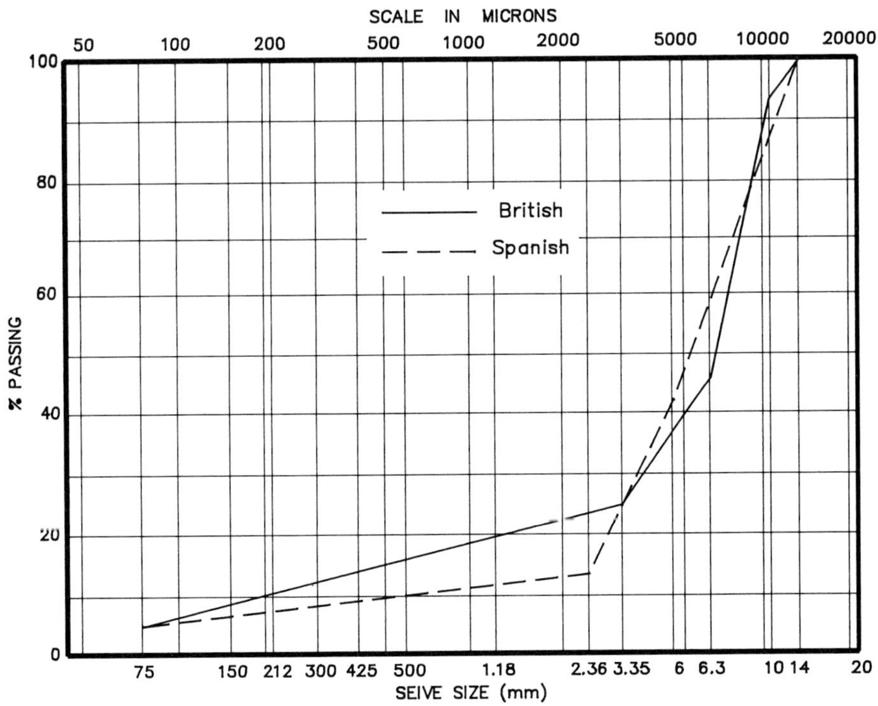

Figure 5. Particle size distribution of the British and Spanish porous asphalt mixes used.

The mixes were compacted in Marshall moulds giving 50 blows per side for each sample. The binder contents used were those suggested in BS 4987 and by the Spanish Specifications, with one binder content on either side of the recommended 5.2% value in BS 4987 to investigate the influence of binder content on the studied parameters. Five samples were prepared for each determination.

Volumetric measurements were made on the prepared samples to calculate the voids contents, for which an industrial cling film of specific gravity of 0.89 was used. The samples were then tested for the elastic modulus in the Repeated Load Indirect Tensile Test (RLIT) at 20°C using the Nottingham Asphalt Tester (NAT) following the procedure in British Standard Draft for Development 213 (BSI 1993). Due to the open structure of the mix, the tip of the LVDT stem was replaced by a 5mm diameter disc for a more consistent contact with the tested samples. The RLIT results are shown in Figure 6 together with voids content. It can be seen that there is a noticeable difference in stiffness between the two British mixes which decreases with increase in binder content. The difference in voids content for the same binder content, however, is considered to be within the experimental accuracy of the measurement. As for the Spanish mix, due to its lower fines content, the voids volume and modulus are higher than those of the corresponding British mix.

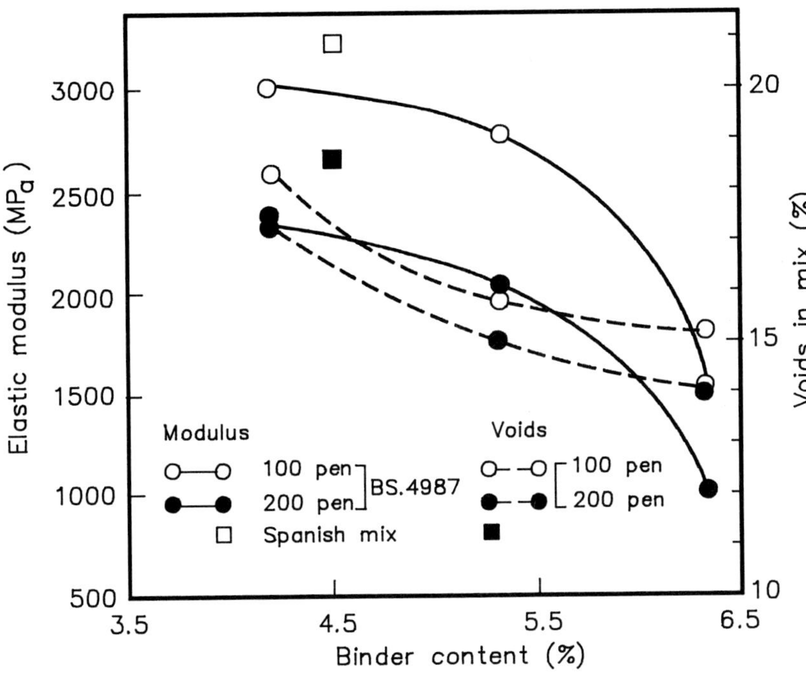

Figure 6. Elastic modulus and voids content of the porous asphalt mixes as a function of binder content/type.

The non-destructive nature of the RLIT enabled the use of the same tested samples for further investigation, thereby avoiding duplication of sample manufacture and lessening sources of variability. As such, it was possible to determine the permeability of all the prepared samples in the perspex set-up shown schematically in Figure 7.

The 30cm high cylinder has markings 5cm apart with the highest being 25cm above the surface of the specimen. Each specimen was placed in the set-up and the time taken for the water level to drop from the 25 to the 20cm mark was noted. This type of measurement is similar in principle to that of the LCS permeameter and the permeameter used by the Swiss Federal Institute of Technology (Isenring et al 1990). If the flow time was greater than 180 seconds the specimen was deemed to have insufficient porosity and the testing of that particular specimen stopped. The permeability measurements gave a large scatter of results reaching up to a standard deviation of 40% of the mean value. However, the objective was to be able to

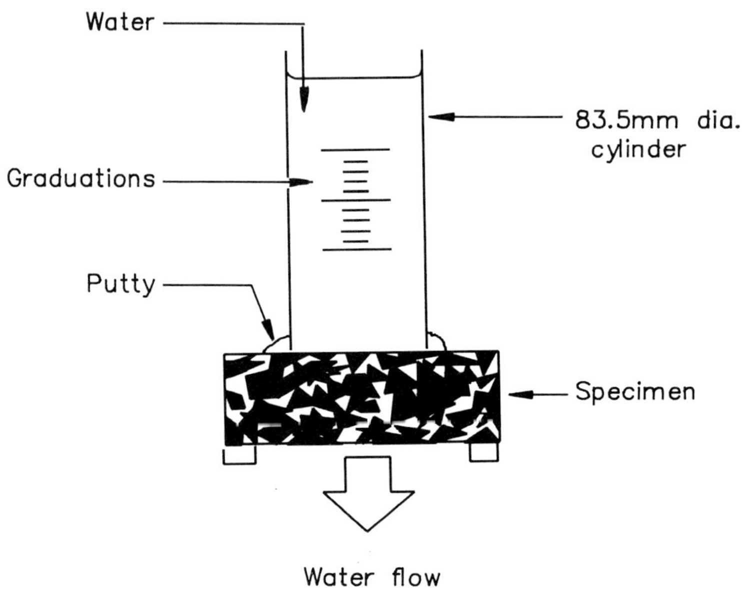

Figure 7. Permeability set-up.

realize the distribution of the voids, not just their volume, and to compare the permeabilities of new and old porous asphalt mixes if the permeameter were to be used in-situ. The permeability results have been correlated with the voids content, as seen in Figure 8, which shows that the lower the binder content the more likely it is to have "permeable" voids.

As referred to earlier, the basic mechanical characteristic which has been taken into account in the Spanish design approach of porous asphalt is the resistance to disintegration as measured in the Cantabro Test. The University of Liverpool has adapted this test using an Impact Box instead of the Los Angeles machine to effect the abrasion action on laboratory prepared samples. This apparatus, shown in Plate 2 is rotated at 30 rpm for 10 minutes after inserting the weighed sample inside, and weight loss upon rattling is measured and expressed as a percentage of the original weight.

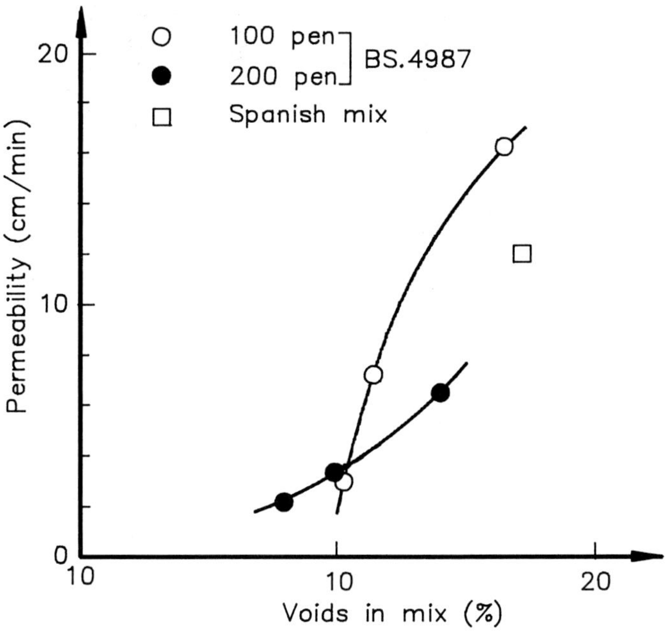

Figure 8. Correlation between permeability and voids content of the porous asphalt mixes.

The same specimens, again, were introduced to the IB after having their permeabilities measured. Figures 9 and 10 show the abrasion loss to increase with decrease in binder content and increase in voids content, almost irrespective of binder type.

All the tests referred to above were carried out at 20°C.

5 Proposed Mix Design

It has been shown in the previous section that the test procedures used in this work, namely the RLIT, permeability and the Cantabro tests address three vitally important parameters which can be used in the design and assessment of porous asphalt. These parameters can be used to evaluate three critical properties of porous asphalts, namely structural support, drainability and resistance to disintegration, upon which the success of the material relies considerably. It is therefore intended here to propose a mix design method for porous asphalt which adopts the above-mentioned tests to arrive at a mix composition that would meet the anticipated service requirements.

In order to ensure that there would be no binder run-off during the transport of porous asphalt, it is recommended to adopt a binder drainage test like the one propounded by TRL (Daines 1992), or like

Plate 2: The Impact Box for the determination of abrasion loss.

the one used in Spain (Pérez Jiménez and Gordillo 1990). Table 2 outlines the proposed mix design procedure for the determination of the Design Binder Content incorporating the referred to properties.

Table 2. Proposed mix design for porous asphalt

Binder Content	Mix Property	Requirement
Maximum	• Binder Run-Off (Binder Drainage Test)	≤ 0.3%
Minimum	• Stiffness Modulus (20°C)	≥ 2000MPa
Maximum	• Drainability a. Voids in Mix b. Permeability	 > 16% > 25m/day
Minimum	• Resistance to Disintegration (Cantabro Test Losses at 20°C)	≤ 25%

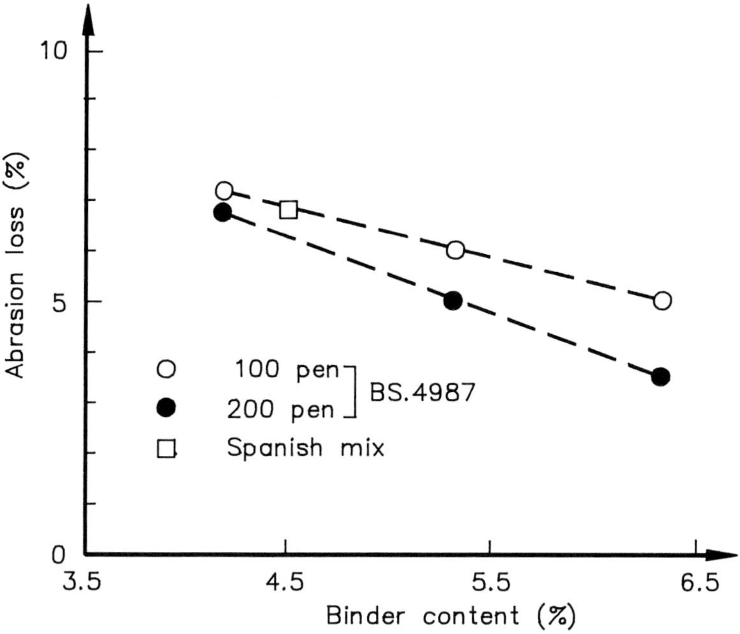

Figure 9. Abrasion loss of the porous asphalt mixes as a function of the binder content/type.

In order to determine the resistance to stripping caused by water, it is recommended to impose a 70% retained modulus after immersion in water at 20°C for 24 hours. Moreover, the Cantabro test can be conducted after the same immersion regime, duplicity of samples notwithstanding, to determine the resistance of porous asphalt mixes to particle losses. This is especially so when it is required to assess the improved adhesiveness brought about by the inclusion of fibres or polymer modified binders.

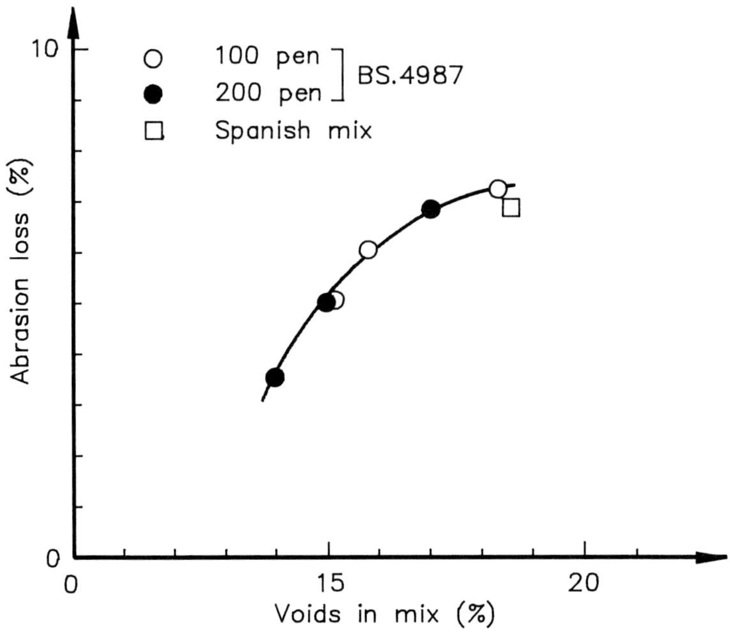

Figure 10. Correlation between abrasion loss and voids content of the porous asphalt mixes.

6 Conclusions

The reviewed literature revealed that, despite their alleged limitations, porous asphalt mixes have been used extensively in continental Europe. There appears to be a consensual agreement among users that the safety and environmental advantages of porous asphalt far outweigh its disadvantages, and thus justify its enhanced use.

In Spain porous asphalts with voids content of less than 20% have varied widely in their behaviour. With heavy traffic, they have closed up after 2 years of use. With medium traffic, however, they have maintained their drainage capacity after 9 years. None of the pavements using this material has shown any serious deterioration. Porous asphalts with voids contents higher than 20% held up very well even under heavy traffic. As is the case with the other mixtures, these have not shown any serioud deteriorations, and after several years' service, they still maintain excellent skid resistance.

The performance of Spanish and British porous asphalt mixes has been assessed in the laboratory with respect to their stiffness moduli, voids/permeability and resistanct to disintegration. The Spanish mix gave a predominantly good performance in all the conducted tests. The British mix has more fines than the Spanish mix, thus leading to less initial voids and lower drainability.

In the light of the adopted tests, i.e. RLIT, permeability and the Cantabro, a mix design method has been proposed for the determination of the Design Binder Content for porous asphalt mixes.

7 References

British Standard Institution (May 1993) Draft for Development : Method for determination of the indirect tensile stiffness modulus of bituminous materials, DD213, **BSI**, 17pp.

Daines, M.E. (1992) Trials of porous asphalt and rolled asphalt on the A38 at Burton **TRRL Research Report** 323, 27pp.

Fabb, T.R.J. (June 1992) The case for the use of porous asphalt in the UK **ICE Seminar**, London, 15pp.

Isenring, T., Köster, H. and Scazziga, I. (1990) Experiences with porous asphalt in Zwitzerland, **Transportation Research Record** 1265, Transportation Research Board, Washington DC, 41-53.

Lefebvre, G. (1993) **Porous Asphalt**, Permanent International Association of Road Congresses (PIARC), Paris, 192pp.

Ministerio de Obras Publicas y Urbanismo (MOPU) (1988) Permeabilidad in situ de pavimentos drenantes mediante el permeametro LCS, **MOPU**, NLT-327/88, Madrid,.

Pérez-Jiménez, F.E. and Gordillo, J. (1990) Optimization of porous mixes through the use of special binders, **Transportation Research Record** 1265, Transportation Research Board, Washington DC, 59-68.

Ruiz, A., Alberola, R., Pérez-Jiménez, F.E. and Sánchez, B. (1990) Porous asphalt mixtures in Spain, **Transportation Research Record** 1265, Transportation Research Board, Washington DC, 87-94.

Sainton, A. (1990) Advantages of asphalt rubber binder for porous asphalt concrete, **Transportation Research Record** 1265, Transportation Research Board, Washington DC, 69-81.

Science and Engineering Research Council (SERC) (April 1993) Porous asphalt road surfacings, **Research Focus** No. 13, 12pp.

Van der Zwan, J.Th., Goeman, Th., Gruis, H.J., Swart, J.H. and Oldenburger, R.H. (1990) Porous asphalt wearing courses in the Netherlands : State of the art review, **Transportation Research Record** 1265, Transportation Research Board, Washington DC, 95-110.

Van Heystraeten, G. and Moraux, C. (1990) Ten years' experience of porous asphalt in Belgium, **Transportation Research Record** 1265, Transportation Research Board, 34-40.

14 FATIGUE CHARACTERISATION OF BITUMINOUS MIXES USING A SIMPLIFIED TEST METHOD

J.M. READ and S.F. BROWN
Department of Civil Engineering, University of Nottingham, Nottingham, UK

Abstract
This paper describes the test method and equipment used to carry out the Indirect Tensile Fatigue Test (ITFT) on bituminous mixtures using the Nottingham Asphalt Tester (NAT). Results of fatigue tests using an ITFT on four typical U.K. bituminous mixtures, are presented to demonstrate the ease of use and capabilities of this simplified test. The work forms part of a major research project dealing with the development of practical test methods to measure the mechanical properties of bituminous materials for pavement construction.

The four mixtures which were tested consisted of a 30/14 Hot Rolled Asphalt (HRA), a Styrene Butadiene Styrene (SBS) polymer modified 30/14 HRA, a 20mm Dense Bitumen Macadam (DBM) and a 28mm DBM 50. The ranking of the materials indicated that the polymer modified HRA performed somewhat better than the unmodified HRA. The two DBM materials were less satisfactory than the HRA's with regard to fatigue characteristics, although both DBM's had a higher stiffness modulus.

The results also indicated that the ITFT appears to be able to characterise the fatigue life of a bituminous mixture by testing a small number of specimens (less than 10) at high temperatures (in excess of 25°C) and at high stress levels (greater than 450 kPa). This means that the fatigue testing time needed to produce an adequate fatigue relationship for a bituminous material could be as little as 2 hours. This is significantly shorter than the time required by other traditional methods, therefore, making the ITFT suitable for routine commercial testing.
Keywords: Bituminous Mixtures, Dense Bitumen Macadam, Equipment Development, Fatigue, Hot Rolled Asphalt, Indirect Tensile Fatigue Test, LINK 'Bitutest' Project, Nottingham Asphalt Tester, Research, Rise-time, Simplified

Performance and Durability of Bituminous Materials. Edited by J.G. Cabrera and J.R. Dixon. © E & FN Spon. Published in 1996 by E & FN Spon, 2-6 Boundary Row, London SE1 8HN. ISBN 0 419 20540 3.

Test Method, Stiffness Modulus, Strain, Stress, University of Nottingham

1 Introduction

1.1 The LINK 'BITUTEST' project

The greater demand on UK roads and the need to reduce costs has lead to the construction industry requiring simplified test methods which measure the mechanical properties of bituminous materials. The Link "BITUTEST" project is making a major contribution to the necessary development work. The project involves collaboration between the Department of Transport, the Science and Engineering Research Council, nine companies (Industrial partners) and fourteen highway authorities (Associate partners). The structure of the project is as shown in Table 1.

Table 1. The structure of the LINK 'BITUTEST' project

Project Leader	University of Nottingham
Public Sector	Department of Transport Science and Engineering Research Council
Industrial Partners	Cooper Research Technology Esso Petroleum Company Limited Foster Yeoman Limited Mobil Oil Company Limited Nynas UK AB Shell Bitumen SWK Pavement Engineering Limited Tarmac Quarry Products Limited Wimpey Asphalt Limited
Associate Partners (all Highway Authorities)	Buckinghamshire Cambridgeshire Cheshire Hereford and Worcester Kent Lincolnshire Norfolk Northamptonshire North Yorkshire Shropshire Suffolk Surrey Warwickshire

The financing of the project is on a 50:50 basis between the public sector and the Industrial partners under the Department of Transport's LINK initiative.

The projects principal aim is the "Development of Simplified Test Methods" with the main areas of study being permanent deformation, fatigue cracking, durability and mixture design. This paper deals with fatigue cracking and the development of the ITFT.

1.2 Fatigue cracking

Fatigue cracking is a significant form of distress in bituminous pavements and, as such, warrants investigation and the development of representative test methods.

The classical theory of fatigue cracking depicts a moving wheel travelling over a bituminous surface (Figure 1). This develops tensile strains at the bottom of the bituminous layer and also in an annulus at the surface of the layer around the wheel. The repetitive application of

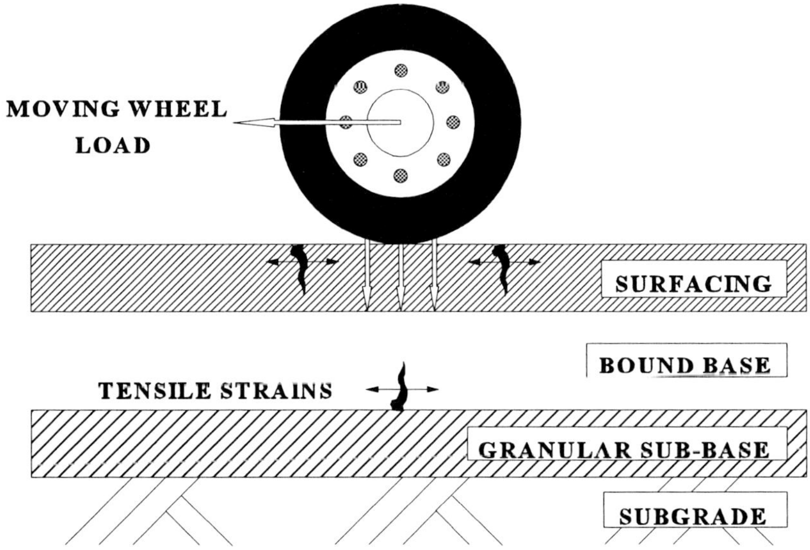

Figure 1. Tensile strains developed by a moving wheel load

these loads eventually causes the coalescence of microcracks to form one large crack (point of crack initiation N_1 load applications) which then propagates to failure (life for crack propagation N_p load applications). Although this is true in an idealised situation, it is often the case that, during construction significant cracks are built into the layer. Because of this or the presence of imperfections in the underlying layers it is likely that the position of the cracks and therefore the time for the cracks to appear at the surface is unpredictable. It is thus necessary to have some indication of both N_1 and N_p from any acceptable test method. The definition of N_1 is achieved by plotting the product of mixture stiffness and number of cycles (R_σ) against number of cycles and this clearly shows the point of crack initiation at the peak of the graph (Figure 2) when testing in the controlled stress mode. This method has been developed by Rowe (1993). Once N_1 has been identified the additional life caused by crack propagation is easily determined.

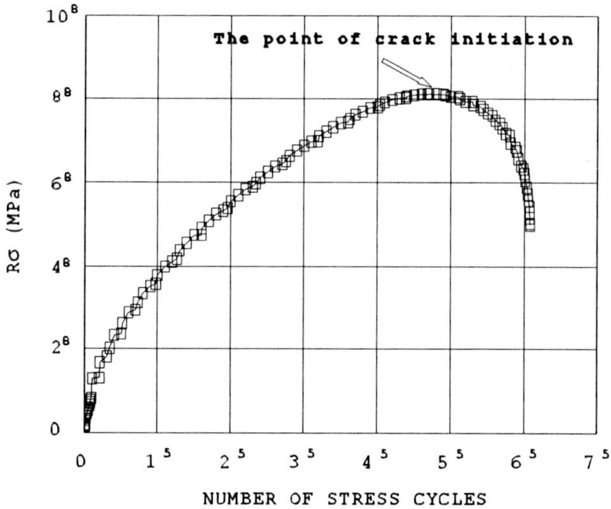

Figure 2. An example graph demonstrating the definition of N_1

2 The Nottingham Asphalt Tester

The Nottingham Asphalt Tester (NAT), shown schematically in Figure 3, is a piece of equipment developed in the mid 1980's at the University of Nottingham. It has been described in detail by Cooper and Brown (1989,1993). It was built originally, as a dedicated piece of equipment, to carry out creep testing but was subsequently found to be very versatile.

The basic NAT consists of a main test frame into which modules are placed in order to carry out a variety of tests. The main test frame has an actuator sited on the top which is capable of applying a load up to approximately 4.3kN through compressed air which is supplied at a pressure of 7 bar. Load is measured with a strain gauged load cell and LVDT's measure the deformation of specimens.

The acquisition of data and the control of the system are carried out using a conventional personal computer connected to a state-of-the-art digital interface and running user friendly software.

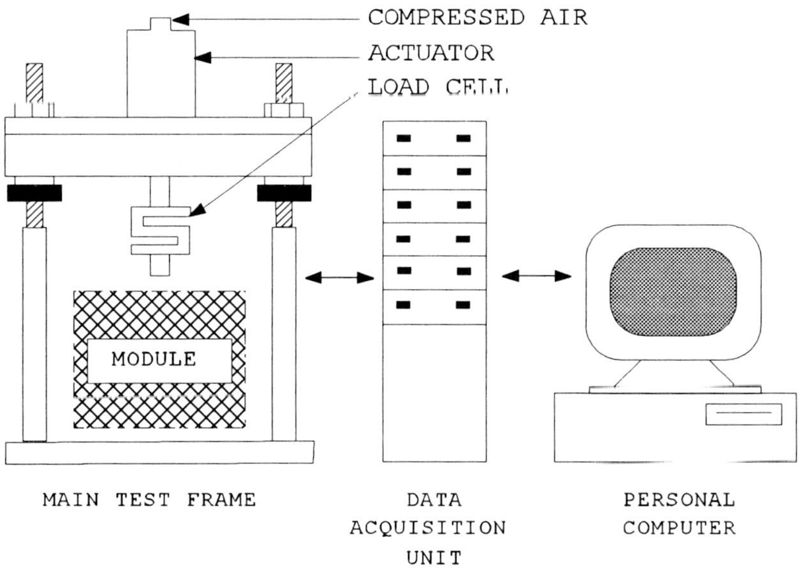

Figure 3. Schematic of the Nottingham Asphalt Tester

A number of different tests can be carried out in the NAT utilising different modules. The tests developed to date are:

1. Stiffness modulus using the repeated load indirect tensile test.
2. Static creep under uniaxial load.
3. Permanent deformation under repeated uniaxial load.
4. Dynamic creep involving pulsed loads with rest periods.
5. Fatigue cracking using the repeated load indirect tensile test

A variation in the software for the stiffness modulus test allows this to comply with the requirements of the ASTM resilient modulus test (ASTM, 1987). All data analysis is carried out using the software supplied with the system and, therefore, ease of use of the apparatus is a key feature for all modes of operation.

3 Test equipment and configuration

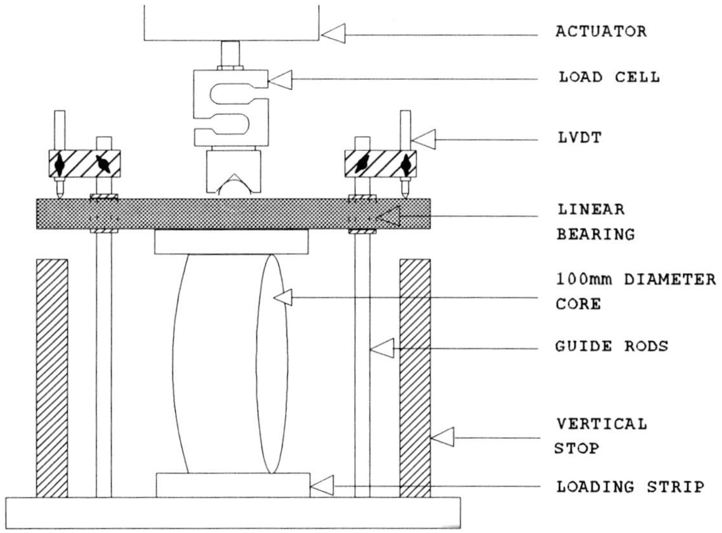

Figure 4. Schematic of the ITFT module for the NAT

The ITFT uses the module shown in Figure 4. The design of this module is based on the indirect tensile stiffness modulus (ITSM) test apparatus as described fully in the British Standard Draft for Development 213 (1993). The main differences are that there is no "yoke" used to mount LVDT's horizontally as at high stresses the specimens fail violently which would cause damage to the LVDT's. The system employed therefore, is to measure the vertical deformation using LVDT's which have a range of 10mm. Two LVDT's are mounted on the guide rods and measure the relative movement of the crosshead. The second major variation from the ITSM test is that linear bearings have been fitted to the crosshead. This is to enable its horizontal position to be maintained throughout testing. The third variation is the addition of vertical stops, which prevent the crosshead from dropping a large distance if a specimen fails suddenly. This is necessary in order to prevent damage to the loading strips which are machined, to very high tolerances, on a radius of 100mm.

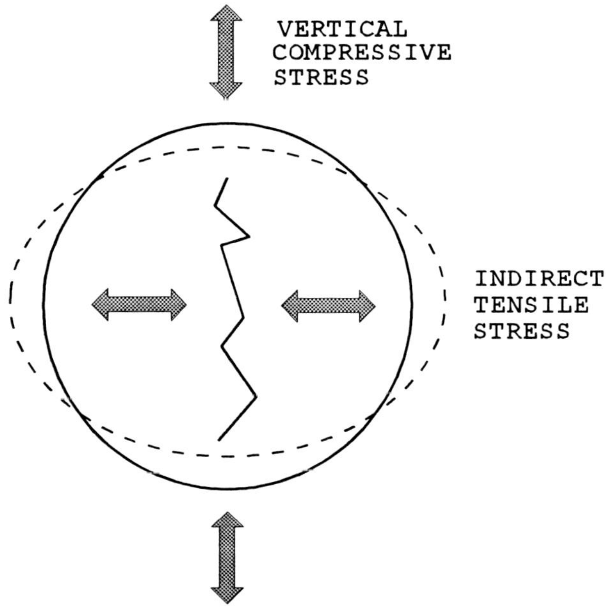

Figure 5. Diagram of the indirect tensile stresses developed by the application of vertical compressive forces

The specimen is loaded diametrally with a vertical compressive force. This generates, indirectly, a tensile stress across the horizontal diameter, hence the name indirect tensile fatigue test (Figure 5).

The magnitude of the compressive force needed is back-calculated from the indirect horizontal stress specified by the operator. The equation used for this back-calculation is:

$$P = \frac{\sigma_{xmax} \pi dt}{2} \qquad (1)$$

P = Vertical compressive force (kN)
σ_{xmax} = Horizontal tensile stress (kPa)
d = Diameter of the specimen (m)
t = Thickness of the specimen (m)

4 The indirect tensile fatigue test method

In order to generate a fatigue line for a particular material a number of specimens need to be obtained which have similar binder contents by volume. The specimens then have to be tested for stiffness as this parameter is required to calculate the strain generated, at the centre of the core, during the fatigue test.

The specimens are then placed in the ITFT module prior to location in the main test frame of the NAT. The software is then run which requires the input of certain key parameters:

1. The filename for data storage
2. The diameter of the core (to the nearest millimetre)
3. The thickness of the core (to the nearest millimetre)
4. The rise-time (in milliseconds)
5. The amount of vertical deformation before the test stops (in millimetres)(used as an indicator of failure)
6. The target horizontal stress, σ_{xmax}(in kiloPascals)

The geometry of the core is important as the size of σ_{xmax} is dependent upon it. The favoured geometry for the ITFT is a diameter of 100mm and a thickness of 40mm. With this geometry and a force of approximately 4.3kN, the maximum tensile stress which can be generated on the horizontal diameter is 600kPa.

The rise-time is defined as the period from the start of the application of load until the load has reached its maximum value. The load pulse is defined as the period from the start of the application of load until the start of the next application of load. A consistent rise-time is necessary as, due to the visco-elastic nature of bituminous materials, the stiffness is time dependent. Therefore, if inconsistent rise-times are employed the stiffness of the specimens vary and, hence, their fatigue characteristics change.

The amount of vertical deformation allowed before the test stops is normally set at 9mm as this ensures that all unmodified materials have reached failure, and that all modified mixes have at least reached their point of crack initiation.

The target horizontal stress is the one key parameter which is varied among any one set of specimens in order to achieve a range of lives and, hence, to generate a fatigue line. The normal method is to test the first core at the maximum achievable σ_{xmax} and to test subsequent specimens at varying lower stresses.

Once all testing is complete the strain generated at the centre of the specimens is calculated and initial strain is plotted against life to failure on a log-log scale. This fatigue relationship can then be compared against other materials to evaluate relative performance.

5 Materials evaluated

The materials evaluated using the ITFT to date include a 30% stone, 14mm nominal aggregate size (30/14) HRA, the same 30/14 HRA modified with SBS, a 20mm DBM and a 28mm DBM 50.

5.1 30/14 Hot Rolled Asphalt

The 30/14 HRA was a laboratory mixed material complying with B.S. 594 (1992) with 7.5% of 50pen bitumen derived from Middle Eastern crude. The material was compacted in a roller compactor, Rowe and Cooper (1988), to a target void content of 3%. The average volumetric proportions are shown in Table 2.

5.2 30/14 Hot Rolled Asphalt (SBS modified)

This material was manufactured to the same specification and in the same manner as the material described in section 5.1. The only variation was the direct substitution of a SBS binder for the unmodified binder. The base bitumen for the modified binder was a 200pen with 7% by mass of SBS added.

5.3 20mm Dense Bitumen Macadam

The 20mm DBM was a laboratory mixed material complying with B.S. 4987 (1993) with 4.7% of 100pen Middle Eastern bitumen. The material was compacted in a roller compactor to a target void content of 7%. The average volumetric proportions are shown in Table 2.

5.4 28mm Dense Bitumen Macadam (50pen bitumen)

The 28mm DBM 50 was a plant-mixed material complying with B.S. 4987 (1993) with 4.0% of 50pen Middle Eastern bitumen. The material was compacted in a large portable mould at the plant and specimens were cored for testing. The average volumetric proportions are shown in Table 2.

All the volumetric proportions were calculated from at least 27 specimens for each mixture.

Table 2. Average volumetric proportions for the four mixtures

Mixture	V_v (%)	VMA (%)	V_B (%)
30/14 HRA	3.7	21.1	17.4
30/14 HRA SBS modified	3.9	21.3	17.4
20mm DBM	5.1	15.6	10.5
28mm DBM 50	11.4	20.3	8.9

V_v = Air Voids (%)
VMA = Voids in the Mineral Aggregate (%)
V_B = Volume of Bitumen (%)

6 Results

For all the results discussed here, the rise-time was 120ms and the entire load pulse was 666ms.

6.1 Commercial use of the ITFT

Figure 6 demonstrates that if the high strain data points are used to obtain the fatigue line of a bituminous material, then the result is almost identical to that of the fatigue line generated using all the data points.

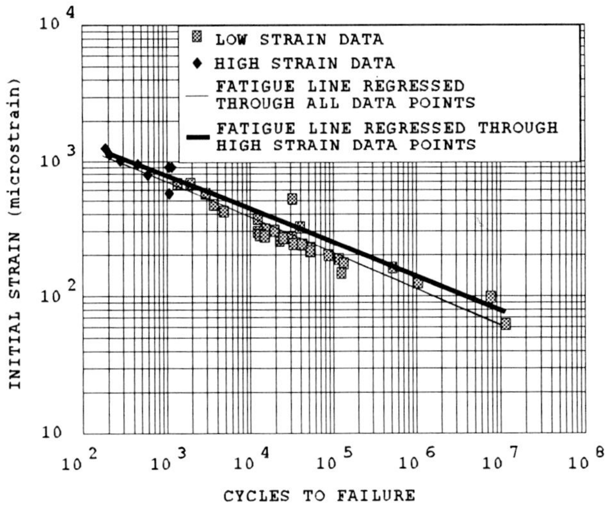

Figure 6. Use of high strain data to determine the fatigue line (30/14)

This result has been substantiated by carrying out the same exercise on the other materials tested in this project.

Using high stress (initial strain) tests makes the ITFT a viable commercial test, as ten specimens can be tested at 20°C in as little as two hours.

6.2 Maximum aggregate size

Due to the preferred geometry of the specimens (diameter 100mm, width 40mm), necessitated by the maximum load which can be applied, mixes with aggregates larger than 20mm were considered to present a potential problem. This proved not to be the case as Figure 7 demonstrates. However, the data in Figure 7 generates a well defined relationship through using a larger number of specimens to overcome the greater scatter in results.

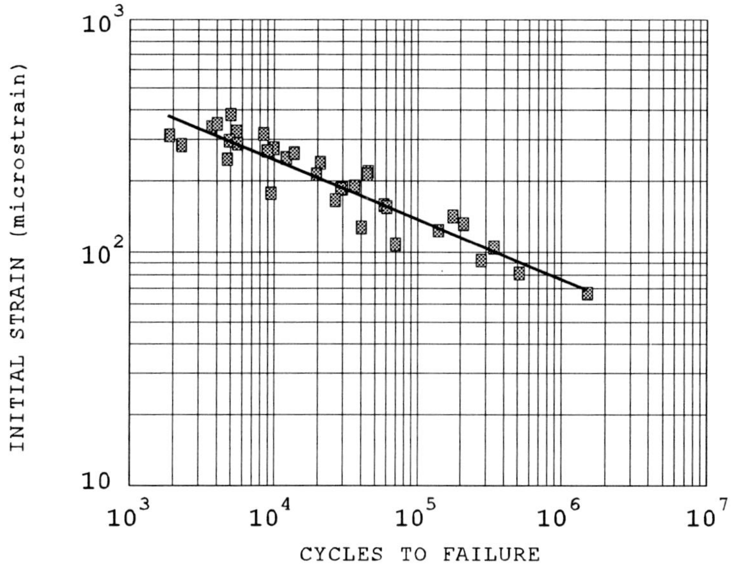

Figure 7. Graph demonstrating the increased scatter found with large aggregate mixtures

6.3 Ranking of the materials

The ranking of the materials tested in this part of the project is as indicated in Figure 8. The 30/14 HRA performs very well, followed by the 20mm DBM and the 28mm DBM 50, both of which have very similar characteristics. The SBS modified 30/14 HRA exhibited the best performance of all the materials. However, as the stiffness of this material is much lower than that of a comparable unmodified mixture, higher strains will develop in service (based upon linear-elastic analysis). This emphasises the need for proper interpretation of fatigue data in the context of pavement design, recognising the interaction between fatigue and stiffness characteristics.

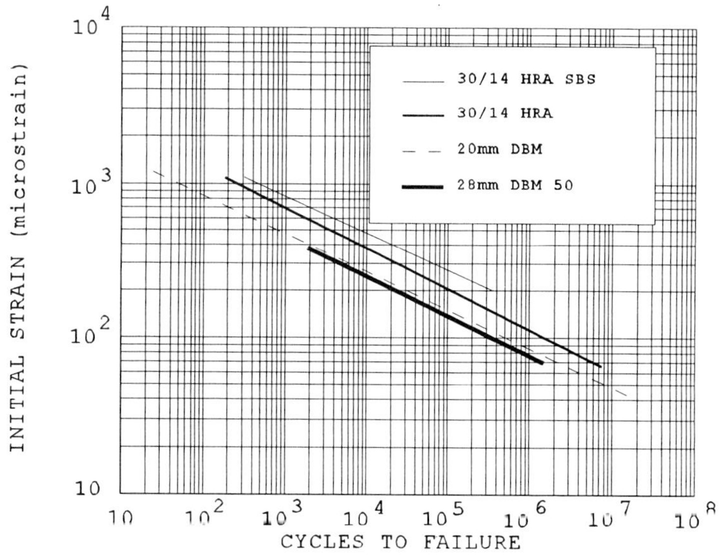

Figure 8. Graph showing the ranking of the four materials discussed in this paper

7 Conclusions

1. The project appears to have gone a long way towards achieving one of its major aims; the development of a simplified fatigue test.

2. The ITFT is a simple, inexpensive, commercially viable piece of apparatus which gives repeatable results.

3. It is possible to characterise the fatigue relationship of a bituminous material in as little as two hours when testing at high stress and temperature. Although it is preferable to have some tests done at low stress, to verify the extrapolation of data from the high stress region.

4. The ITFT test method and configuration is valid for testing large stone mixes. These generate greater scatter in the data, which leads to the need for more

tests to be carried out on any one material to get a statistically significant result.

5. The ranking of the four mixtures which were tested is as follows:

   1. 30/14 HRA SBS modified
   2. 30/14 HRA
   3. 20mm DBM
   4. 28mm DBM 50

6. The relative performance in a pavement will be a function of both the fatigue characteristics and the stiffness.

Acknowledgements

The support of the Department of Transport, Science and Engineering Research Council and the LINK industrial partners (Table 1) is gratefully acknowledged. The authors would also like to acknowledge the contributions to this work made by the associate partners (Table 1).

References

ASTM D 4123 - 82 (reapproved 1987), "Indirect tension test for resilient modulus of bituminous mixtures".

British Standards Institute (1993), "Method for Determination of the Indirect Tensile Stiffness Modulus of Bituminous Materials", Draft for Development 213.

British Standard Institute (1992), "Hot rolled asphalt for roads and other paved areas", BS 594:Part 1.

British Standard Institute (1993), "Coated Macadams for Roads and Other Paved Areas", BS 4987:Part 1.

Cooper K.E. and Brown S.F. (1989), "Development of a simple apparatus for the measurement of the mechanical properties of asphalt mixes", Proc. Eurobitume Symp., Madrid, pp 494-498.

Cooper K.E. and Brown S.F. (1993), "Assessment of the mechanical properties of asphaltic mixes on a routine basis using simple equipment", Proc. Eurobitume Congress, Stockholm, Volume 1B, pp 872-876.

Rowe, G.M. (1993), "Performance of Asphalt Mixtures in the Trapezoidal Fatigue Test", Asphalt Paving Technology, Volume 62.

Rowe G.M. and Cooper K.E. (1988), "A practical approach to the evaluation of bituminous mix properties for the structural design of asphalt pavements", Asphalt Paving Technology, Volume 57, pp 484-501.

15 FACTORS AFFECTING THE DURABILITY OF BITUMINOUS PAVING MIXTURES

T.V. SCHOLZ and S.F. BROWN
Department of Civil Engineering, University of Nottingham, Nottingham, UK

Abstract
The durability of bituminous mixtures is defined as their resistance to damage caused by environmental factors. Assuming that a pavement layer is constructed according to specification, it is generally agreed that the two primary factors affecting durability are embrittlement of the bitumen due to age hardening and damage due to moisture but a number of other parameters are also considered to be of importance. The principal mechanisms which cause age hardening of and water damage to bituminous paving mixtures are identified and described.
Keywords: Durability, Bituminous mixtures, Water sensitivity, Moisture damage, Ageing, Age hardening

1 Introduction

1.1 Definition of Durability

A product which is durable is one which is able to exist for a long period of time without significant deterioration. The factors which affect the durability of bituminous mixtures, under this definition, would include all factors which contribute to deterioration. However, the highway industry generally restricts the term durability to those effects which are related to the environment; namely moisture and ageing. For example, Whiteoak (1990) states that, 'Durability can be defined as the ability to maintain satisfactory rheology, cohesion and adhesion in long-term service.' The Asphalt Institute (1988), however, refers only to water when discussing the durability of bituminous mixtures. Terrel and Al-Swailmi (1992) state that, 'Environmental factors such as temperature, air, and water can have a profound effect on the durability of asphalt concrete mixtures.' In addition, Terrel and Shute (1989) indicate that, 'Environmental factors, traffic and time are the factors which need to be accounted for in the development of test procedures to simulate the field. Environmen-

Performance and Durability of Bituminous Materials. Edited by J.G. Cabrera and J.R. Dixon. © E & FN Spon. Published in 1996 by E & FN Spon, 2-6 Boundary Row, London SE1 8HN. ISBN 0 419 20540 3.

tal factors include; moisture from precipitation or groundwater sources, temperature fluctuations (including freeze-thaw conditions) as well as aging of the [bitumen]. The effect of traffic could also be considered as an external influence or environment.' Further, Terrel and Al-Swailmi (1992) concluded that repeated loading (i.e., simulation of traffic loading) is a very important variable to be included in water conditioning protocols. Similarly, in an earlier study, Lottman (1971) found that heavy traffic volume appeared to increase the rate of damage due to moisture more effectively than climatic extremes of precipitation and temperature.

The above lack of consensus regarding the factors which influence the durability of bituminous paving materials makes an accurate definition difficult. Clearly, there is general consensus that water and ageing affect durability but uncertainty as to whether traffic loading should be included. Lottman (1982), Tunnicliff and Root (1984), and Terrel and Al-Swailmi (1992) have incorporated the effects of temperature variation in the procedures they have developed and Terrel and Al-Swailmi include repeated loading to simulate the effects of traffic. For the purposes of this paper the following definition will be used:

> Durability as it applies to bituminous paving materials is defined as the ability of the materials in the bituminous pavement structure to resist the effects of water, ageing, and temperature variations, in the context of a given amount of traffic loading, without significant deterioration for an extended period.

1.2 Problem Statement

Highway engineers attach considerable importance to the durability of bituminous paving mixtures as the costs of maintenance and rehabilitation of pavement structures that do not survive their design life can be substantial. While the principal failure mechanisms resulting from traffic loading are cracking and permanent deformation, adverse environmental effects can accelerate the deterioration process. Many factors such as the composition of the bitumen, the type and grading of aggregate, the interaction between bitumen and aggregate, bitumen content, mixture permeability, construction practices and climate affect the durability of bituminous mixtures. Assuming that a pavement layer is constructed according to specifications, it is generally agreed that the two primary factors that affect the durability of the mixture are embrittlement of the bitumen due to age hardening and damage due to moisture.

It is well known that bitumen becomes stiffer (increases in viscosity) during the mixing and construction process as well as while the pavement is in service. This hardening, often referred to as ageing or age hardening, is manifested in the stiffening of the bituminous layer(s) which, to some extent, is beneficial. However, if the hardening of the bitumen is excessive, the mixture can become brittle and crack, resulting in partial or significant failure of the bound layer. Excessive hardening of bitumen can also result in decreased adhesion between the bitumen and aggregate (Traxler, 1963), often resulting in loss of material at the layer surface.

Damage due to moisture can also significantly influence the durability of bituminous mixtures. It is generally agreed that there are two mechanisms by which moisture can degrade the structural integrity of the bitumen-aggregate matrix (Kennedy, 1985; Terrel and Al-Swailmi, 1992):

a) loss of cohesion (strength) and stiffness of the bitumen;
b) failure of the adhesion (or bond) between the bitumen and the aggregate, often referred to as stripping.

Both mechanisms of water damage result in a weaker pavement layer. In addition, stripping can result in loss of material and severe stripping can deteriorate the bituminous mixture to a virtually cohesionless state (Lottman, 1982).

1.3 Purpose

The purpose of this paper is to provide a synthesis of selected literature regarding age hardening and water sensitivity of bituminous mixtures with emphasis on their relation to durability. More specifically, this paper provides a synopsis of the factors which affect durability and, in particular, the mechanisms and consequences of age hardening and water sensitivity.

2 Properties of Bitumen

The durability of bitumen and, thus, bitumen-aggregate mixtures, is largely determined by the physical properties of the bitumen, which in turn is determined by its chemical composition. A brief review of bitumen chemistry is, therefore, required prior to discussing the factors which affect the durability of bituminous mixtures.

2.1 Source of Bitumen

Bitumen is the product of the distillation of crude oil and is produced from the residue of the distillation process. It, thus, contains the components of the crude oil having the greatest molecular weight. Crude oil is thought to originate from the sedimentation of large quantities of organic and vegetable matter, together with mud and rock fragments, on the ocean floor. These were converted into hydrocarbons by heat from the earth's crust and pressure applied by the upper layers of sediment, possibly aided by bacteria and radioactive bombardment (Whiteoak, 1990). It is not known how long the process took but it has been reported to be on the order of millions of years (Petersen, 1984; Halstead, 1985; Whiteoak, 1990). Due to the way in which crude oil originated and the large number of such deposits throughout the world, the physical and chemical properties of the crudes vary widely as do the properties of bitumens produced from these crudes.

2.2 Elemental Composition of Bitumen

Bitumen consists of a complex mixture of organic molecules which vary widely in composition. The molecules contain, primarily, hydrogen and carbon, referred to as hydrocarbons but most contain one or more heteroatoms (nitrogen, sulphur, and oxygen) and trace amounts of metals, primarily vanadium, nickel and iron. Petersen (1984) notes, 'Because the heteroatoms impart functionality and polarity to the molecules, their presence may make a disproportionately large contribution to the differences in physical properties among [bitumens] from different sources.'

Although elemental composition is important to note, it provides little information regarding how the atoms are assembled into molecules or what types of molecular structures are present in the bitumen, knowledge of which is necessary for a fundamental understanding of how composition affects physical properties and chemical reactivity (Petersen, 1984).

2.3 Molecular Structure

The way in which the elements are incorporated into molecules and the type of molecular structure present is far more important than the total amounts of each element present in bitumen. Because of the way the source of bitumen was derived from living organisms, it is not surprising that the molecular structure of the components of bitumen are highly diverse. A full discussion of the various types of organic compounds found in bitumens is well beyond the scope of this paper but it is important to note the three ways in which carbon atoms are linked to one another as follows:

a) Aliphatic or paraffinic types—carbon atoms are linked in straight or branched chains;
b) Naphthenic—carbon atoms are linked in simple or complex (condensed) saturated rings, where "saturated" means that the highest possible hydrogen to carbon ratio is present;
c) Aromatic—materials characterised by the presence of one or more especially stable six-atom rings (e.g., benzene, toluene, etc.).

Bitumen incorporates all three types of compounds in widely varying concentrations and makeup.

2.4 Fractional Composition of Bitumen

Because the number of molecules with different chemical structures is astronomically large, chemists have not seriously attempted to separate and identify all the different molecules in bitumen (Petersen, 1984; Halstead, 1985). Instead, various techniques have been developed to separate bitumen into less complex and more homogenous fractions. An attempt will not be made to summarise the techniques as the literature abounds with descriptions of these (e.g., Petersen, 1984; Halstead, 1985; Bell, 1989). However, it is important to note that the techniques divide the bitumen into groups or generic fractions based on molecular size, chemical reactivity and/or polarity and that the different separation techniques lead to fractions having different characteristics (i.e., the fractions from one separation technique differ from those using another technique with regard to chemical and physical characteristics). Because Corbett's separation method (Corbett, 1969) has probably found the widest use for research purposes the subsequent discussion with regard to the fractional composition of bitumen will be based on this technique. This will also avoid confusion.

It is generally agreed that bitumen is composed of asphaltenes and maltenes. The asphaltenes are considered to be the most complex fraction containing the molecules with the highest polarity and tendency to interact and associate. Asphaltenes, consisting largely of hydrocarbons and some heteroatoms, are brittle solids when isolated. It is generally believed that the asphaltenes are primarily responsible for bitumen viscosity (Rostler and White, 1959; Corbett, 1970; Halstead, 1985). Whiteoak (1990) states that, 'The asphaltene content has a large effect on the rheological characteristics of a bitumen. Increasing the asphaltene content produces a harder bitumen with a lower penetration, higher softening point and consequently higher viscosity.'

The asphaltenes, of high molecular weight, are dispersed or dissolved in a lower molecular weight oily medium referred to, by most authors, as maltenes. The maltenes are made up of saturates, naphthene aromatics and polar aromatics. The saturate fraction is a viscous oil lacking polar chemical functional groups. The molecules of this fraction are non-polar and may contain saturated normal and branched-chain (i.e., aliphatic) hydrocarbons, saturated cyclic hydrocarbons and, in addition to sulphur, possibly a small amount of mono-ring aromatic hydrocarbons. The naphthene aromatic fraction is a viscous liquid which constitutes the major proportion of the dispersion medium for the asphaltenes. This fraction may contain condensed non-aromatic and aromatic ring systems and, possibly, sulphur and the heteroatoms oxygen and nitrogen. The polar aromatic fraction, comprising highly condensed aromatic ring systems and functional groups containing heteroatoms, serve as the peptisers or dispersing agents for the asphaltenes. This fraction is highly polar giving it strong adhesion characteristics.

Each of the components of bitumen contain many different chemical compounds that coexist in neat bitumen as a homogenous mixture which is made possible by the interaction of the various components with one another to form a balanced and compatible system. It is the balance of the components which give bitumen its unique viscoelastic properties. Imbalance or incompatibility amongst the components, as sometimes manifested by component phase separation, leads to undesirable properties (Petersen, 1984).

2.5 Functionality and Polarity

As previously mentioned, the heteroatoms impart functionality and polarity to the molecules present in bitumens. Although present in small quantities, the heteroatoms significantly affect the physical properties and performance characteristics of bitumens. Functionality refers to the way in which molecules in bitumen interact with each other as well as with the molecules and/or surfaces of other materials (e.g., aggregate). Polarity refers to the way in which the electrochemical forces in the molecules are imbalanced, producing a dipole. Polar compounds, or dipoles, have the characteristic that similar charges repel and opposite charges attract one another, a characteristic which produces interactions that strongly influence physical properties, particularly flow characteristics.

3 Ageing of Bituminous Mixtures

3.1 Factors Affecting Ageing

Age hardening of bitumen occurs as a result of compositional changes in the bitumen. For example, ageing due to oxidation, the primary cause of bitumen hardening, is believed to result primarily from the introduction of oxygen-containing chemical functionalities which, due to their polar nature, greatly increase molecular interaction forces, thus increasing viscosity (Petersen, 1990). The changes that occur as a result of age hardening are, as yet, not well understood, primarily due to bitumen being a rather complex mixture of organic molecules that vary widely in composition. No two crudes, and there are nearly 1500 (Whiteoak, 1990), are exactly alike. However, many researchers have investigated age hardening of bitumens and bituminous mixtures and have provided significant advances toward a better understanding of the mechanisms involved.

3.2 Mechanisms of Age Hardening

Traxler (1963) identifies 15 effects which may influence the chemical, rheological, and adhesion characteristics of bitumen as shown in Table 1. He provides experimental data for some of the effects but notes that some of those listed have not been given experimental consideration. He also notes that the effects are not necessarily given in order of importance and that time, temperature and film thickness are factors in all of the effects.

Petersen (1984) states that, 'Durability is determined by the physical properties of the [bitumen], which in turn are determined directly by chemical composition. An understanding of the chemical factors affecting physical properties is thus fundamental to an understanding of the factors that control [bitumen] durability.' He identifies three composition-related factors which govern the changes that could cause hardening of bitumen in pavements as follows:

a) Loss of the oily components of bitumen by volatility or absorption by porous aggregates;
b) Changes in chemical composition of bitumen molecules from reaction with atmospheric oxygen;
c) Molecular structuring that produces thixotropic effects (steric hardening).

Of the three factors listed, he identifies reaction with atmospheric oxygen as probably being the major and best understood cause of age hardening. In pavements where bitumen exists in thin films exposed to atmospheric oxygen, rapid and irreversible oxidation occurs resulting in the formation of polar, strongly interacting, oxygen-containing chemical functional groups that greatly increase viscosity and alter complex flow properties, a phenomenon which often leads to embrittlement of the bitumen and ultimately pavement failure. The chemical functional groups formed on oxidative ageing include sulphoxides, anhydrides, carboxylic acids and ketones. Table 2 presents data from tests conducted on four bitumens from different crudes that had been aged under identical conditions. The data indicates that ketones and sulfoxides are the major oxidation products while anhydrides and carboxylic acids are formed in smaller amounts. Table 3 presents data which shows that the concentration of ketones formed on oxidative ageing are greatest in the asphaltene and polar aromatic fractions; smaller concentrations are found in the naphthene aromatic (shown as aromatic in Table 2) and saturate fractions. Petersen (1984) explains that because the polar aromatic and asphaltene fractions are known to contain the highest concentrations of aromatic ring systems, they have the highest content of hydocarbon types sensitive to air oxidation. It must be stressed, however, that oxidative ageing requires the presence of oxygen. Thus, in pavements having

Table 1. Effects which may reduce the binding properties of bitumen (After Traxler, 1963).

	Influenced by					Occurs	
Effects	Time	Heat	Oxygen	Sun-light	Beta & Gamma Rays	At Surface	In Mass
1. Oxidation (in dark)	✓	✓	✓			✓	
2. Photo-oxidation (direct light)	✓	✓	✓	✓		✓	
3. Volatilisation	✓	✓				✓	✓
4. Photo-oxidation (reflected light)	✓	✓	✓	✓		✓	
5. Photo-chemical (direct light)	✓	✓		✓		✓	
6. Photo-chemical (reflected light)	✓	✓		✓		✓	✓
7. Polymerisation	✓	✓				✓	✓
8. Development of internal structure (Thixotropy)	✓					✓	
9. Exudation of oil (Syneresis)	✓	✓				✓	
10. Changes by nuclear energy	✓	✓			✓	✓	✓
11. Action by water	✓	✓	✓	✓		✓	✓
12. Absorption by solid	✓	✓				✓	✓
13. Adsorption of components at solid surface	✓	✓				✓	
14. Chemical reactions or catalytic effects	✓	✓				✓	✓
15. Microbiological deterioration	✓	✓	✓			✓	✓

Table 2. Chemical functional groups formed in bitumens during oxidative ageing (After Plancher et. al., 1976).

	Concentration (moles per litre)				
Bitumen	Ketone	Anhydride	Carboxlic Acid[a]	Sulphoxide	Average Hardening Index[b]
B-2959	0.50	0.014	0.008	0.30	38.0
B-3036	0.55	0.015	0.005	0.29	27.0
B-3051	0.58	0.020	0.009	0.29	132.0
B-3602	0.77	0.043	0.005	0.18	30.0

Note: Column oxidation (Davis and Petersen, 1966), 130°C, 24 hours, 15 micron film.
[a]Naturally occurring acids have been subtracted from reported value.
[b]Ratio of viscosity after oxidative ageing to viscosity before.

Table 3. Carbonyl functional groups formed in Wilmington bitumen fractions during column oxidation (After Petersen, et. al., 1974).

	Concentration (moles per litre)		
Fraction	Ketone	Anhydride	Carboxlic Acid
Saturate	0.045	0.010	Trace
Aromatic	0.32	0.017	—[a]
Polar Aromatic	1.48	0.088	—[a]
Asphaltene	1.82	0.080	ND[b]
Whole Bitumen	1.02	0.052	0.007

[a]Some acids lost on alumina column during component fractionation.
[b]Not determined.

very low air voids (or more correctly, very low permeability) oxidative ageing is not likely to significantly affect the rheological properties of the pavement. For example, Vallerga and Halstead (1971) found that for pavements with less than 2% air voids, field ageing during 11 to 13 years of service, subsequent to hardening occurring during mixing, transport and laydown, appeared to be negligible.

Molecular structuring, a slow and largely reversible phenomenon which appears to occur concurrently and synergistically with oxidative ageing, can produce significant changes in the flow properties of a bitumen without changing its chemical composition and may be a significant factor contributing to embrittlement of the bitumen. Petersen stresses, however, that this phenomenon is difficult to quantify as the recovery processes (i.e., use of solvents, heat, and mechanical working to obtain neat bitumen from bituminous mixtures) destroys most or all of the structuring.

The loss of volatile components (i.e., the nonpolar saturate or oily fraction of bitumen) occurs during the mixing, storage, transport, and laydown of the mixture (i.e., while the bitumen is in a thin film at an elevated temperature) as well as due to absorption of the polar components by porous aggregate. Petersen states that, 'With current specifications and construction practices, volatility is probably not a significant contributor to pavement hardening.' Similarly, Whiteoak (1990) states that, 'Penetration grade bitumens are relatively involatile and therefore the amount of hardening resulting from loss of volatiles is usually fairly small.' The absorption of the polar components by porous aggregate results in compositional changes in the bitumen which may significantly affect its properties and ageing characteristics. Traxler (1963) suggests that chemical reactions or catalytic effects at the bitumen-aggregate interface may, under certain situations, change the properties of the bitumen enough to affect its durability in service.

Petersen also recognizes that environmental factors, particularly water, can seriously affect the performance and durability of bituminous paving materials. However, although damage due to water may be related to bitumen composition and adsorption of bitumen components onto aggregate surfaces, it is primarily an interfacial phenomenon.

3.3 Consequences of Age Hardening

Age hardening of the binder in bituminous mixtures is the result of compositional changes causing an increase in viscosity in the bitumen. Figure 1 illustrates the increase in viscosity over time for several different bitumens while in service (extracted from data from Zube & Skog, 1969). Similar findings have been reported by other researchers (e.g., Kemp & Sherman, 1984; Kandhal & Koehler, 1984; Shah, 1978; Culley, 1969). The viscosities of the bitumens increase roughly two orders of magnitude during nearly 10 years of service. It should be noted, however, that the steric hardening (molecular structuring) component may not be represented in the data shown as the viscosities were determined on recovered bitumen from the mixtures which probably destroyed the effect and, therefore, the values shown are probably somewhat lower than that of the in-situ bitumens.

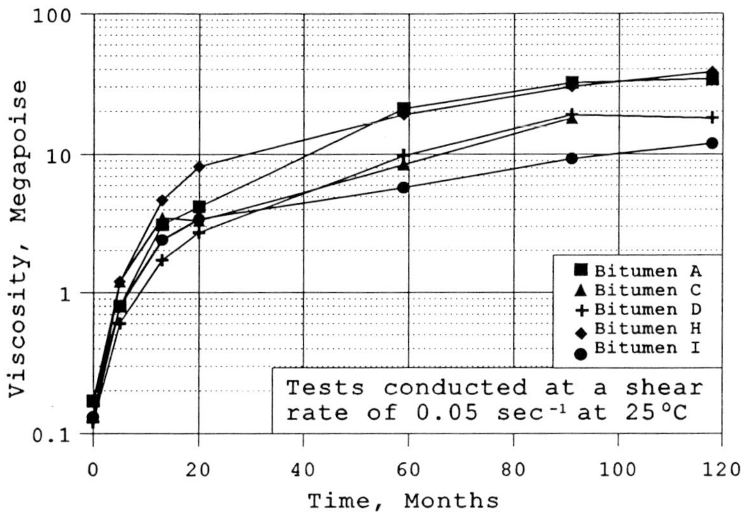

Figure 1. Viscosity change of several bitumens during service in pavements (After Zube & Skog, 1969).

Excessive age hardening can result in a brittle bitumen which contributes to various forms of cracking in the bituminous mixture. Cracking generally occurs in the form of fatigue, thermal, or reflective cracking. Fatigue cracking is the result of an accumulation of damage, arising from repeated or fluctuating stresses (i.e., traffic loading), which eventually leads to fracture. Thermal cracking is the result of thermally-induced tensile stresses which exceed the tensile strength of the bitumen. Thermal cracking can occur as a result of the mixture temperature falling below some limiting value and/or as the result of an accumulation of permanent tensile strain arising from repeated or fluctuating thermal stresses. Reflective cracking occurs in mixtures which overlay existing roadways that are cracked. The cracks in the overlay appear directly above cracks in the existing roadway, hence the term 'reflective.' Reflective cracking generally occurs as a result of stresses developed in the overlay via differential movement of the portions of the existing roadway immediately adjacent to a crack.

Age hardening reduces, through embrittlement of the bitumen, the ability of the bituminous mixture to support traffic- and thermally-induced stresses and strains. Hence, age hardened bitumen has a reduced ability to flow, by virtue of increased stiffness, under the influence of external loading. This reduction in the flow characteristics of the bitumen directly affects its vulnerability to cracking. For example, Figure 2 (Vinson et. al., 1992) shows that oven-aged bituminous mixtures have a higher fracture temperature than do unaged mixtures, as determined by the thermal stress restrained specimen test where a 50 × 50 × 250 mm specimen is held at constant length while its temperature is reduced at a constant rate until fracture occurs.

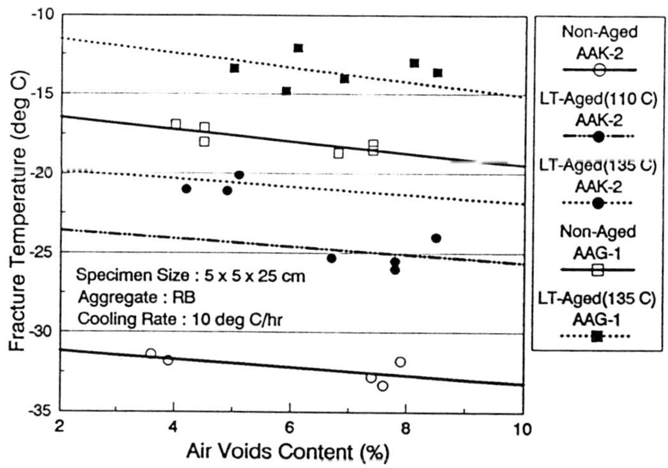

Figure 2. Effect of age conditioning on fracture temperature (After Vinson et. al., 1992).

4 Water Damage to Bituminous Mixtures

4.1 Factors Affecting Water Sensitivity

Although many factors contribute to the degradation of bituminous mixtures, moisture appears to play a major role. In general, water can reduce the stiffness or strength of the bitumen-aggregate matrix and/or cause the bond between bitumen and aggregate to fail, both potentially resulting in significant distress to the pavement. The mechanisms of damage due to moisture are not clearly understood. However, many researchers have investigated moisture sensitivity of bituminous mixtures and have provided significant advances toward a better understanding of the mechanisms involved.

4.2 Mechanisms of Moisture Damage

It is generally agreed that moisture can degrade the integrity of bituminous mixtures in two ways (Kennedy, 1985; Terrel and Shute, 1989):

a) By causing a reduction in the cohesive strength and stiffness of the mixture, characterised by softening;
b) By causing failure of the adhesion (or bond) between bitumen and aggregate, referred to as stripping.

Lottman (1982) provides a more comprehensive list of the moisture damage mechanisms that cause stripping and mixture softening as follows:

a) Pore pressure of water in the mixture voids due to wheel-loading repetitions; thermal expansion-contraction differences produced by ice formation, temperature cycling above freezing, freeze-thaw, and thermal shock; or a combination of these factors;
b) [Bitumen] removal by water in the mixture at moderate to higher temperatures;
c) Water-vapour interaction with the [bitumen]-filler mastic and larger aggregate interfaces;
d) Water interaction with clay minerals in the aggregate fines.

Of the mechanisms identified above, stripping has been given, by far, the greatest attention. It has traditionally been thought that stripping is related to rupture of the adhesive bond at the bitumen-aggregate interface, a complex phenomenon involving physical and chemical properties of both the bitumen and the aggregate, with the properties of the aggregate surface playing an important role in determining the adhesive properties of the bitumen-aggregate bond (Petersen, et. al., 1982). However, recent research has shown that cohesive failures within the bulk aggregate or bitumen (or both), rather than separation at the bitumen-aggregate interface, are a major mechanism where stripping occurs (Jamieson, et. al., 1993).

Work by Curtis et. al. (1991) on adhesion and adsorption characteristics of bitumen-aggregate systems showed that the adsorptive behaviour of bitumen and bitumen model components on aggregates is highly specific and particularly influenced by the aggregate surface chemistry; the chemistry of the bitumen has less influence. The polar components of bitumen adhere to active sites on the aggregate surface through chemical bonding, electrostatic forces, hydrogen bonding and Van der Waals interactions.

Surfaces rich in alkaline earth metals are less likely to be susceptible to adhesive debonding than are surfaces rich in alkali metals (Jamieson, et. al., 1993). The susceptibility of siliceous aggregates to stripping may be associated with the presence of water soluble cations and aluminosiclicates where the mechanism of water stripping is probably (Scott, 1978):

a) The dissolution of water soluble salts;
b) The dissolution of silica resulting from the high pH environment generated by solubilisation of the alkaline earth cations;
c) Electrostatic repulsion between the negatively charged aggregate and anionic components of the bitumen at the interface;
d) Dissolution of soaps formed between acid anions on the bitumen surface and alkali metal cations on the aggregate surface.

Thus, it can be seen that the adhesion and debonding characteristics of a bitumen-aggregate system cannot be determined by the generic aggregate type but must be determined by the physical and chemical nature of the surface with which the bitumen comes in contact.

4.3 Consequences of Moisture Damage

Damage due to moisture occurs in various forms and degrees of severity. As alluded to earlier, the primary consequence of moisture damage is that of stripping, characterised by failure of the bitumen-aggregate bond. Stripping is often initially manifested in localised areas where the bitumen has migrated to the surface of the bituminous layer, referred to as flushing or bleeding. This migration of bitumen results in an unstable matrix in the lower portions of the bituminous layer which can lead to permanent deformation in the form of rutting and/or shoving as well as the development of potholes and cracking under the action of traffic loading. Subsequent intrusion of water into these localised water-damaged areas, coupled with traffic loading, further degrades the structural integrity of the pavement layer and, possibly, underlying layers, which if not repaired could lead to substantial localised failure of the pavement structure. Stripping can also result in ravelling which is characterised by loss of material at the surface of the bituminous layer.

The other major consequence of moisture damage is that of a reduction of stiffness in the bituminous layer which decreases the load spreading capabilities of the pavement. Under the action of traffic loading, a pavement with reduced stiffness due to water damage is prone to rutting as a result of increased stresses and strains in the underlying layers. Loss of strength in the bitumen-aggregate matrix may also encourage stripping (Kennedy, 1985).

5 Concluding Summary

A thorough review of the extensive literature devoted to durability of bituminous mixtures has exhibited the following major points:

a) Age hardening of bitumen can adversely affect the durability characteristics of bituminous mixtures.

b) Atmospheric oxidation is the principal cause of age hardening, resulting in increased viscosity and reduced penetration of the binder.

c) There is clear evidence that the aggregate plays a significant role in the way the binder in bituminous mixtures hardens over time.

d) Steric hardening may significantly contribute to reduced durability characteristics of bituminous mixtures.

e) Distress due to age hardening is usually manifested in cracking of the bound layer(s).

f) Moisture can also adversely affect the durability characteristics of bituminous mixtures.

g) Moisture damage can be manifested in loss of adhesion between the bitumen and aggregate and/or loss of cohesion in the bitumen-filler mastic.

h) Loss of adhesion is presently believed to result from failure in the aggregate, failure in the bulk bitumen or a combination of the two.

i) The adhesion and adsorption characteristics of bitumen-aggregate systems is dependent more on the aggregate surface chemistry than the composition of the bitumen. Thus, the adhesion and debonding characteristics of bitumen-aggregate systems must be determined by the physical and chemical nature of the surface with which the bitumen comes into contact.

6 Acknowledgements

The Department of Transport, the Science and Engineering Research Council and the LINK BITUTEST Industrial Partners are gratefully acknowledged for their financial sponsorship. The authors would also like to thank the Worshipful Company of Paviours, the County Surveyors Society and the Rees Jeffreys Road Fund for additional financial support.

7 References

Bell, C.A. (1989) Summary Report on Aging of Asphalt-Aggregate Systems, **SHRP-A/IR-89-004**, Strategic Highway Research Program, National Research Council, Washington, D.C.

Corbett, L.C. (1969) Composition of Asphalt Based on Generic Fractionation Using Solvent Deashaltening, Elution-Adsorption Chromatography, and Densimetric Characterization in **Analytical Chemistry**, Vol. 41, pp. 576-579.

Corbett, L.C. (1970) Relationship Between Composition and Physical Properties of Asphalt in **Proceedings, Association of Asphalt Paving Technologists**, Vol. 39, pp. 481-491.

Culley, R.W. (1969) Relationships Between Hardening of Asphalt Cements and Transverse Cracking of Pavements in Saskatchewan **Proceedings, Association of Asphalt Paving Technologists**, Vol.38, pp. 629-659.

Curtis, C.W., Ensley, K., and Epps, J. (1991) Fundamental Properties of Asphalt-Aggregate Interactions Including Adhesion and Adsortion, **Final Report, SHRP A-003B**, Strategic Highway Research Program, Washington, D.C.

Davis, T.C. and Petersen, J.C. (1966) An Adaptation of Inverse Gas-Liquid Chromotography to Asphalt Oxidation Studies in **Analytical Chemistry**, Vol. 38, pp. 1938-1940.

Halstead, W.J. (1985) Relation of Asphalt Chemistry to Physical Properties and Specifications in **Proceedings, Association of Asphalt Paving Technologists**, Vol 54, pp. 91-117.

Jamieson, I.L., Jones, D.R., and Moulthrop, J.S. (January 1993) Advances in the Understanding of Binder-Aggregate Adhesion and Resistance to Stripping in **Highways and Transport**, pp. 6-19.

Kandhal, P.S. and Koehler, W.C. (1984) Significant Studies on Asphalt Durability: Pennsylvania Experience in **Transportation Research Record 999**, Transportation Research Board, Washington D.C., pp. 41-50.

Kemp, G.R. and Sherman, G.B. (1984) Significant Studies on Asphalt Durability: California Experience in **Transportation Research Record 999**, Transportation Research Board, Washington D.C., pp. 36-41.

Kennedy, T.W. (1985) Prevention of Water Damage in Asphalt Mixtures in **STP 899**, American Society for Testing and Materials, Philadelphia, pp. 119-133.

Lottman, R.P. (1971) The Moisture Mechanism that Causes Asphalt Stripping in Asphaltic Pavement Mixtures, **Final**

Report, Department of Civil Engineering, University of Idaho, Moscow, Idaho.

Lottman, R.P. (1982) Laboratory Test Method for Predicting Moisture-Induced Damage to Asphalt Concrete in **Transportation Research Record 843**, Transportation Research Board, Washington, D.C., pp. 88-95.

Mix Design Methods for Asphalt Concrete and Other Hot-Mix Types, Manual Series No.2 (MS-2) (1988) The Asphalt Institute, Lexington, Kentucky.

Petersen, J.C. Barbour, F.A. and Dorrence, S.M. (1974) Catalysis of Asphalt Oxidation by Mineral Aggregate Surfaces and Asphalt Components in **Proceedings, Association of Asphalt Paving Technologists**, Vol. 43, pp. 162-177.

Petersen, J.C., Plancher, H., Ensley, E.K., Venable, R.L., and Miyake, G. (1982) Chemistry of Asphalt-Aggregate Interaction: Relationship with Pavement Moisture-Damage Prediction Test in **Transportation Research Record 843**, Transportation Research Board, Washington, D.C., pp. 95-104.

Petersen, J.C. (1984) Chemical Composition of Asphalt as Related to Asphalt Durability: State of the Art in **Transportation Research Record 999**, Transportation Research Board, Washington D.C., pp. 13-30.

Petersen, J.C. (1990) Effects of Physical Factors on Asphalt Oxidative Aging, paper submitted for presentation at the **ASCE Materials Engineering Congress 90** in their session of Durability and Durability Tests for Asphalts, Denver, Colorado.

Plancher, H., Green, E.L. and Petersen, J.C. (1976) Reduction of Oxidative Hardening of Asphalts by Treatment with Hydrated Lime—A Mechanistic Study in **Proceedings, Association of Asphalt Paving Technologists**, Vol. 45, pp. 1-24.

Rostler, F.S. and White, R.M. (1959) Influence of the Chemical Composition of Asphalts on Performance, Particularly Durability, **Special Technical Publication 277**, American Society for Testing and Materials, pp. 68-84.

Scott, J.A.N. (1978) Adhesion and Disbonding of Asphalt Used in Highway Construction and Maintenance in **Proceedings, Association of Asphalt Paving Technologists**, Vol. 47, pp. 19-48.

Shah, S.C. (1978) Asphalt Cement Consistency in **Transportation Research Record 695**, Transportation Research Board, Washington, D.C., pp. 1-7.

Terrel, R.L. and Shute, J.W. (1989) Summary Report on Water Sensitivity, **SHRP-A/IR-89-003**, Strategic Highway Research Program, National Research Council, Washington, D.C.

Terrel, R.L. and Al-Swailmi, S. (1992) Final Report: Water Sensitivity of Asphalt-Aggregate Mixtures Test Development, **Final Subtask Report: C.5, TM-OSU-A-003A-92-22**,

Strategic Highway Research Program, National Research Council, Washington, D.C.
Traxler, R.N. (1963) Durability of Asphalt Cements in **Proceedings, Association of Asphalt Paving Technologists**, Vol. 32, pp. 44-63.
Tunnicliff, D.G. and Root, R.E. (1984) Use of Antistripping Additives in Asphaltic Concrete Mixtures in **NCHRP 274**, Transportation Research Board, Washington, D.C.
Vallerga, B.A. and Halstead, W.J. (1971) Effects of Field Aging on Fundamental Properties of Paving Asphalts in **Highway Research Record 361**, Transportation Research Board, Washington, D.C., pp. 71-92.
Vinson, T.S., Jackson, N.M., and Jung, D.(1992) Thermal Cracking Resistance of Asphalt Concrete: an Experimental Approach in **Proceedings, 7th International Conference on Asphalt Pavements**, Vol. 3, pp. 395-409.
Whiteoak, D.(1990) **The Shell Bitumen Handbook**, Surrey, UK.
Zube, E. and Skog, J. (1969) Final Report on the Zaca-Wigmore Asphalt Road Test in **Proceedings, Association of Asphalt Paving Technologists**, Vol. 38, pp. 1-38.

16 A CRITICAL APPRAISAL OF RECYCLING UK ROADS

A.D. GILL
Colas Limited, Crawley, Sussex, UK
A.R. WOODSIDE and W.D.H. WOODWARD
Department of Civil Engineering and Transport, University of Ulster, Carrickfergus, Northern Ireland

Abstract
This paper examines the various methods which are presently available in the United Kingdom to enable the recycling of bituminous road materials. It outlines the merits and disadvantages of each system ranging from 'Hot mix off-site recycling at a central plant' to the 'Deep cold in-situ recycling' process. The processes are defined and described and the potential of each operation is assessed by the authors. The paper suggests that in the future a very definite bias will be given towards the use of secondary materials in the sub-bases and road-bases and highlights how a substantial saving of primary aggregates may be made if their place can be taken by secondary materials. The enhancement of secondary materials is considered and the paper concludes by suggesting possible future developments in the light of impending European Standards.
Keywords: Conserving Resources, Proven Performance, Cost Effective, Environmentally Friendly.

1 Introduction

Never has the highway industry in the United Kingdom been faced with a greater range of conflicting pressures. Government predictions highlight the need for more roads while the growing "environmental awareness lobby" is calling for restrictions on quarrying and questioning the need for the proposed investment in highway construction.

At present, approximately 320 million tonnes of aggregate are used in the U.K. road building industry every year. This is expected to increase to between 370 million tonnes and 440 million tonnes by the year 2011.

More importantly to meet this demand, imports of aggregate are expected to soar from their present level of 4 million tonnes per year to over 100 million tonnes by the year 2011.

This has created a problem as to where the necessary supplies of aggregate are to come from. To answer this question it may be possible to offer recycling as an acceptable alternative for the future of highway construction in the United Kingdom.

This paper considers the potential for using recycling as a viable option in highway construction. The basic construction and material requirements are outlined first.

Performance and Durability of Bituminous Materials. Edited by J.G. Cabrera and J.R. Dixon. © E & FN Spon. Published in 1996 by E & FN Spon, 2–6 Boundary Row, London SE1 8HN. ISBN 0 419 20540 3.

Sources of materials are then discussed, followed by the factors which need to be considered if recycling is to see future growth in the United Kingdom.

The various methods which are presently available are reviewed, ranging from the traditional view of recycling as shallow in-situ cold-mix and off site/hot-mix, to the more recent environmentally acceptable alternatives such as in-situ/hot-mix and deep-cold/in-situ methods.

Finally, the future of recycling is considered in view of the comments proposed.

2 The layered structure of a road

The structure of a road is made up of a number of layers. Aggregates are required at all levels but both the quality and the cost of the materials used generally increases from the bottom towards the top. This means that specification requirements for the wearing course or topmost layer, are considerably greater than for the bottom capping and sub-base layers.

By building in layers, a very wide range of constructional materials can be used. From a practical point this relates to reductions in cost if abundant local low quality materials can be used to provide the large amount required for the lower layers. Although the quantity of materials is less with higher layers their total cost may be greater as they may have to be transported considerable distances should suitable local supplies not exist.

For each layer, different specification requirements are needed as the different layers perform different functions. This ranges from the ability to withstand the polishing and attrition caused by trafficking to the distribution of stresses in the lower layers, i.e. it is a case of "horses for courses".

In this context, it would be very attractive should secondary and other types of recycled materials be shown to perform to the same standard as traditional sources now in use.

3 Sources of construction material in the United Kingdom

The term construction material has been used to include the wide range of materials now seen to be available for road construction. They include:-

3.1 Primary or virgin sources of aggregate produced specifically for road construction

These are the traditional sources of aggregate, and have been selected over time as being suitable to give the necessary levels of in service performance. However, they have a problem in their distribution which is dependent on the geology of the United Kingdom. This means that there are few sources of the required "hard rock" to be found in South-east England where demand is greatest. Rather, aggregates that have been used tend to be found further to the North and West, for example, in the Peak District and North Wales. Indeed, a number of Northern Ireland quarries now find it profitable to export aggregate, both high quality surfacing chippings and lower quality sub-base material, into South-east England.

3.2 Low quality aggregates

These are sources which the highways industry may have tried in the past and have been shown not to possess the necessary levels of performance when used in road construction. As a result, such materials may now only be used as filler or as sub-base.

3.3 Secondary materials

These have been created as a by-product of another industry. Examples include slate waste from North Wales, colliery waste from the coal mining areas, china clay sand from Cornwall, fly ash created by electricity production and slag from the steel industry.

3.4 Processed demolition waste

This is an industry of growing importance, especially in parts of Europe where natural resources may be even more scarce than in the U.K. Many quarries are now recycling materials such as concrete rubble from building demolition.

This practise is being enforced with levies imposed against the use of quarried virgin aggregate. In its favour, the Dutch have for example found that when used as sub-base material, a recementation process occurs which serves to increase the strength of the layer.

Current research being carried out at the University of Ulster in conjunction with Building Research Establishment is assessing the potential of such materials.

3.5 Surface planings and bituminous/unbound aggregate dug-out from pre-existing highway construction

This offers the greatest potential for recycling as the raw material is already on-site and only requires a suitable type of recycling process to produce a material which will perform in use. As a result, the existing road is being used as a horizontal quarry.

4 Factors to consider about recycling

Although it is possible to say that there are many hundred's of millions of tonnes of potentially recyclable material available, the fact of its existence does not automatically warrant their use in a highway's construction. Other factors need to be taken into account before what is perceived as the environmentally acceptable alternative of recycling is adopted by the industry. Some of these factors will now be considered.

4.1 Location

Whilst one can quote mountains of slate waste in North Wales, such material still has to be transported considerable distances to where demand for bulk aggregate is greatest, i.e. predominantly the South-east of England. In practise this implies that greatest use of this type of material is restricted to a localised market.

In contrast, there are many miles of the British road network now undergoing advanced stages of structural or surfacing failure. Recent figures suggest that the number of roads with less than one years life expectancy may be over 15% of the total road network. It is these failed roads which could provide abundant on-site sources of material which favour in-situ recycling processes.

4.2 Transportation costs

These are a major factor to consider in the future use of low-grade and secondary materials. Other than for localised use, transportation over distance only becomes attractive once the value of the product is raised, such as in the case of high skid-resistant surfacing aggregate.

4.3 Traditional experience

The use of non-traditional sources of aggregate, such as those proposed for recycling, do not have the long term track record of traditionally used sources of aggregate.

This aspect of going into the unknown, is one which the British roads industry has traditionally been loath to risk.

4.4 Long term performance

Whether due to pressure from Europe, environmentalists, or growing concern about rapidly depleting sources of aggregate, governmental bodies such as the Department of Transport and the Department of the Environment are now investing considerably into research about the use of these materials so that the necessary in service experience is obtained.

4.5 Material durability

If aggregate is to be used in the context of the modern road, then it must be expected to possess adequate long term performance properties.

There is no point in using recycled materials in a situation where in service conditions cause premature structural failure. Consider the example of recycling a bituminous material which had prematurely failed due to aggregate durability or soundness reasons. Unless detected by adequate screening and testing procedures, the resulting structure is simply premature failure waiting to occur.

4.6 Modern environmental pressures

This is a veritable minefield with the quarrying and highways industries placed in the middle of conflict between those who forecast the need for increased growth and those who would impose greater restrictions on all future growth. Hard rock aggregate has traditionally been quarried in what are now public areas, such as the Peak District National Park and so future development is under much debate.

An alternative has been to go away from the public and develop the remote super-quarry such as that at Glensanda off the West coast of Scotland where the in-side of a mountain is quite literally being quarried away. Having a coastal site, handling and transportation costs may be minimised.

5 Uses of recycled materials in highway construction

There are two main uses for recycled materials:-

5.1 Capping and sub-base materials

As the specification requirements for these materials are quite low, there is a great potential market for such recycled materials, either on their own or in combination with primary materials such as virgin crushed rock aggregate.

5.2 Roadbase and surfacing materials

As the specification requirements for these layers are much higher, this necessitates the raw recycled material to possess a higher level of performance. Candidates for this type of use are surfacing planings which should contain a relatively high quality aggregate. However, as mentioned previously, unsound aggregate is a problem that must be considered.

6 Roadbase and surfacing recycling processes

The use of recycling in these layers usually, but not always, requires the material to be bound with bitumen or cement. It is possible to categorise the different types of recycling process as used for roadbase and surfacing layers as follows:-

Hot-mix/off-site
Cold-mix/off-site
Shallow hot-mix/in-situ
Shallow cold-mix/in-situ
Deep cold-mix/in-situ.

Generally, in terms of cost, the in-situ processes are to be favoured as they do not require extra transport, handling and processing - if, it can be shown that the in-service performance is comparable to the traditional hot-mix/off-site process.

Each of these process types will now be defined and discussed. The potential of each will also be given.

6.1 Hot-mix/off-site processes

This is the traditional type of recycling of which there is still a limited knowledge in the United Kingdom. In this process, existing materials are removed by planing, transported to a hot-mix plant and then reprocessed with virgin aggregate and bitumen to comply with specification requirements for hot-mix materials such as Hot Rolled Asphalt.

In Northern Ireland, this method has been used by the Department of Environment Roads Service, for a number of motorway and dual-carriageway re-surfacing contracts. The first was in 1988 on the M1 motorway and involved the use of Hot Rolled Asphalt planings applied to virgin aggregate and bitumen. It was found that the recycling mixing process required careful control to provide a satisfactory end-product. Initial trials proved that mixes containing up to 50%

recycled material could be used successfully.

However, the future of this type of recycling in Northern Ireland is limited due to the abundance of cheap, high quality virgin aggregates. But, with the use of a modern, mobile hot-mix plant on-site, this method may prove to be a viable option in mainland Britain where surface aggregate costs are significant.

As regards its future, the Department of Transport currently only permit the use of mixes with up to 10% although levels of 50% are used elsewhere in Europe. In the long-term, European environmental and health concerns may not allow this type of recycling process to reach its full potential, but may however favour the growth of cold-mix processes.

6.2 **Cold-mix/off-site processes**

This is similar to hot-mix/off-site in that it involves recycling at a central plant. The exception is that the process involves the use of cold mixing with either one or a combination of foamed bitumen, bitumen emulsion, cement and lime.

As regards the only operational plant in the United Kingdom, this is now operated by R J Maxwells and has to date concentrated on the use of foamed bitumen stabilisation. This, when mixed with bituminous planings produces a material suitable for roadbase use. As such this process has significant potential within the U. K marketplace. However, it has also been trialed as a permanent re-instatement material under the U.K's HAUC regulations which may further extend its market potential.

6.3 **Shallow hot-mix/in-situ processes**

This may be termed a surface re-generation process for existing wearing course materials such as Hot Rolled Asphalt.

The process first involves heating the roads surface layer, scarifying followed by reshaping and then reinforced by a thin overlay of new asphalt. Typically this is 20 - 25mm thick and is heat welded to the old material. The total depth of treatment is about 50mm with cost savings of 15 - 20% claimed. Examples of this type of process include that known as Repave.

Due to the size of plant involved, this process has tended to be restricted to major roads. But, as trunk roads and motorways account for only about 4% of the United Kingdom's total road network, the processes potential expansion must be restricted. However the process is beginning to show favour in Eastern Europe.

6.4 **Shallow cold-mix/in-situ processes**

This process, commonly known as Retread, has been in service for over 50 years in this country. It was originally introduced as a relatively cheap method of repairing badly damaged roads during the war.

The fact that it has survived as long, is testimony to its value within the road maintenance industry in this country.

Retread involves firstly the scarifying and reshaping of an existing road or footway surface. Once completed, virgin aggregate may be added to reprofile the road surface, alternatively excess aggregate may be removed.

After the desired profile has been achieved, bitumen emulsion is applied using a spray tanker. This is harrowed in to the full 75mm depth of the retread layer to ensure an even mix. This is then followed by compaction of the layer.

Finally a surface dressing is applied using between 3mm - 14mm chippings to give adequate texture depth to the surface.

Depending upon the type of emulsion used in the final dressing, and if the site is to be subsequently overlaid, a further surface dressing may be required in 9-12 months to finally seal the surface.

This method of in-situ cold recycling is appropriate for the rejuvenation or reshaping of residential and generally lightly trafficked roads. The Retread process has been shown to be a cost effective alternative to planing out and adding a new overlay. Giving a claimed cost saving of between 25 - 35%.

As it uses a "cold emulsion" it also has the advantage of being attractive from both a Health and Safety viewpoint as well as to the environment.

6.5 In depth cold-mix/in-situ processes

As the name implies, this process treats the road to a far greater depth than does the shallow cold-mix/in-situ process. This type of process can recycle an existing road surface to a depth ranging from 150mm to 300mm.

This process involves pulverising the existing road surface to a depth of up to 300mm. This material is then compacted and reshaped.

Excess material is removed at this stage. Once the desired profile has been achieved the material is rotovated again, during which time bitumen emulsion, foamed bitumen and/or cement will be added in pre-determined quantities and thoroughly mixed throughout the layer.

Again the layer is compacted and shaped before being sealed with sprayed bitumen emulsion and sealing grit. Typically this layer is then overlain with some other material to provide a new running surface.

As well as an enhanced speed of operation the in-depth recycling process typically offers a cost saving of up to 40% and an energy saving of up to 90% when compared to traditional methods, since the existing road is being used as a horizontal quarry.

Its principal advantage is that is very flexible depending upon what is being recycled. Due to its significant financial and environmental benefits, and the pressure now being placed upon local authorities to recycle, this type of process must have potential in the future. Recent research at the University of Ulster has shown this material was capable of out performing "virgin" material.

7 Conclusions

To be critical about the future of recycling in the United Kingdom, it is possible to conclude:-

Modern roads require materials that can provide the necessary levels of long term performance.

The high demands for surfacing aggregate are generally too great for secondary recycled materials.

Both the Department of Transport and the Department of the Environment desire to see increased growth in the use of recycled materials for structural and/or foundation layers of existing and proposed roads due to the quantity of materials required.

Hot-mix recycling is already acknowledged by the Department of Transport as a viable option, but has problems such as extra cost and quality control.

In the debate of hot versus cold processes, cold mixes are more attractive due to their environmental acceptability.

There would appear to be good, long term prospects for the in-situ cold recycling processes.

Off-site processes still involve considerably more energy expending activities - such as road excavation, planing and importation of the recycled materials.

The authors wish to acknowledge the help and co-operation received from their colleagues in the preparation of this paper and would emphasise that the views expressed are entirely their own and are not necessarily those of the University of Ulster or Colas Limited.

17 A REPEATED LOAD COMPRESSION TEST FOR ASSESSING THE RESISTANCE OF BITUMINOUS MIXES TO PERMANENT DEFORMATION

J.M. GIBB and S.F. BROWN
Department of Civil Engineering, University of Nottingham, Nottingham, UK

Abstract
The paper describes the use of a repeated load uniaxial compression test for assesssing the permanent deformation resistance of bituminous paving mixtures. The test equipment, procedures, specimen preparation and test configuration are described in some detail. The merits of using a repeated load test are demonstrated and results of tests on asphaltic concretes which show good correspondence with performance in wheeltracking experiments are presented. The case for the potential use of such a test system in designing bituminous mixtures to make best economic use of resources and as an instrument for performance specification is presented.The work forms part of a major project concerned with the development of simplified test methods to measure the mechanical properties of bituminous mixtures for use in a practical construction environment.
Keywords : Bituminous Mixtures, Permanent Deformation, Repeated Loading, Laboratory Testing

1 Introduction

The mechanical response of a bituminous material to the application of an external load is, in part, viscous. That is to say, an element of the deformation induced by the application of the load is time dependent and irrecoverable. Under repeated applications of load, the permanent deformation continues to accumulate. The consequence of this , in the case of bituminous paving mixtures subject to vehicular traffic, is that ruts develop in the wheelpaths. If severe, these ruts have a major effect on the serviceability of the pavement and maintenance is then required to restore the profile.

The performance of a bituminous mixture is strongly dependent on the volumetric proportions of aggregate, binder and air voids. These

Performance and Durability of Bituminous Materials. Edited by J.G. Cabrera and J.R. Dixon. © E & FN Spon. Published in 1996 by E & FN Spon, 2–6 Boundary Row, London SE1 8HN. ISBN 0 419 20540 3.

proportions may be specified or, to a degree, surrogated by the use of recipe specifications as is current practice in the UK (eg. BSI, 1993). However, resistance to permanent deformation is dependent not only on the relative proportions of the constituent materials of a mix but also on their nature and mechanical characteristics. The shape, surface texture and grading of the aggregate particles and the grade and rheological properties of the binder all have an effect. It is essential, therefore, that mechanical testing for assessment of resistance to permanent deformation should form a part of mixture design and evaluation

2 The Creep Test

In the 1970's the Shell Organisation developed the use of a static unconfined uniaxial compression test, the Creep Test, for assessing the permanent deformation resistance of bituminous materials (Hills, 1973). This test gained wide acceptance, principally due to ease of specimen preparation, simplicity of test procedure and low cost of test equipment. The only requirements for the test specimen were that it should be prismatic with flat and parallel ends normal to the axis of the specimen. The test procedure required only the application of a constant stress to the specimen for one hour and measurement of the resultant deformation. The test equipment was, therefore, rather simple and early versions of the equipment (de Hilster and van de Loo, 1977) generally applied the load as a dead weight via a mechanical lever arm.

Shell developed a rut prediction procedure based on the Creep Test, but it was found that the method under-predicted rut depths measured in trial pavements (Hills et al, 1974) and, therefore, they subsequently introduced an adjustment factor, derived empirically for each mix type, to account for the effects of dynamic loading. Since the introduction of the Creep Test, further concerns have arisen over the use of static loading. Permanent deformations observed in the Creep Test are largely associated with viscous flow within the binder film. Under repeated loading, larger strains may occur due to the effect of the pulsed loading on the aggregate skeleton. There is also evidence to suggest that the static test does not reflect the improved performance of modifiers which enhance the elastic recovery properties of a material (Valkering et al 1990), whereas this can be demonstrated under repeated loading conditions.

3 The Nottingham Asphalt Tester - NAT

Hitherto, the facility to carry out repeated loading tests had generally been limited to major research laboratories because of the cost and complexity of the equipment required. Research carried out at the University of Nottingham during the 1980's into the application of pneumatic load and digital control systems led to the development of the Nottingham Asphalt Tester (NAT) (Cooper and Brown, 1989 and 1993). In this system, the supply of compressed air to a pneumatic actuator is governed by a solenoid valve. The operation of this valve is controlled by an IBM PC compatible microcomputer via a digital to analogue converter. The load applied to a specimen and the resultant deformation are monitored by a strain gauged load cell and Linear Variable Differential Transformers (LVDT's) respectively, and outputs from these devices are acquired by the computer through an analogue-digital interface. The pneumatic actuator is capable of applying a load of up to 4.2kN and transient deformations as low as 1 micron can be recorded.

The use of microcomputer control and data acquisition allows, through the use of suitable software, both considerable flexibility in the application of load and also the facility to acquire data continuously and automatically throughout a test. The advantage of pneumatic systems is their low cost compared to servo-hydraulic apparatus, though precision in control of load application is diminished. The NAT has been designed to accommodate a number of different modules for measuring the properties of bituminous materials using cylindrical test specimens and its success in providing a means of routine evaluation of properties can be judged by the fact that there are currently over 30 units in use with Highway Authorities and material suppliers in the UK alone.

Figure 1 shows, schematically, the configuration of the equipment for permanent deformation testing. Permanent axial deformation is monitored with LVDT's attached directly to the top of the upper loading platen. The test is carried out on cylindrical (moulded or cored) specimens, the ends of which are prepared by trimming with a diamond tipped saw.

4 The Effect of Repeated Loading

The NAT may be used to perform the static Creep Test, though in view of the limitations and concerns associated with this test, discussed in Section 2 above, a repeated loading test is normally used. In this test, which is more representative of vehicular loading effects, stress pulses of 100kPa are applied for a duration of one second, with an interval of one second between load pulses. The test is generally run for 3600 cycles, giving an accumulated loading time of one hour. This test is known as the Repeated Load Axial (RLA) test.

To demonstrate the effect of repeated loading, a comparison was made between the static Creep Test and the RLA test. Both methods were used on an asphaltic concrete manufactured from 20mm nominal size crushed granite. Figure 2 shows the results in the form of strain-time plots for the static tests, while the results for the RLA tests are shown on

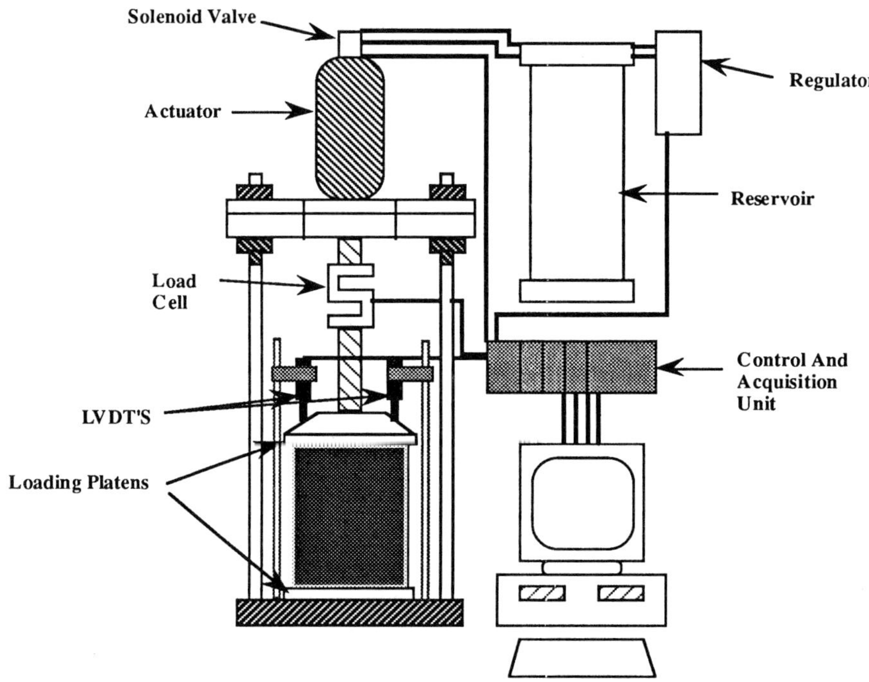

Fig. 1. The Nottingham Asphalt Tester

Figure 3. The void contents of the specimens were in the range 3-7% for both series of tests. All tests were carried out at 40°C.

These plots clearly show a significant difference in the response of the material to the different types of loading , with considerably greater deformation having occurred in the specimens subjected to repeated loading. This is most probably due to the fact that the RLA test represents a harsher examination of the aggregate structure in the material than does the Creep Test.

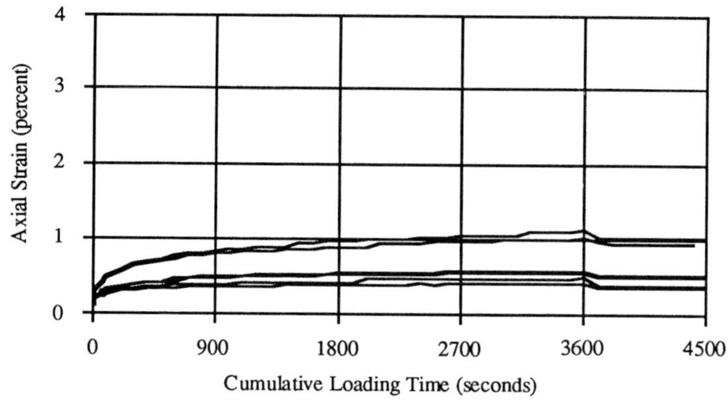

Fig. 2. Creep Test Plots : 20mm Asphaltic Concrete

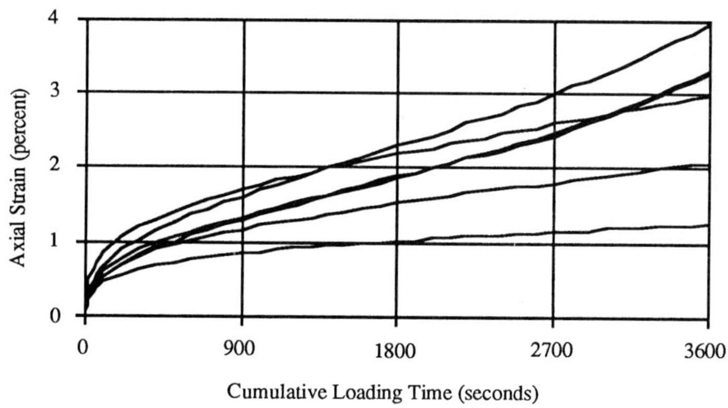

Fig. 3. RLA Test Plots : 20mm Asphaltic Concrete

5 Comparison With Wheeltrack Testing

A test programme to assess the permanent deformation resistance of asphaltic concretes using the Creep Test, RLA test and a wheeltracking test was carried out on 8 mixes made up from the combinations of two aggregates, two binder types and two binder contents. The aggregates were a crushed granite and a dredged, uncrushed chert. Both aggregate and binder data are given in Table 1 and their reference codes are shown in Table 2. The two binder contents for each mixture, shown in Table 3, were determined by Marshall and Hveem designs on the particular bitumen-aggregate combinations. The aggregate grading is shown in Figure 4.

Table 1. General Characteristics Of Mixture Constituents

Aggregate		Description	
Watsonville granite		Rough angular, crushed rock	
Texas Chert		Smooth rounded gravel	
Bitumen	**Pen**	**S.Pt(°C)**	**P.I.**
Valley	53	47	-1.34
Boscan	70	48	-0.53

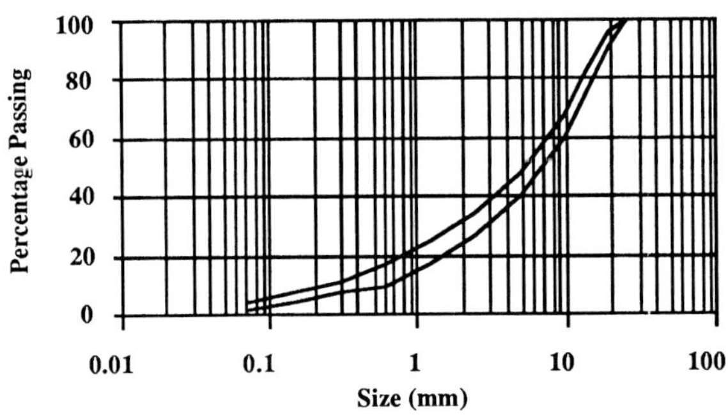

Fig. 4. Grading Envelope For 20mm Asphaltic Concrete

The wheel-tracking test was an adaptation of the DD184 method (BSI, 1990). The equipment is shown schematically in Figure 5.

Table 2 . Variables Evaluated In Test Programme

Variable	Description	Reference Code
Bitumen	Boscan	B
Type	Valley	V
Binder	Optimum	O
Content	High	1
Aggregate	Texas Chert	T
Type	Watsonville Granite	W

The specimens for the wheeltracking test were slabs measuring 404mm by 280mm in plan with a depth of 75mm. Each slab was subjected to 5000 wheel passes at a temperature of 40°C. Wheel loads of both 554N and 809N were used, giving contact pressures of 650kPa and 950kPa respectively. Performance was assessed by strain rate in the Creep and RLA tests and deformation rate in the wheeltracking test. The ranking of the different mixes by each of the three tests is shown in Figure 6.

Table 3 . Binder Contents Used In Test Specimen Manufacture

Mix Reference (see Table 2)	Binder Content (% by mass of mix)
VOT	4.1
VIT	4.8
BOT	4.3
BIT	5.0
VOW	4.9
V1W	5.5
BOW	5.1
B1W	5.7

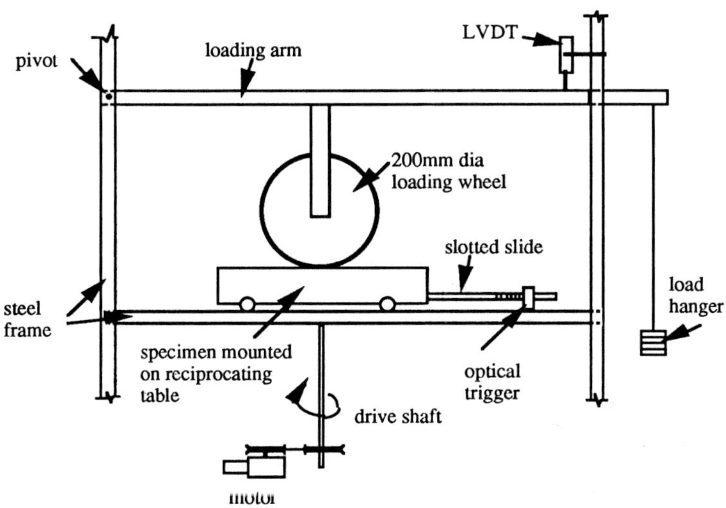

Fig. 5. The Wheeltracking Test

It can be seen that there is quite good correspondence between the RLA and wheeltracking tests, but that the ranking produced by the creep test is markedly different.

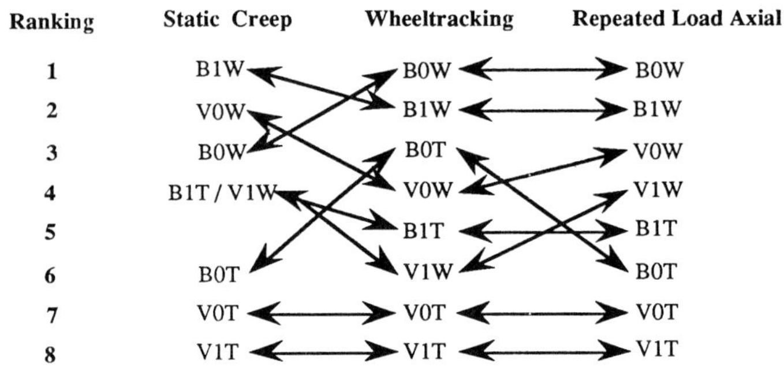

Fig. 6. Comparison Of Mix Performance

6 Application of the RLA Test

Mix design is an important application for the RLA test, since it provides a means for quantifying the effect on permanent deformation resistance of varying mix proportions . Figures 7 and 8 show the effect of varying binder content on rolled asphalt wearing courses with aggregates from two different sources. The test parameter used here is the mean strain rate which was obtained by averaging the strain rate, calculated in increments, over the whole duration of the test.

Rolled asphalt wearing course is a gap graded material with a high proportion of filler and 50 pen bitumen which is used extensively on UK roads. At present, the material is either specified by recipe or optimisation of binder content is performed using a modified version of the Marshall procedure. The design binder contents for the materials in Figures 7 and 8 were 8.1% and 7.3% respectively. The indication from the RLA results is that these binder contents are at the upper limit of the optimum range for permanent deformation resistance. The plots also show that below a certain threshold level, deformation resistance is relatively insensitive to binder content and this provides a "window" to determine the design binder content based on consideration of other properties such as workability, durability and resistance to fatigue cracking. These plots also provide an indication of the likely tolerance of the material to variations in binder content during production.

The RLA test also has a role in "end product" testing. Figures 9 and 10 show data for cores taken from a 20mm nominal size continuously graded bituminous basecourse, made with 200 pen bitumen.

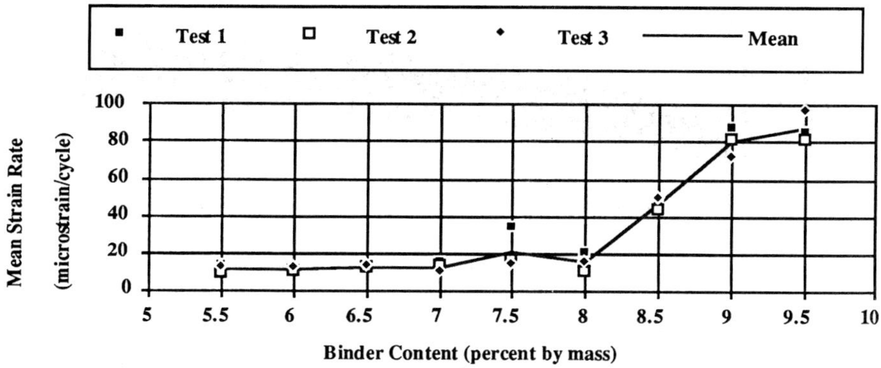

Fig. 7. Effect Of Binder Content On Mean Strain Rate : 30/14 Hot Rolled Asphalt : Aggregate "A"

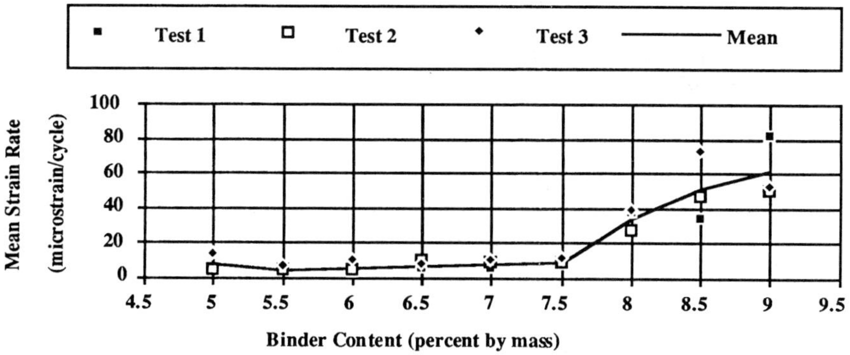

Fig. 8. Effect Of Binder Content On Mean Strain Rate: 30/14 Hot Rolled Asphalt : Aggregate "B"

Void contents of the specimens for which the test results are shown in Figure 9 were lower than those in Figure 10. It is evident that the material with lower void content has the poorer resistance to deformation. This finding probably reflects incorrect volumetric proportioning of the material which has resulted in an excess volume of binder at high levels of compaction.Whatever the cause, this example illustrates how the RLA test can be employed to obtain data on the performance potential of a material which would not be revealed by compositional analysis.

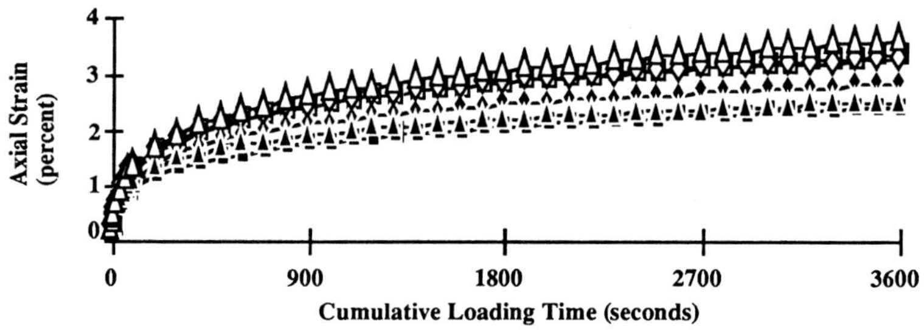

Fig. 9. RLA Test Results For 20mm DBM, 200 Pen, Void Content Range 3.1 - 3.5%

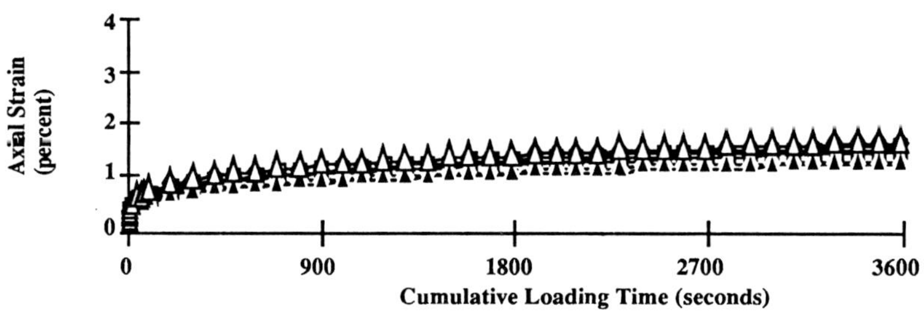

Fig. 10. RLA Test Results For 20mm DBM, 200 Pen, Void Content Range 4.8 -5.3%

7 Conclusions

1. The development of the Nottingham Asphalt Tester has provided the facility to carry out testing on bituminous materials using repeated loading and with automatic data acquisition on a routine basis.
2. The Repeated Load Axial test retains the simplicity of test procedure and specimen preparation of the Creep Test.
3.The Repeated Load Axial test gives different permanent deformation results from the Creep Test.
4.The Repeated Load Axial test ranks material more closely to the wheel-tracking test than does the static Creep Test.
5. The Repeated Load Axial test has the potential to be used for mixture design and evaluation.

8 References

British Standards Institution (1993) Coated macadams for roads and other paved areas. **BS 4987,** Part 1

British Standards Institution (1990). Method for the determination of the wheel tracking rate of cores of bituminous wearing courses. **DD 184**

Cooper, K.E. and Brown, S.F. (1993). Assessment of the mechanical properties of asphaltic mixes on a routine basis using simple equipment. **Proceedings, Eurobitume Congress,** Stockholm, Vol 1B pp 872-876

Cooper, K.E. and Brown, S.F. (1989). Development of apparatus for repeated loading in creep and indirect tension. **Proceedings, Eurobitume Symposium,** Madrid

de Hilster,E. and van de Loo, P.J.(1978) The Creep Test: Influence of test parameters. **Proceedings, Colloquium 77 Plastic Deformability of Bituminous Mixes,** pp 173-215, Zurich

Hills, J.F (1973) The creep of asphalt mixes. **Journal of the Institute of Petroleum, Vol 59 No. 570,** pp247-262

Hills,J.F. Brien,D. and van de Loo, P.J. (1974). The correlation of rutting and creep tests on asphalt mixes. **Institute of Petroleum, Paper IP 74-001**

Valkering, C.P., Lancon, D.J.L., de Hilster, E. and Stoker, D.A. (1990). Rutting resistance of asphalt mixes containing non-conventional and polymer modified binders. **Proceedings, Association of Asphalt Paving Technologists,** Vol 59, pp 590-609.

18 THE USE OF THE WHEEL TRACKING TEST FOR WEARING COURSE DESIGN AND PERFORMANCE EVALUATION

I.D. WALSH
Engineering Services Laboratory, Kent County Council, Aylesford, Kent, UK

1 Introduction

The Wheel Tracking Rate (WTR) test was developed by TRRL in 1977 to simulate the in-service rutting of hot rolled asphalt. The test has been published as a British Standard Draft for Development DD184.[1]

Briefly, the test involved subjecting a 50mm thick slab of material, either manufactured in the laboratory, or using a 200 mm diameter core taken from the wearing course after laying, to a rolling wheel load, which traverses the specimen at a constant temperature, normally 45°C, and standard load. The test measures rutting under the wheel over a period of time.

The test can also be used for design purposes using plant processed dry aggregates, mixed with binder and compacted in the laboratory to known density using a vibrating hammer. The slabs, if 300mm square can be used to obtain two determinations, therefore 3 slabs are necessary for the six determinations required by DD 184. For the purposes of evaluation these are made up at the same time as the Marshall design check. The paper discusses this design method and compares it to the Marshall method described in BS 594[2] based on 14 plants, a range of binder contents and primarily onc sand source.

TRRL developed a relationship based on work on the A30 Winchester[3] which related the results from the test to the number of commercial vehiclcs travclling at 60 mph required to form a 10mm rut at the end of a 20 year life, as follows:

$$\text{Max WTR} = \frac{14{,}000}{C_v + 100}$$

Where: WTR is wheel tracking rate in mm/hr
C_v equals number of commercial vehicles in the lane per day.

Thus for a typical heavily trafficked motorway slow lane (4500 cv/lane/day) WTR of 3mm for a max might be required, a typical lightly trafficked high speed road (500cv/lane/day) 23mm/hr max would apply.

However, the effect of loading time on deformation is significant, ie as speeds fall, binder modulus decreases; therefore for a given mixture, rut resistance decreases and the required WTR should decrease.

Performance and Durability of Bituminous Materials. Edited by J.G. Cabrera and J.R. Dixon. © E & FN Spon. Published in 1996 by E & FN Spon, 2–6 Boundary Row, London SE1 8HN. ISBN 0 419 20540 3.

The relationship between loading rate and deformation as linear[4] within the loading rate and working temperatures typical of UK highways for a material such as HRA where properties are determined by the binder rather than aggregate skeleton.

Kent County Council apply a speed factor of $^x/_{60}$ to correct the WTR for the effect of actual vehicle speed where x is measured in miles per hour for typical commercial vehicle flow. BS 7533[5] recommends that, for canalised traffic, traffic volumes should be increased by a factor of 3 if traffic wheels are constrained within the same path; this occurs at bus stops, where pedestrian refuges have been inserted and on narrow contraflow lanes. It has been recognised by Szatkowski when he was preparing his recommendations[9] for heavily trafficked roads given in Table 1, Columns 2 and 3 below.

Table 1. Relationship: traffic, WTR and stability (Refs 9,10,22,25).

TRAFFIC Intensity cv/l/day	(Max WTR) mm/hr	Stability sand asphalt (min) kN		Stability 30% Stone (min) kN	
			incl tol.		EVA modified
<750	32	-	-	-	
750	16	1.5	2.1	2.7	1.8
1500	8	2.1	3.1	4.0	2.7
3000	4	3.1	4.6	6.0	4.0
6000	2	4.6	6.5	8.5	5.7
6000+	<2	>6.5	>6.5	>8.5	>5.7
6000 HCT	<1	>6.5	>7.5	>9.5	>6.5

HCT = heavy canalised traffic

incl tol. refers to the increased stability required to ensure that the minimum value is achieved at the extremes of the tolerances on composition permitted by the British Standard.

The effect of speed and canalisation, for example, for a typical urban street with speeds around 20mph and 300 cv/lane/day, gives a min WTR requirement of 4.7mm/hr and on motorway contraflow and slip road contracts with 4500 cv/lane/day at 30mph nominal, 0.5 mm/hr WTR is required.

The formula has been found to compare reasonably well with in-service performance for sand based hot rolled asphalts on high speed [3] and other roads.[20] However, on A303 Mere[24] a best-fit WTR for the current criterion of 0.5mm/yr was 3.8mm/hr or 5.9mm/hr for a rate of 0.65mm/yr ie 13mm formation over 20 yr life. The design formula specified 12.7mm, (9.3mm with the speed factor). On this contract the WTR formula would not give an adequately rut resistant surfacing.

The WTR test is useful up to 10mm/hr WTR, the maximum usually obtained with HRA and 50 pen bitumen binder. It cannot sensibly be used for WTR in excess of 15mm/hr. The test method therefore is particularly applicable for the design of wearing courses for high traffic intensity, slow speed situations. It should be noted that contraflow during construction is often the worst case and may require special surfacing especially if these works take place in Summer.

Work by Choyce and Woolley (1988)[6], Carswell (1987)[26] and Colwill and Carswell (1990)[22] with Ethyl Vinyl Acetate (EVA) polymer and by Denning and Carswell (1981) using styrene-butadiene-styrene (SBS)[7] polymer has shown the extremely beneficial effects on WTR of modifying bitumen with polymers. 5% EVA in a binder not only improves WTR by a factor of 0.3 but makes the mixture less sensitive to binder content, with little change over the permitted tolerance. Polymer modification can reduce the temperature susceptibility of the bitumen[6][22] and therefore probably probably also improve the low temperature characteristics, this may help to reduce the propensity of HRA wearing course to crack in winter. HRA mixtures containing Trinidad Lake Asphalt (TLA) have been found to have improved deformation resistance at slow speeds compared with unmodified bitumen.[8]

In designing wearing course mixtures for considerable numbers of heavy vehicles, slow traffic, or highly stressed sites, the engineer may specify a particular polymer modified mixture or may decide to give the contractor a choice of any of the binders given in Table 2 and specify a maximum WTR; mixtures with the same WTR should give approximately the same rut resistance performance allowing the economics of the various options to be determined by the contractor.

The formula for wheel tracking rate given above has been correlated with the performance in-service of a 40 mm thick HRA surfacing containing 30% coarse aggregate. There is at this time insufficient evidence available as to whether or not this formula may also be used for DBM type mixtures where the method of deformation is not one of plastic flow of the bitumen binder/filler/sand matrix during hot weather, but is the reorientation of the particles as the softer bitumen ceases to be a glue and becomes a lubricant.

Table 2. Comparison of wheel tracking rates of HRA with 50 pen bitumen, EVA and SBS modified binder and 50/50 TLA pen bitumen blend.

Binder	Pen	WTR at 45°C (mm/h)
50 pen	56	3.2
70 pen + 5% EVA (45/33)	57	1.0
200 pen + 7% SBS	84	0.7
50/50 TLA + 50pen bitumen	24	1.6

2 Test method for manufacture of slabs

The Marshall design is carried out in accordance with BS 598 Part 107[12] and the target binder content obtained. The mix density is also measured.

Sufficient dry material of known particle density and bitumen is weighed out, heated and mixed with the target binder content to produce a slab 305 x 305 x

50mm thick. The material is compacted hot with a vibratory hammer using a 100mm by 140mm rectangular foot to achieve approximately 98% of the Marshall density. The density to be used is obtained from the mix density curve.

In some cases additional slabs are made with less or more binder content, generally at target + 0.3 and + 0.6%. The Marshall stability of these slabs is obtained from the stability/binder content curve plotted as part of the design mix process.

Historically a pair of slabs is made at each binder content. One slab can be used to produce two Wheel Tracking Rate determinations. The mean of the four determinations gives the laboratory Wheel Tracking Rate. This is less than the 6 required by BS DD184 because of the perceived improved precision of this test method compared with field cores and confirmed by recent precision trials.

WTR rate is affected by a number of factors considered individually below, binder content, and binder penetration being the most significant. However, fine aggregate and filler content and type mix density (air voids) and stone content are also relevant. Single cores/slabs, can have their WTR distorted by individual pieces of aggregate at the centre of the specimen; this is particularly true for higher Wheel Tracking rates exceeding, say 10mm. The effect of plant conditioning on aggregates may also need consideration when deciding if the laboratory specimens need to be made from such material or whether samples of stockpile material would suffice. Where plants normally produce crushed rock fines macadams or sand/rock fines blends are proposed, the effect of contamination or variation in fine aggregate percentages needs consideration. When measuring stiffness modulus, it is also known that the age of the sample, ie time that the bitumen and aggregate have been in contact, affects the result. Since there is a correlation between WTR and stiffness modulus, it is probably prudent to eliminate this factor by the test protocol. Most of these factors are also relevant to the Marshall Test procedure.

3 WTR and Marshall stability

HRA wearing course is commonly designed by the Marshall design method in BS 594 to obtain the target binder content of, normally 50 pen, bitumen in the mixture, by measuring the stability and flow under the standard test conditions, following compaction by 50 blows of the hammer. Work by Szatkowski[9] and Jacobs[10] related Marshall stability of sand asphalt specimens to WTR, suggesting a clear relationship was possible. This is shown in Table 1. This work supplemented by the work of Choyce[25] on the additional stability provided by the coarse aggregate, suggested a factor of 1.3 x should be applied. The requirements were published in BS 594[2] Table B1 for 30% stone content HRA laid 40mm thick. Work on other nominal stone sizes and thickness of material has not been published.

Work by Lees[12] has recommended a minimum flow of 2mm is required to exclude excessively brittle mixtures. The table demonstrates that compositional tolerances have been partially included within the current recommendations. Increasing thickness without increasing stone content is likely to improve workability, increase chip embedment and retention in cold weather, but reduce rut resistance.

The WTR test method on prepared slabs described above has enabled the relationship between Marshall stability and WTR to be established in KCC laboratory for one sand source (Charing) and 30% stone content. This is shown in Figure 1 and as expected, WTR falls as stability increases. There is a significant scatter in the results but an equation of the form WTR = 11.6 - 1.39(stability) has a correlation coefficient of 0.80. This is improved slightly to 0.82 with an exponential regression WTR = 158 x $e^{-0.65\ \text{stability}}$. When compared to Marshall Quotient (stability/flow), an exponential regression of the form WTR = 15 x $e^{-1.2\ MQ}$ has a correlation coefficient of 0.85.

Figure 1 also gives the relationship found by Jacobs(3) at Winchester, by Daines(24) on A303 Mere (1989) and Choyce(25) between WTR on cores from the road and Marshall stability. The last two were on crushed rock fines material. It suggests that either the local Kent Charing sand gives a better WTR than its Marshall stability would suggest or laboratory slabs give lower WTR than cores. As shown in Para 11, the latter is the case; the 'corrected' line also being given on Figure 1.

4 Effect of binder content

The effect of binder content on WTR for three sources of sand is given in Figure 2. It demonstrates:

a) that increased binder content results in increased WTR;

b) different sands reflect this to a differing degree, Snodland sand being more sensitive (3.0mm change for 0.5% increase in binder content) compared to Charing sand (2mm change for 1% increase);

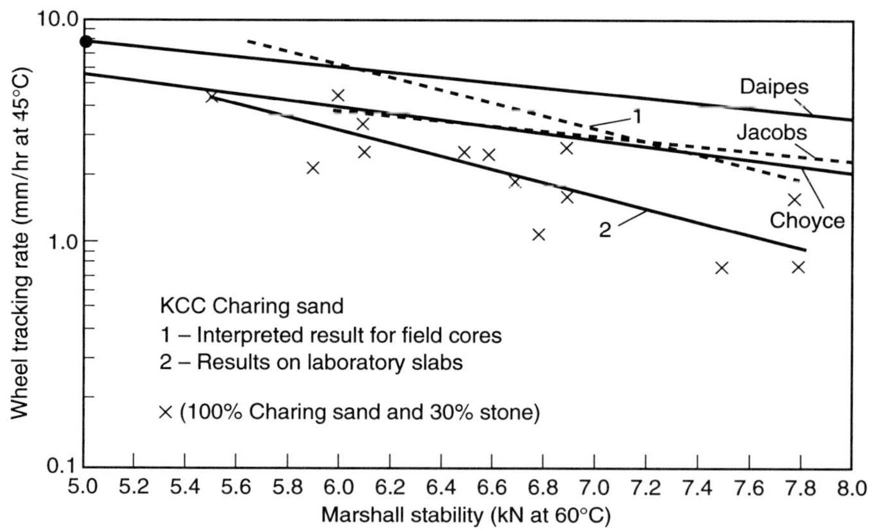

Fig. 1 Relationship between Marshall stability and WTR for Charing sand and 30% stone content, and for other materials.

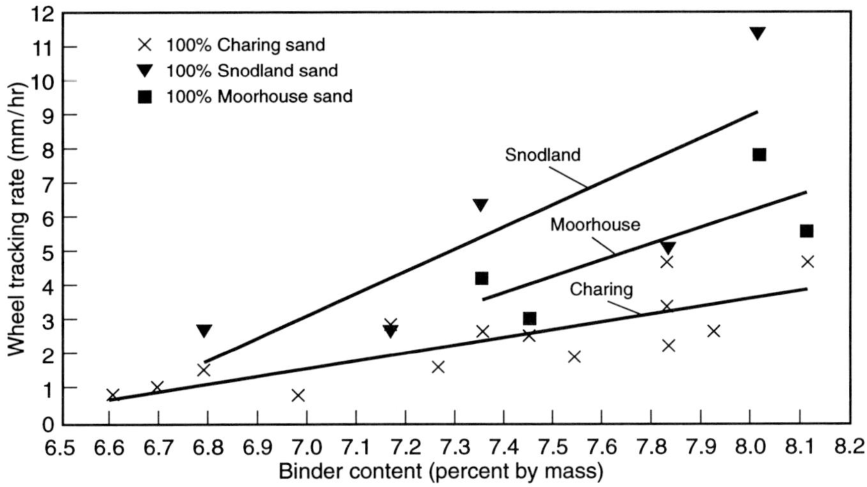

Fig. 2. Effect of binder content on WTR for three sources of sand.

c) Charing sand, with a target binder content of 7.4% or less (max binder content 7.7%) is suitable for 30mph traffic up to 2280 cv/l/day, whilst Snodland sand, a similar pit sand from a source only 30 km away and of the same geological type, would only be suitable for traffic levels of 900 cv/l/day. Both of these are however values satisfactory for roads with above average traffic.

At higher binder contents, there is a slight increase in the scatter of results of 4 determinations on 2 slabs. Using Charing sand and 50 pen bitumen the range was 25% at target +0.6%, whilst at target the range was 21%. This is probably due to binder migration during the compaction process leading to an uneven distribution through the specimen.

Work(22)(26) has shown that EVA modification reduces binder content sensitivity as shown in Figure 3.

The relationship of WTR to binder content does not show a clear maximum/minimum value, therefore it cannot be used for determining an optimum binder content and other methods should be used.

5 Effect of binder penetration

The effect of binder penetration is important because whilst laboratory work is carried out with binder of penetration within the range 40 - 60 pen, and generally close to 50 pen, when cores are taken from the field, much lower penetrations are likely.

Work done as part of the DoT Remix Recycling(13) study also showed that

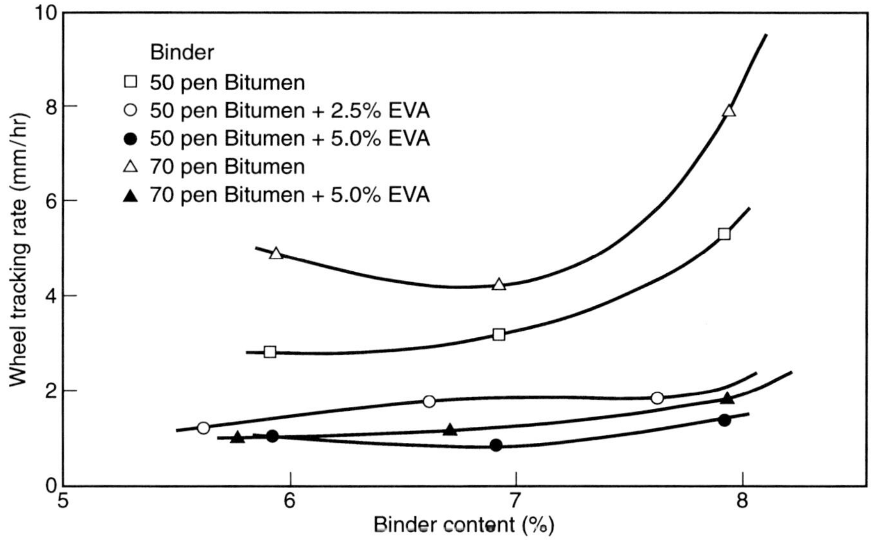

Fig. 3. Wheel tracking rates of rolled asphalt (30% stone content) with various binders using Bristol Channel fine aggregate.

reheating bulk samples in order to make, for example, Marshall stability moulds or WTR slabs, has a significant effect on the penetration of bitumen. Whilst unheated cores from two sites had average penetrations of 49, & 70, reheated and recovered bulk samples had an average penetration of 24 and 33, a 2.1 fold reduction. Reheating as method of sample preparation is therefore not recommended.

From the A614 contract the relationship given in Figure 4 demonstrates this effect and shows that a variation within the permitted range of BS 3690 gives a change in WTR from 1.65 to 3.42 on this mixture.

Figure 5 shows that it is necessary to consider the penetration and WTR on specific cores. This figure gives WTR on a group of cores vs the recovered penetration from a bulk specimens taken from the same area. This averaging masks the correlation.

Typically, binder in HRA recovered in the field has a pen of around 35 therefore one would expect that the WTR on cores from site would be lower by 50% than slabs at 50 pen as a result of this factor, but density also plays a part.

6 Effect of filler content

Laboratory specimens are made up at 10% filler content, field mixtures can vary by ± 2% Work by Szatkowski(9) given in Figure 6 demonstrated that increasing filler content above the designed value adversely affects WTR, particularly in combination with high binder content.

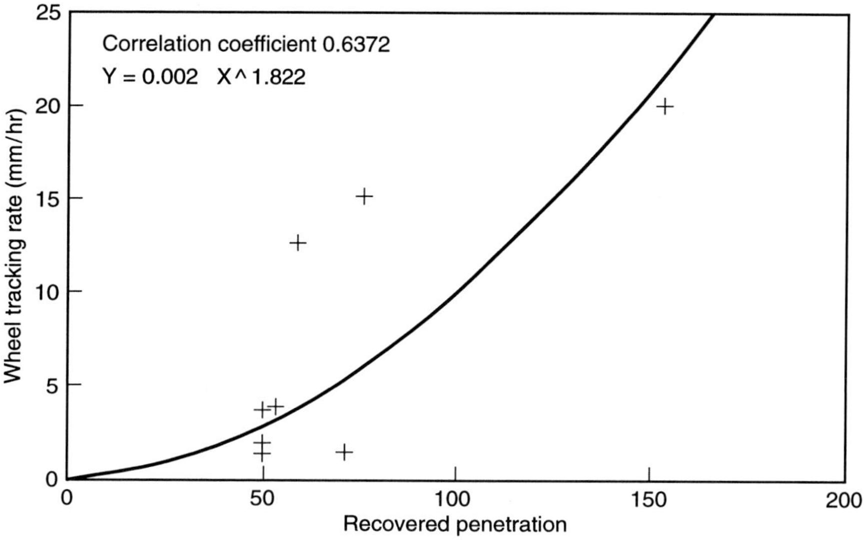

Fig. 4. Relationship between WTR and recovered penetration on remix material from A614 Nottingham – core specific.

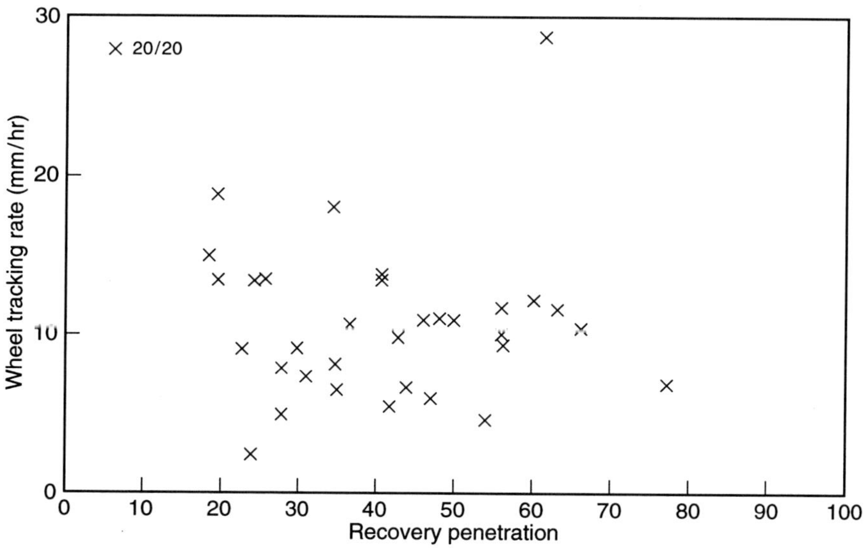

Fig. 5. Relationship between WTR and recovered penetration on remix material – non-specific summary.

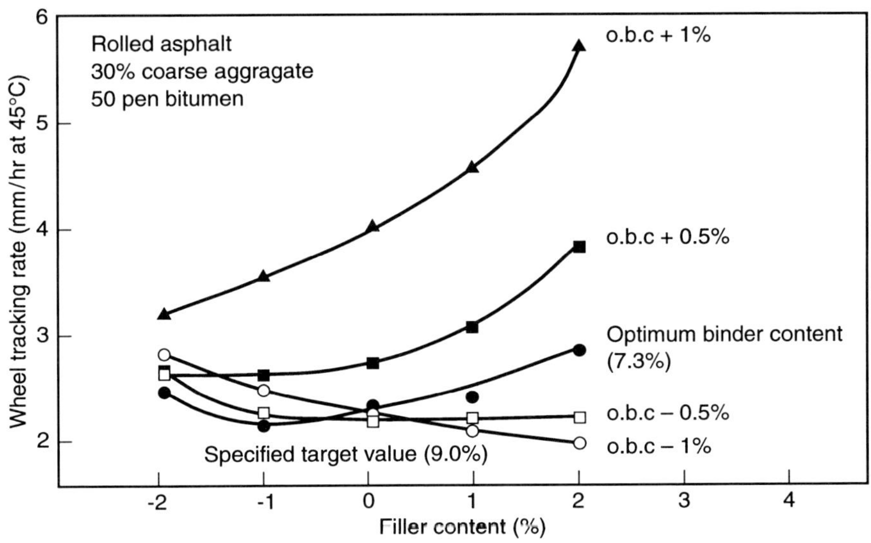

Fig. 6. Effect of binder and filler contents on resistance to deformation.

Kavussi and Lees [23] found that not only was filler content but also filler type was relevant for performance initially and also affected ageing.

7 Effect of mix density

When making up slabs in the laboratory the air voids are typically within the range 3.5 to 4.2%, but slabs with higher binder content, (target + 0.6) can have voids as low as 1.2% whilst those with binder content at target -0.3% can be as high as 6.2%. The effect of air voids on laboratory slabs from one supplier (R) has found a trend towards lower WTR for lower voids (Figure 7). This relationship is not apparent for other suppliers or if the results from more than one supplier are aggregated, even if the fine aggregate is the same, with a trend of lower WTR for higher voids apparent. This is due to the overriding effect of the binder content particularly for the mix with higher WTR and higher binder content from supplier B.

The standardisation trial[21] carried out to assist in ascertaining precision identified that a variation in density between specimens made some contribution to the repeatability of the test method with two mixes but not with two others. One mix in particular generating a 0.5mm/hr change on 10mm for 80 kg/m^3 density change (3% change in air voids). Since it is very possible to make up slabs to ± 1% this suggests the effect can be overcome at the design stage.

As discussed in Para 11, specimen density may be the cause of differences between cores and laboratory-made slabs. All cores were taken soon after laying but Daines[24] pointed to the effect of voids on ageing suggesting an optimal void content of 3-4% for durability.

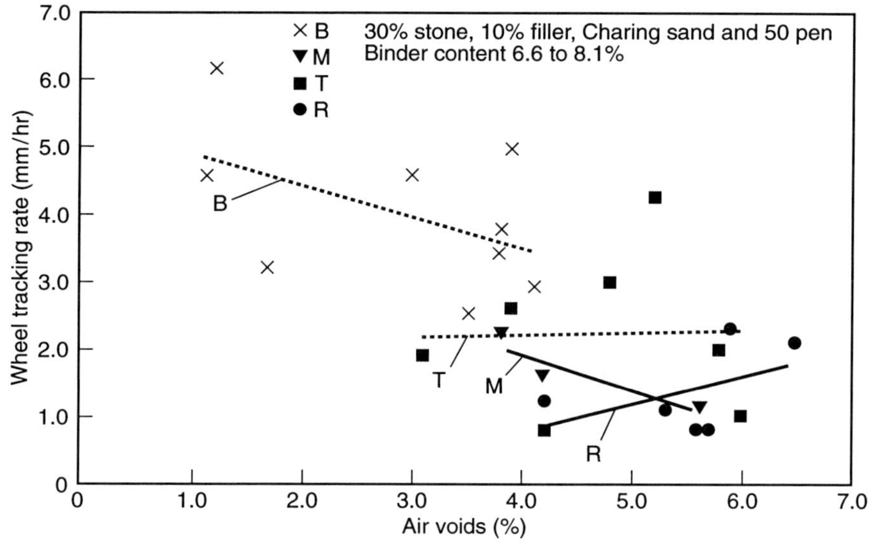

Fig. 7. Effect of air voids on wheel tracking rate.

The Asphalt Institute[14] recommends a range of voids of 3 to 5% at the mix design stage. The County Surveyors' Society[15] recommends 4% ± 2% for 30% stone asphalt wearing courses in situ, in the absence of specific information.

The minimum voids level is specified to allow for the further densification of the mixture under traffic without creating instability in the mixture; the maximum value specified is to ensure adequate durability by reducing air and moisture migration through the mixtures, though this is also significantly affected by interconnections between voids as much as their quantity. Shklarsky and Kimshi[16] and Lees and Yap[17] showed different permeabilities can exist for identical air voids depending upon types of mix, binder content and degree and type of compaction.

In UK practice, a typical HRA, of the rear of a paver, has about 10% voids. Further compaction occurs as the rollers press the chippings into the mat to achieve the required texture depth. As work relating to the 1991 Remix Contracts summarised in Table 3 shows, compaction of hot rolled asphalt wearing courses to achieve void contents below 6% is not usually a problem. This is usually achieved by 3 point deadweight rollers or tandem drum rollers, with or without vibration, with the exception of the edges of the mat, where a combination of wind chill, overchipping and lack of rolling can lead to joint fretting problems.

Table 3. Air voids outside the range 4 ± 2% (13).

<2%	>6%	Total Cores
4	17	146

It was noted, that 12 of the 17 results with high air voids occurred on the two schemes where the weather was cold, leading to rapid mat cooling. Texture depths were high in parts and the mixtures' willingness to accept chippings, as measured by its 'mouldability' in the Wheel Tracking Rate test was very stiff.

Jacobs[3] found a clear relationship between long term texture depth and WTR on A33 Winchester Bypass and ongoing monitoring of the Remix recycling sites[13] will provide further data on this relationship.

8 Effect of stone content

On the basis that for say 40% stone content, the predominant mechanism is still plastic flow of the binder/filler matrix, there is no reason to believe that the WTR test is not valid. The effect of increased stone content on laboratory slabs above this level has not been evaluated.

Table 1 demonstrates that increasing stability results from increasing stone content. Work in Kent suggested that, at the 7kN, level increasing from 30% to 40% stone content increased stability by 1kN. This relationship is used so that if 40% stone content is necessary for 50mm thick layers[18], KCC designers increase the specified stability by this amount. The stone content/binder correction data on BS 598:Part 102:1989 is used to correct the target binder content so that mix designs need only be done at 30% stone content.

9 Effect of plant processing

Figure 8 demonstrates the effect of different asphalt plants on the WTR/Binder content relationship. It shows that at a target binder content of say 7%, the type of plant can introduce a range from 1.3mm to 1.9mm, or at a higher binder content of say 7.8% (usually target +0.6%) a range from 2.1mm to 3.4mm. It is postulated that this effect relates to the effectiveness of the dust extraction system at the plant removing material smaller than 63 micron, plant B being different from the others. Some evidence is coming available from a manufacturer who uses both drum and batch plants that the effect of automatic dust recycling may also affect WTR. However, all the plants considered here produce a material with a WTR falling within a narrow band and it may be that for a given sand all WTR results are well below a design requirement, in which case obtaining plant processed material for design purposes may not be necessary.

10 Effect of types of fine aggregate

It has been shown that, in order to get a completely uncontaminated sand sample from a plant making macadam with crushed rock fines, upwards of 200T may need to be passed through the plant to remove all traces of contamination. However, work by Earland at TRL[19] has shown that a sand/crushed rock blend has little effect on WTR as shown in Figure 9 where at OBC of 6.5%, a 75/25 sand/basalt mix had a WTR of 2.1mm/hr compared to 100% sand with WTR of 2.9 mm/hr. This has also been observed in Kent with an 80/20, sand/Arklow CRF mix where the blend had higher WTR at the same binder content than unblended fines. However, from another supplier, with a plant giving almost identical WTR performance, a 25% addition of crushed rock fines to Charing sand reduced WTR from 1.6mm to 0.9mm and made the mix less sensitive to higher binder contents. Each potential blend must therefore be tested.

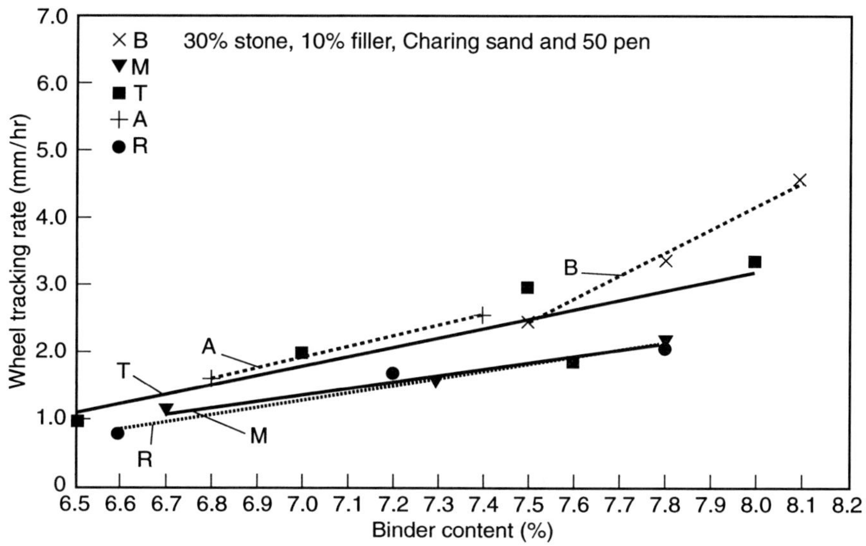

Fig. 8. Effect of sand content on relationship between wheel tracking rate and binder content.

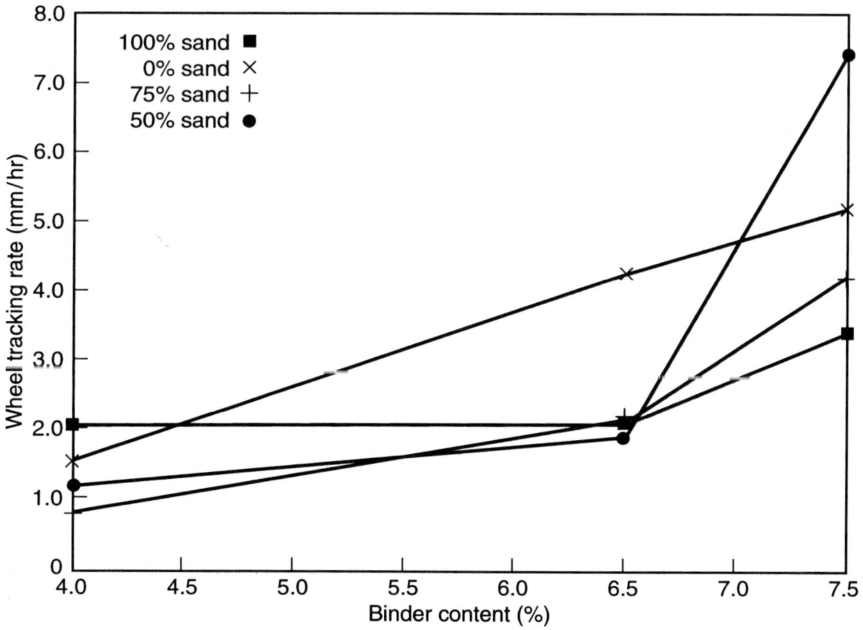

Fig. 9. Effect of different asphalt plants on WTR/binder content relationship.

A number of manufacturers believe blending fine aggregate sources can be beneficial for Marshall stability. An example of this is with a sand/slag blend, where it was possible to measure slag content accurately by magnet and relate to Marshall stability. Where blends are used it is therefore necessary to consider the tolerance of the cold feed bin conveyor system and carry out a sensitivity analysis for these different fine aggregate sources.

11 Relationship between slabs and field cores

To date cores have only been obtained from a small number of contracts. Resident Engineer staff are unwilling to have 6 No. 200 dia cores taken from their wearing course especially since there is unlikely to be any HRA available to backfill them so that cold-lay macadam has to be used. It is also true that on most roads in Kent the WTR requirement is in excess of 5mm and often in excess of 10mm. It is only on particular sites as described in Para 1 that WTR as low as 2mm are required. Given that our most common sand will achieve below 5mm there is little incentive to check every contract.

The relationship between slabs and cores, based on 6 contracts and Charing sand is given in Figure 10A and 10B. It can be seen that laboratory slabs generally give lower WTR than cores from site. This is contrary to the prediction from binder penetration discussed in Para 5 above. However, if the results from slabs made up at target +0.6% binder, the maximum permitted, are considered, a better correlation is possible. This relationship may not hold for other sands which have a different WTR/binder content relationship. It may be observed that the results which deviate greatest from the 1:1 relationship were those from a mix with very low target binder content. Whilst the vibrating hammer in the laboratory can readily overcome this to achieve air voids of 4%, in the field it is probable that these stiff mixes would have a lower density than this and hence higher WTR. It is not straightforward to measure air voids of HRA wearing courses because of the presence of precoated chippings. More data on this needs to be collected. Making slabs at 6% air voids, the maximum likely in the field, may be difficult practically with a vibrating hammer and foot. An alternative for this sand source is to introduce a 2 x correction factor based on a 'best fit' line, ie at target binder content laboratory slabs will have half the WTR of cores taken from the road. This relationship, when plotted on Figure 1, makes this sand conform to data from elsewhere.

Work on the A2 East Cliff Viaduct, Dover, which was specified using the WTR test[21] with a maximum of 2mm required, demonstrated a good correlation between cores and slabs and, using a 35 pen TLA blend, the road pavement continues to perform in accordance with the design formula, including a speed factor. Previous Marshall stability design mixtures had rutted within 3 years.

12 Precision of WTR test

A standardisation experiment was conducted by TRL in September 1992.[21] This was carried out on cores taken from a trial strip containing 50 pen, 100 pen and 200 pen bitumen. Eleven laboratories reported data, however, particularly with the 200 pen material, excessively high WTR were obtained (in excess of 15mm) so that non-standard extrapolation was carried out by some laboratories. WTR in excess of 10mm are likely to cause problems with homogeneity of the sample and interaction of 20mm aggregate particles in the rutted area. Some laboratories also used "best fit by eye" methods of interpretation using small scale graphs.

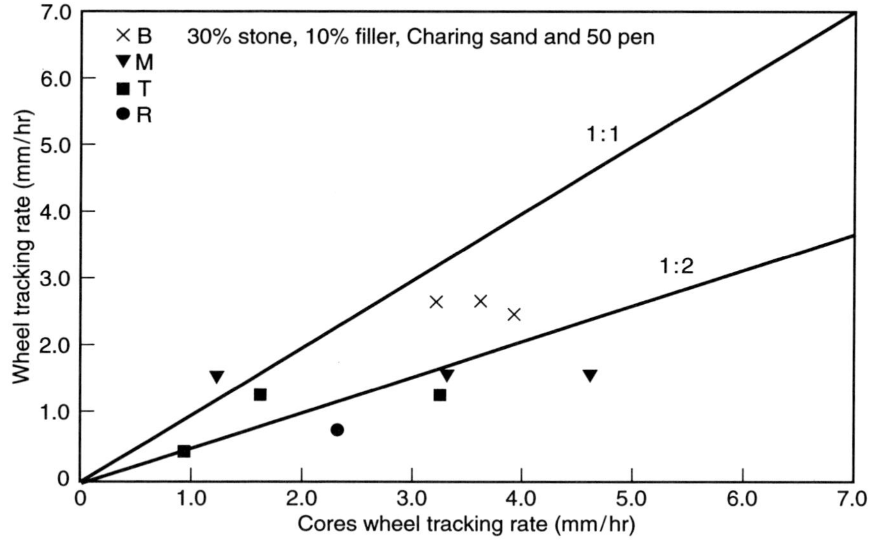

(a) Slabs at target binder content %

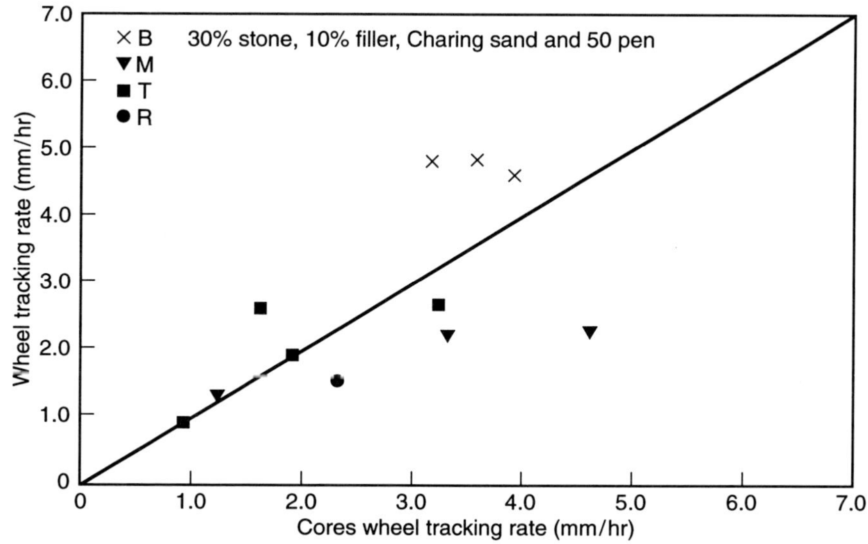

(b) Slabs at target binder content +0.6%

Fig. 10. Relationship between WTR for slabs and cores.

As a result of this work, a further study into precision using laboratory prepared specimens, to largely eliminate density and composition effects is in progress. It indicates that the repeatability has improved, the reproducibility cannot readily be compared with the previous experiment because of the different level of rutting

The work in Kent on slabs suggests that the mean difference between two slabs, each having two determinations, is 0.32 mm (at 95% confidence level) for WTR in the range 0.8 to 4.8 mm/hr. The mean difference between two determination on one slab was found to be 11%, ie ± 0.3mm/hr on 4mm/hr.

13 Conclusions

1. The manufacture of slabs in the laboratory is straightforward and produces specimens which are suitable for use in a mix design process.

2. It is necessary to use plant processed dry materials, as different plants affect the dust content significantly, if the mixture, as designed, will be close to the permitted maximum.

3. Slabs should be tested as soon as possible after manufacture to reduce the effect of time-related binder penetration changes.

4. Slabs made at target binder content had approximately half the WTR of cores taken from the road, slabs made up at target + 0.6% binder content achieve results similar to those from cores taken in the field. These are empirical relationships from work on one sand source. Field cores will have WTR reduced by having lower pen bitumen, but increased as the density will be less than the density used for slab manufacture, especially with mixes with a low target binder content.

5. The WTR/binder content relationship is such that it cannot be used alone for mix design. Its value is for checking the performance of mixtures.

6. EVA and SBS are effective at reducing WTR and making the mix less sensitive to temperature and binder content fluctuations. SBS is almost certainly necessary if 0.5mm/hr WTR is required for specific conditions, including those pertaining during construction.

7. The WTR / Marshall stability relationship will vary between fine aggregate sources, permitting or excluding some sources if WTR is used in the design of lower WTR/high stability mixes.

8. By the inclusion of the speed factor, KCC have been satisfied that the 'Jacobs' formula provides an adequately rut resistant road, however other evidence suggests the formula is not conservative enough and more data should be obtained.

14 Acknowledgement

I would like to thank Ian Valentine and the team at our Aylesford Laboratory for producing the data in this report.

References

1. BRITISH STANDARDS INSTITUTION (1990). Method of Determination of the Wheel Tracking Rate of cores of bituminous wearing courses Draft for Development DD184. BSI London.
2. BRITISH STANDARDS INSTITUTION (1992). Hot Rolled Asphalt for roads and other paved areas; Specification for Constituent materials and asphalt mixtures BS 594 Part 1.
3. JACOBS FA (1983). A30 Winchester Bypass, the performance of rolled asphalts using the Marshall Test TRRL Report LR 1082.Transport Research Laboratory.
4. NUNN ME (1985). Prediction of permanent deformation in bituminous pavement layers TRRL Research Report RR26.Transport Research Laboratory.
5. BRITISH STANDARDS INSTITUTION(1992). Structural design of Pavements constructed with Clay and Concrete pavers. BS 7533. BSI London.
6. CHOYCE PW and WOOLLEY KG (1988). EVA Modified Binders. Highways Vol 56 (January).
7. DENNING J and CARSWELL J (1981). Improvements in Rolled Asphalt Surfacings by the addition of organic polymers, TRRL Report LR 989.Transport Research Laboratory Crowthorne.
8. Otto H (1992). Wirtschaftlichkeit von Asphaltbauweisen 10th Int. Conf Trinidad Lake Asphalt, Berne.
9. SZATKOWSKI WS (1980). Rolled Asphalt Wearing Courses with high resistance to deformation Proc Conference on performance of rolled asphalt surfacings, ICE London.
10. JACOBS FA (1983). A22 Winchester By-Pass - The Performance of Rolled Asphalts designed by the Marshall Test TRRL Report LR 1082 Transport Research Laboratory.
11. BRITISH STANDARDS INSTITUTION (1990) Method of test for the determination of the composition of design wearing course rolled asphalt. BS 598:Pt 107. BSI.
12. LEES G (1978). Asphalt Mix Design for Optimum Structural and Tyre Interaction Purposes 6 Int. Conf on Structural Design of Asphalt pavements Ann Arbor Michigan.
13. KENT COUNTY COUNCIL (1993). Remix Recycling, Final Interpretative Report for DoT on Field Trials Project Report 295/8, Engineering Services Laboratory, Aylesford.
14. THE ASPHALT INSTITUTE (1988). Mix design methods for asphalt concrete and other hot mix types Manual Series No.2.
15. COUNTY SURVEYORS' SOCIETY (1992). Highway Authorities Standard Tender Documents, Notes for Guidance 1 and 2 Hot Rolled Asphalt and Macadam.
16. SCHLARSKY E and KIMCHI A (1962). Influence of voids bitumen and filler contends on permeability of sand-asphalt mixtures Highways Research Board Bulletin 358.
17. LEES G and YAP PK (1967). Voids, permeability relationship in bituminous mixtures Department of Transportation. University of Birmingham Research Journal No.2.
18. KENT COUNTY COUNCIL (1988). Road Pavement Design Technical Memorandum M88/1 Engineering Services Laboratory, Aylesford.
19. EARLAND MG (1991). The effect of compositional variations of HRA wearing course on laboratory performance. TRL limited circulation working paper. WP/MC/24.
20. KENT COUNTY COUNCIL (1990) A2 East Cliff Viaduct Dover Materials Report, Engineering Services Laboratory, Aylesford.
21. SYM R (1992). Wheel Tracking test precision (standardisation) trial: Statisticians report. Transport Research Laboratory.
22. COLWILL DM & CARSWELL J.(1990) The role of polymers in upgradig low stability aggregates 1st Int Symp on Highway surfacings Univ of Ulster.
23. KAVUSSI A & LEES G (1989) An accelerated weathering technique to assess the hardening of bitumen and filler-bitumen mixes Conf on Durability and Perf of Bit. High. Mats. Hatfield Poly.
24. DAINES ME (1992). The performance of hot rolled asphalt containing crushed rock fines, A303 Mere. TRL Research Report, RR298, Transport Research Laboratory.
25. CHOYCE PW (1980). Relationships between wheel tracking and Marshall test results for 30% stone content rolled asphalt Jnl IAT No.38 (Sept).
26. CARSWELL J (1989). The effect of EVA modified bitumens on rolled asphalts containing different fine aggregates. TRRL Research Report RR122 Transport Research Laboratory.

19 QUALITY CONTROL DURING CONSTRUCTION OF BITUMINOUS MIXTURES USING A SIMPLE AIR PERMEABILITY TEST

J.G. CABRERA and T.Q.M. HASSAN
Civil Engineering Materials Unit, Department of Civil Engineering, University of Leeds, Leeds, UK

Abstract

This paper presents an evaluation of a simple air permeameter for use in the laboratory during the design stage or in the field for control of the quality and uniformity of pavements during the construction stage.

The values of air permeability are correlated statistically with the engineering parameters which describe the structural properties of a bituminous mixture and with parameters related to its composition. The statistical analysis indicates that permeability is strongly affected by binder content and that porosity of a mix can be predicted by using the measure of air permeability. The statistical analysis gives a ranking of the variables which affect permeability.

The test is easy to carry out; it takes a very short time to obtain readings for calculation of permeability and it is highly repeatable. The apparatus can be used in the laboratory and also during construction in the field.

1 Introduction

Controlling the quality and uniformity of hot bituminous mixtures during the construction of pavements leads to long service life and adequate performance. The traditional method of quality control by non-destructive methods involves measuring the density of in-situ compacted materials and the calculation of their porosity from the knowledge of their theoretical composition.

It is now recognised that permeability is a parameter which affects directly the performance and durability of materials; however, methods which have an acceptable repeatability and which are easy to carry out during routine quality control are not available except in research laboratories.

Performance and Durability of Bituminous Materials. Edited by J.G. Cabrera and J.R. Dixon.
Published in 1996 by E & FN Spon, 2–6 Boundary Row, London SE1 8HN. ISBN 0 419 20540 3.

This paper describes an air permeameter which has been designed for use in the laboratory during the design stage or for control of the quality and uniformity of the bituminous layers in a pavement during construction.

Statistical analysis is carried out to assess the influence of compositional parameters and engineering properties of hot bituminous mixes on air permeability values.

2 Background Information

Methods for measuring permeability of bituminous materials have been proposed as earlier as 1953[1]. Many studies and designs of equipment appeared during the 1960s, these were devoted to study the nature of liquid and gas flow through bituminous materials and to explore possible correlations with the parameters characterising a bituminous mix. For example McLaughin and Goetz[2] suggested that air permeability measurements give a better indication of durability than for example measurements of porosity. Ellis and Schmidt[3] indicated that the porosity of a mixture is not necessarily proportional to permeability when the variations occur due to particle size distribution of the aggregates. Other studies explored the effect of compaction characteristics on air permeability[4] and concluded that there was certain relationship between density and permeability of bituminous mixtures and that the air permeability test proposed was reproducible providing that the data is corrected for variations of temperature since these affect the viscosity of the flowing air. Mullen studying bituminous pavement samples concluded that pavements with less than 6% porosity are impermeable and perform better than pavements with higher porosity. Davies and Walker[5] studied a variety of asphalt mixes and proposed a classification of materials according to their permeability values. This classification is shown in Table 1.

Table 1. Classification of bituminous mixtures in terms of permeability

Permeability K, cm^2	Bituminous Mixture Class
10^{-14} - 10^{-11}	Impervious
10^{-11} - 10^{-9}	Partially impervious
10^{-9} - 10^{-7}	Poor drainage
10^{-7} - 10^{-6}	Fair drainage
10^{-6} 10^{-3}	Good drainage

Jacob[6] studied the properties of hot rolled asphalt and asphaltic concrete compacted at various levels. His conclusions show that porosity is a good indicator of the degree of compaction but that it does not provide a direct relation to the permeability of the material. His results also show that air permeability is very sensitive to changes in the degree of compaction of asphaltic concrete but not so to changes in degree of compaction of hot rolled asphalt. Akeroyd et al[7] showed that the structure of pores of hot rolled asphalt is more likely to ensure impermeability than the structure of pores of continuous graded mixes. Asphaltic concretes with a porosity greater than 5% are permeable to air and water, whereas porosity of more than 10% can result in impermeable gap graded mixes (hot rolled asphalt).

Difficulties of interpretation regarding the influence of various engineering and compositional parameters on air permeability arise from the close interdependence between some of these parameters. Therefore the most fruitful manner of studying their effect on permeability is to use statistical techniques which may then lead to the formulation of numerical relationships for prediction purposes.

3 Apparatus and test procedure

The Leeds air permeameter (LAP) designed at Leeds is based on the same principle as the apparatus designed by Ellis and Schmidt[3]. It is very simple to construct and operate; it is in fact a "low technology instrument" capable of being constructed, maintained and operated without support or facilities. It is used as part of the routine of laboratory design in Leeds and it has been tested for use in the field. Fig. 1(a) shows the apparatus ready for testing on a pavement. Fig. 1(b) shows the permeameter cup after removal from the position of the test next to two rubber rings of difference external diameter which are used to control the air path length during the field test. Fig. 2 shows the same apparatus ready for test in the laboratory.

The field apparatus consists of:

1. a steel cup which is made to penetrate at least 4 mm into the pavement and therefore has sharpened edges;
2. a water container of sufficient volume to carry out at least 12 tests before recharge of water;
3. a manometer to control differential pressure;
4. graduated cylinders of 50 ml, 250 ml and 500 ml;
5. rubber rings of constant internal diameter (100 mm) and variable external diameter (120 mm, 200 mm and 300 mm);
6. stopwatch;
7. grease-kaolin mixture of high viscosity to seal the steel cup and rings to the pavement surface; and
8. brushes: one to clean the pavement and one to apply the grease-kaolin mixture.

The laboratory apparatus (Fig. 2) requires also a cup but instead of the rubber rings a mould and two rubber membranes are used to seal the specimen. The mould and specimen are placed standing over a perforated base so that air can flow through the specimen under the differential pressure created by discharging water from the container.

The procedure to carry out a test is very simple and it is as follows:

1. <u>In the field</u>: after brushing the surface covering the test area the cup is inserted into the pavement by applying pressure on it. The rubber ring is placed in position and the cup and ring are sealed using the high viscosity grease-kaolin mixture. <u>In the laboratory</u>: the two rubber membranes are placed inside the mould and folded over the top and bottom edges of the mould. This is to ensure the sealing between the

Fig. 1(a). Apparatus ready for field use.

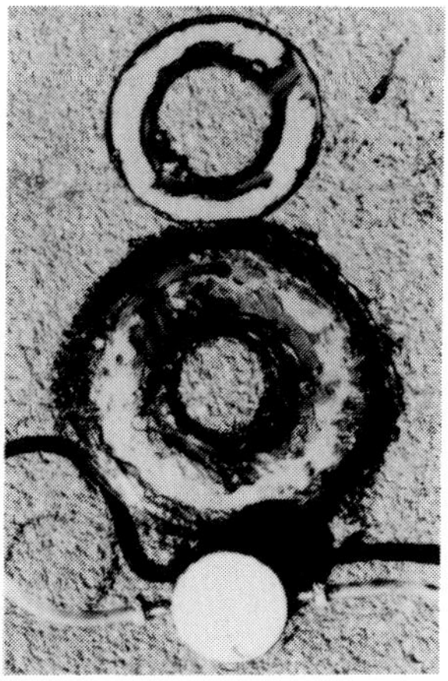

Fig. 1(b). Permeameter cup after removal from the test position, next to two rubber rings of different external diameters.

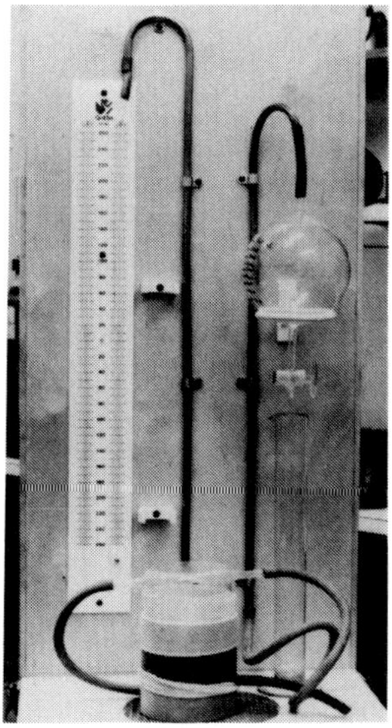

Fig. 2. Laboratory apparatus using mould and two rubber membranes to seal specimen.

specimen, the mould and the cup. The air trapped between the membranes and the mould is removed by suction through a pipe fixed to the middle of the mould. A very thin layer of silicon grease is applied around the cup, one membrane is folded over the cup and the other over the mould to secure sealing the specimen. The mould assembly is then placed over the perforated base.

2. A pressure difference is obtained by opening the valve of the water container. After the pressure has stabilised the time taken for 50 ml of water flowing from the container into a graduated cylinder is registered. This procedure is repeated three times or more to obtain a representative flowing time. A pressure differential of 50 mm of water is recommended for routine measurements. However, if the time of flow is short the pressure differential can be reduced to 38 mm or 25 mm of water.

The air permeability is obtained from the relation given by Wycoff[8] as follows:

$$K = u.V.L/(A.dP.dT) \quad (1)$$

where:		
K	=	Permeability, cm^2
V	=	Volume of water passing, cm^3
u	=	Viscosity of air, dyne sec.cm^2
L	=	Specimen height or length of path given by rubber ring used, cm
A	=	Cross-sectional area of specimen of steel cup, cm^2
dP	=	Differential pressure, dynes per cm^2
dT	=	Time taken for water with volume V to flow from the container to the graduated cylinder, sec.

For specimens with a 100 mm diameter air viscosity of 1.853×10^{-4} dyne sec/cm^2 at 20 degree C and volume of fluid equal to 50 ml equation [1] becomes:

$$K = 1.143 \times 10^{-4} L/dP.dt \quad (2)$$

The differential pressure dP is measured in cm; therefore the value of K becomes:

$$K = 1.143 \times 10^{-4}.L/980.7 \times dP.dT \quad (3)$$

$$K = 1.165 \times 10^{-7}.L/dP.dT \quad (4)$$

To convert K into coefficient of permeability k in cm/sec at normal temperature the following relation is used:

$$k = K \times 10^5 \quad (5)$$

4 Materials and mix composition

4.1 Material components

The experiments carried out in this investigation involved only hot rolled asphalt (HRA). This was designed according to the appropriate British Standards[9]. The

HRA was suitable for wearing course, its maximum particle size was 14 mm with a stone content of 33%.

The coarse aggregate was crushed granite, the sand was a processed "Bunter" sandstone, and the filler consisted of limestone for the standard control mix and four different pulverised fuel ashes for the other mixes. The filler content was constant for all the mixes and equal to 10%.

The binder used was of two viscosities, ie 68 pen and 116 pen.

4.2 Preparation of mixes

The mix components were heated in the laboratory at 160 degree C +/- 10 degree C and mixed in the appropriate proportions in a thermostatically controlled twin-vertical paddle mixer. The mixing process involved mixing the mineral aggregates and filler for 60 seconds and then introducing the binder and continuing the mixing for two minutes. The loose hot mix was then transferred to the mould (100 mm diameter and approximately 50 mm height) for compaction.

Compaction was carried out at two energy levels and using two different modes of compaction, i.e. Marshall hammer 50 blows and 75 blows and Gyratory Testing Machine (GTM) at 0.7 MPa and 1.4 MPa pressure, 1 degree angle of gyration and 30 revolutions.

The GTM is used in the Leeds Design Method (LDM)[10] as an alternative to the Marshall hammer compactor recommended in the BS method. The GTM compactor produces a shearing action in the specimen by a gyratory motion of a steel mould while pressure is maintained at each end against steel loading plungers whose faces remain parallel to each other. This "kneading action" which is a combination of shearing and compressive action is closer to the compaction action produced by a roller in the field and therefore produces an aggregate mineral structure which has random orientation unlike the oriented structure resulting from the Marshal compaction.

Specimens compacted were then cooled down and extruded from the mould. They were stored until testing at laboratory room temperature.

4.3 Testing programme

The programme of tests was designed to provide data for analysis of various compositional and engineering parameters on the air permeability of hot rolled asphalt and also to provide comparative information on the laboratory and field measurements. The study consisted of preparing HRA mixes containing:

1. five types of filler;
2. five bitumen contents at 0.5% interval; and
3. two types of bitumen (68 pen and 116 pen)

The mixes were compacted using:

1. two modes of compaction (Marshall hammer and GTM); and
2. two compaction energy levels (light and heavy compaction).

The tests carried out included those used in the LDM[10]. The dependent variable was in all cases the air permeability.

A limited field programme was carried out in a section of the M62 motorway between Leeds and Manchester while resurfacing was taking place. The HRA used for resurfacing was of similar characteristics to the HRA used in the laboratory study except that the HRA in the field included "chipping" which were spread to a mean value of 14 Kg/m^2.

5 Analysis of laboratory results

The optimum binder content for the mixes containing the two types of binder and compacted by the two modes and two energy levels gave results well within the expected values for a mix of this type. The values of stability, flow and porosity were in general within the requirements for motorway standard except for mix containing the DR filler for which the values were in general lower and reached only the requirement for medium traffic roads.

The analysis of variance carried out on the results showed that most of the independent variables tested have influenced the value of permeability except the "filler type" variable which did not have a measurable effect. Because of the many variables involved it was difficult to obtain general relationships of statistical validity, therefore the analysis was carried out using stepwise regression. The variables which did not affect the correlation were eliminated. The variables which had similar effect were grouped together in order to make the correlation as general as possible. From the analysis it was concluded that porosity and binder content were the most influencing parameters with relation to permeability. The relationships were of the following form:

$$K = a + b.P^2 \tag{6}$$

$$K = a - b.P + c.P^2 \tag{7}$$

where: K = Air permeability, $cm^2 \times 10^{-10}$
P = Porosity %
a,b,c = Constants

and

$$K = a - bB + cB^2 \tag{8}$$

where: B = Bitumen content %

The significance of this relationships was assessed by the multiple correlation coefficient value and this was in every case equal or greater than 0.80 indicating that 80% or more of the variation in permeability is explained by the bitumen content (eq. 8) or by the variation in porosity (eqs. 6 & 7).

The effect of compaction energy is indirectly considered in equations 6 and 7 since energy affects porosity. The influence of mode of compaction is very important as is shown in Fig. 3 where the mixes compacted by the marshall hammer exhibit far lower permeabilities than those compacted by the GTM. This is due to the oriented nature of the aggregate matrix which arises from the unidirectional nature of the mode of

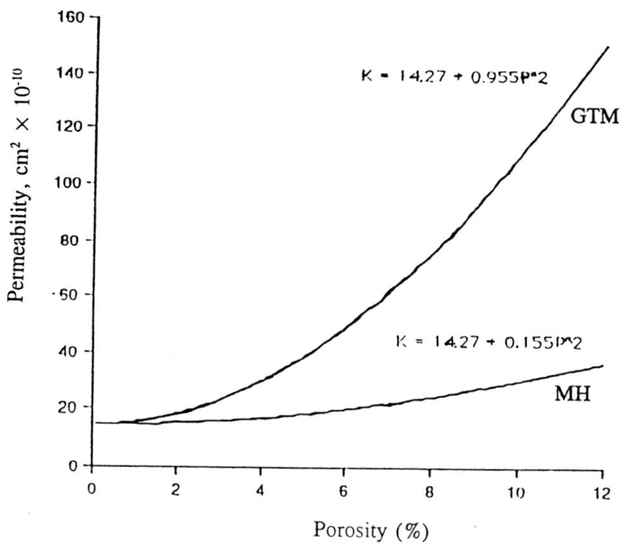

Fig. 3. Effect of mode of compaction on the permeability–porosity relationship.

compaction provided by the marshall hammer and has important bearing when assessing laboratory results against field results.

Fig. 4 shows an example of the relationship for the HRA made with EG filler. At very low porosity values arising from the large volume of mineral aggregates and filler, the permeability is high because the available pores are interconnected continuous pores due to the lack of binder. As the porosity increases due to the replacement of aggregate for a lighter relative density binder, the permeability decreases up to an "optimum porosity" value above which the permeability increases.

Fig. 5 shows a typical relationship for the standard LS filler mix. The permeability decreases as the binder content increases up to an "optimum binder content" above which the permeability increases. This trend is in effect confirmation of the trend between porosity and permeability.

6 Analysis of field results

Field air permeability values were obtained from measurements made on the wearing course of the fast lane and overtaking lane of the eastbound carriageway of the M62 Leeds/Manchester motorway and on the wearing course of the hard shoulder of the westbound carriageway. Altogether 36 points were surveyed and three results were obtained on each point.

Measurements of site density were also carried out next to the permeability measuring points. Cores were obtained from the same area of the permeability measurement for evaluation of porosity and other engineering properties. Laboratory permeability values were also obtained from these cores.

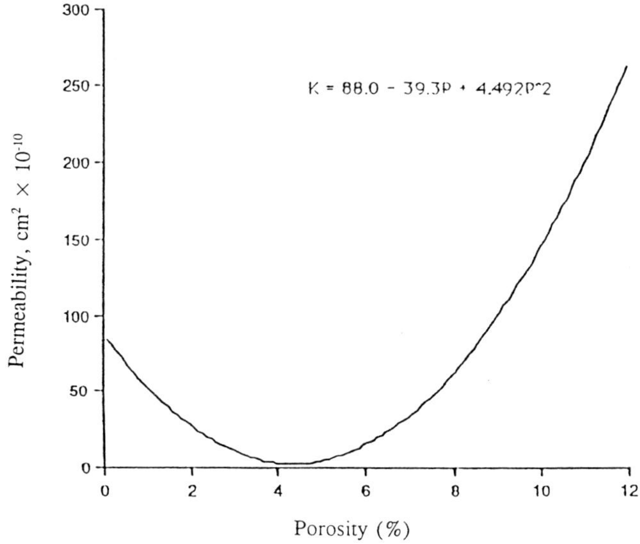

Fig. 4. Permeability–porosity relationship.

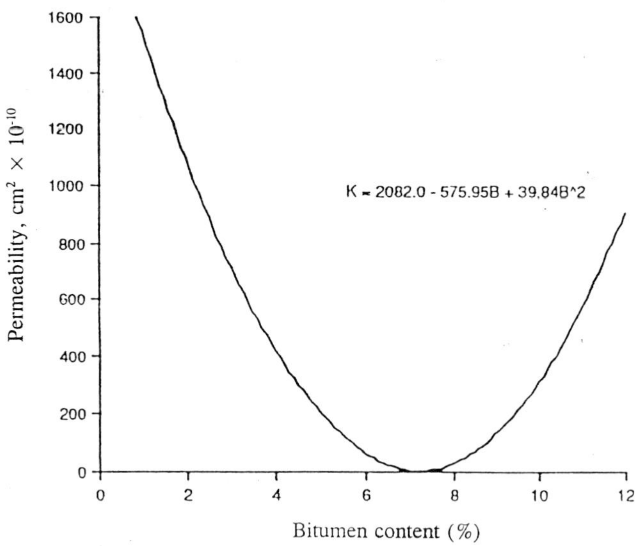

Fig. 5. Permeability–bitumen content relationship.

The results show that field permeability is repeatable for 60% of the measurements at 5% coefficient of variation and for 80% of the measurements at 10% coefficient of variation. Therefore the field permeameter is acceptable in terms of repeatability.

Field permeability values correlated well with the laboratory values of the cores obtained from site. The field values were in all cases higher than the laboratory values. During the analysis it was found that using rubber rings of different external diameter did not influence the permeability indicating that air does not enter from the surface but that the flow of water is influenced by the air reserve inside the pavement. This is then the cause of higher values.

The measurements then should be carried out without the rubber ring since the path of flow is non existing as in the laboratory. The correlation between laboratory and field measurements gave a correlation coefficient r = 0.995 and the relationship was of the following form:

$$\text{Log } kf = a + b.kl \tag{9}$$

where:

kf	=	Field coefficient of permeability, cm/sec
kl	=	Laboratory coefficient of permeability, cm/sec
a,b	=	Coefficients

Correlations with binder content or porosity were very poor. It is believed that the lack of statistical relationships arise from the inclusion of the "chipping" which alters the porosity value mainly in the upper part of the specimen. Most of the field porosity values were between 2% to 4%; however the permeability values varied between 10^{-3} and 10^{-2} cm/sec which is obviously a far larger variation. Thus it seems that the field permeability test is a more sensitive test regarding the quality and uniformity of the bituminous mix in place.

7 Conclusions

The major conclusions of this study are:

1. Porosity and binder content are the major parameters influencing the permeability of hot rolled asphalt. The relationships are polynomials of first or second order and have high statistical significance. As such they show that there is an optimum binder content and an optimum porosity value for minimum air permeability.

2. Laboratory measurements of permeability are an effective means of detecting changes in mix composition and constructional parameters like compaction level and mode of compaction.

3. The field air permeability test has acceptable repeatability and it is easy to carry out. However the values obtained by this test are invariably higher than the values of the laboratory test. This is due to the fact that there is no fixed length path for the air to move but rather there is a large reservoir of air within the pores of the pavement.

4. The sensitivity of the test to variation of composition or compaction energy and modes make it desirable as a test for quality and uniformity control. Specifying a value of maximum permeability for acceptance purposes achieves the double purpose of controlling uniformity and ensuring performance which is not the case when specifying prescribed mixes.

8 References

1. Gilber P and Keyser JH. (1973) *A study of currently used methods for determining the permeability of bituminous mixtures.* Jour Test and Eval, vol 1, No 6, p 484.
2. McLaughlin, JF and Goetz WH. (1955) *Permeability, voids content and durability of bituminous concrete.* Proc Highway Res Board, vol 34, p 274.
3. Ellis WH and Schmidt RJ. (1960) *A method for measuring the air permeability of asphalt concrete pavements.* ASTM Special Technical Pub. no 294, p 85.
4. Warner DB and Moavenzadeh F. (1964) *Permeability-compaction characteristics of bituminous mixtures.* Proc ASTM Tech Papers, vol 64, p 981.
5. Davies JR and Walker RN. (1969) *An investigation on the permeability of asphalt mixes.* Ontario Joint Highway Res Programme. No RR145. Dept of Highway, Ontario.
6. Jacob F. (1977) *Properties of rolled asphalt and asphaltic concrete at different states of compaction.* TRRL Report. SR 288.
7. Akeroyd FML, Hoban T and Chipperfield EH. (1978) *Mix design to resist rutting in rolled asphalt wearing course.* Eurobitume Seminar (London), p 99.
8. Wycoff RD, Botset HG, Muskat M and Reed DW. (1933) *The measurement of the permeability of porous media for homogeneous fluids.* The Review of Scientific Instruments, vol 4, New Series, p 394.
9. British Standard Institution, (1985) *Hot rolled asphalt for roads and other paved areas.* BSI London. BS 594: Part 2.
10. Cabrera JG. (1994) *Hot bituminous mixtures: Design for performance.* Presented to this Symposium, Leeds.

20 BITUMINOUS TESTING IN EUROPE

B. ALEY
Bedfordshire County Council, Bedford, Bedfordshire, UK

1. UK PANEL

Panel B510/1/WG2 provides input to TG2 on "Sampling and examination of Bituminous Mixtures".

The panel has seven permanent members which are supplemented by individual experts as required.

2. BASIC TESTS COVERED BY UK

Originally thirty basic test procedures were identified as being required and were divided between the EC member states representatives whose tasks were to:-

- Identify costs, both of equipment and time involved in performing the procedure.
- Summarise each group
- Issue to other EC Member states
- Formulate drafts.

The UK got secretaryship for nine proposed Euro Norms:-

Ref Number	Identification
1.0a	Sampling
1.0b	Sample Preparation
1.1	Binder Content (Quantity)
1.3a	Bitumen Recovery - Rotary Evaporator
1.3b	Bitumen Recovery - Fractionating Column
1.10	Mix Temperature
1.11	Moisture Content
1.13a	Texture Depth - Patch
1.13b	Texture Depth - Laser

Performance and Durability of Bituminous Materials. Edited by J.G. Cabrera and J.R. Dixon. © E & FN Spon.
Published in 1996 by E & FN Spon, 2–6 Boundary Row, London SE1 8HN. ISBN 0 419 20540 3.

Of these nine, the two dealing with Texture Depth were later to pass to WG5 who deal with "Surface Assessment".

A decision was taken to allow alternatives for each of the Basic Operations involved in determining Binder Content (1.1) and this has resulted in a rather complex EN which allows a number of alternative combinations of the following nine operations as shown in Figure 1 attached.

Number	Identification
B1	Hot Extractor (Paper Filter)
B2	Hot Extractor (Wire Mesh Filter)
B3	Soxhlet Extractor
B4	Bottle Rotation Machine
B5	Centrifugal Filter Press
B6	Continuous Flow Centrifuge
B7	Pressure Filter
B8	Bucket Type Centrifuge
B9	Binder Recovery Apparatus for use with Bucket Type Centrifuge.

Assessing the precision associated with each of the routes through the procedures will be a mammoth task.

3. OTHER BASIC TESTS

Other basic tests being processed to Euro Norms by other member states include:-

Ref Number		Identification
1.4		Mix Density
1.5		Specimen Density
	(a)	Hydrostatic
	(b)	Measurement
	(c)	Gamma Rays
1.6		Voids Content (Calculation from 1.4 & 1.5)
1.7		Compaction Degree
1.8		Compactability

1.9		Water Sensitivity
	(a)	Specimen
	(b)	Loose Mix
1.12		Segregation Sensitivity
1.14		Abrasion/Spiked Tyres
1.15		Abrasion/Porous Asphalt
1.16		Binder Drainage/Porous Asphalt
1.17		Permeability/Porous Asphalt
1.18		Marshall Test
1.21		Wheel Tracking Test
1.22		Indirect Tensile Test

4. FUNDAMENTAL TESTS

Once the basic test methods have been "europeanised" work will start on the following fundamental tests:-

Ref Number	Identification
2.1	Fatigue
2.2	Permanent Deformation
2.3	Dynamic Stiffness
2.4	Low Temperature Cracking
2.5	Layer Adhesion
2.6	Layer Thickness

5. IMPACT OF EUROPEAN SPECIFICATIONS

The theme of this symposium is "Performance and Durability of Bituminous Mixtures" but of course innovations or comparisons require test methods.

My general impression of test writing activity in Europe is that there is a lot of striving towards standardisation but this will sometimes go against quality and some tests that are being processed will fall when the time and finance is available to assess their precision.

On the brighter side it will result in the standardisation of a number of tests which were previously only used in research establishments thus, hopefully providing better correlations between results throughout the European Community.

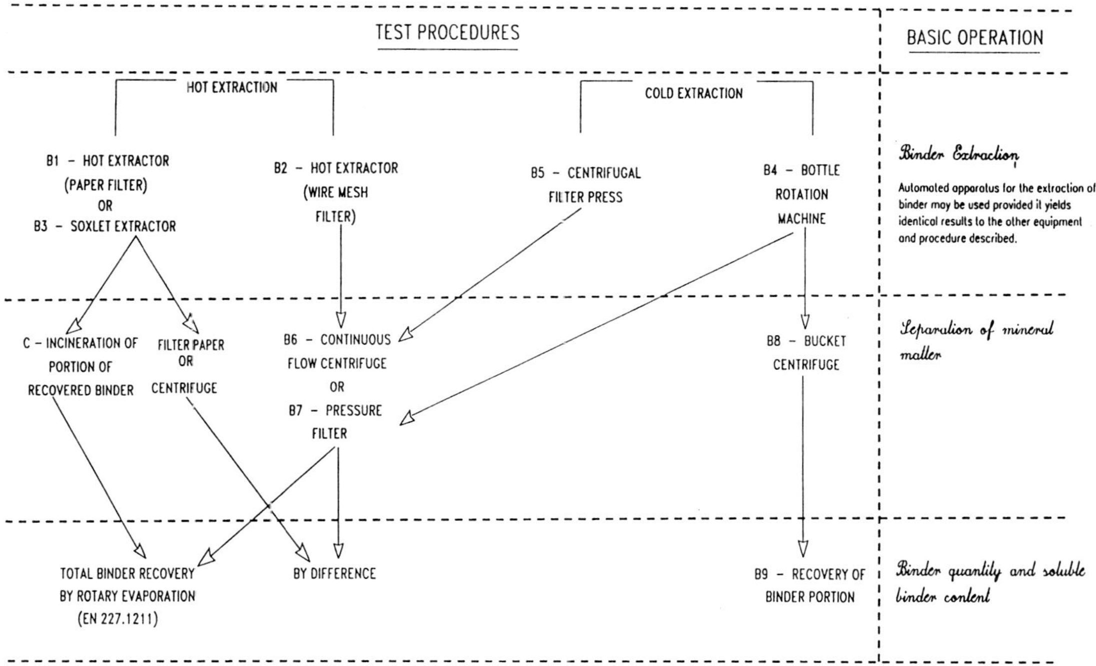

Figure 1. Alternative Procedures for Determination of Binder Content

21 EUROPEAN STANDARDISATION IN THE SHADOW OF THE CONSTRUCTION PRODUCTS DIRECTIVE

C.A. LOVEDAY
Tarmac Quarry Products Ltd, Ettingshall, Wolverhampton, UK

When we consider the frantic activity involved in the preparation of European Standards we should step back a little and think why it is all happening and what is driving it. We should not assume that the European Standards will be used in the same way as our familiar old British Standards. They will not. This is because they form part of a framework of Regulations which are required by the European Community in order to guarantee that the free market works.

The developments stem from our membership of the European Community and will significantly change responsibilities for the demonstration of product conformity. In essence, the 'rules' set to ensure a free market between the member states of the EC will apply to all transactions in our domestic construction market and will modify the traditional system we are familiar with.

Thus, just because we are not intending to supply HRA to Madrid does not mean that we shall not be affected by European harmonization. The changes stem from the EC Construction Products Directive (CPD) which was implemented into UK law on 27th of June 1991. The CPD is an ambitious and far-reaching Directive which will affect all producers, exporters, suppliers and retailers of construction products and the construction industry generally. Its purpose is to promote free movement of construction products by removing barriers to trade. The Directive does this by setting up the machinery for the production of new European Standards; then, providing that a product complies with the relevant Standard, its free movement within the EC cannot be hindered by a member state (or by an engineer operating within a member state). In the process of harmonization of Standards, special priority has been given to construction products because of the size of the industry and the amount of public expenditure involved.

In the legislation it has been recognised that harmonized Standards on their own are not enough to ensure freedom from barriers to trade: there must also be a recognized way for producers to prove compliance of their products with the harmonized Standards. Without such a provision, each country, or each contract, could set up its own 'approvals procedure' and this would in itself constitute a barrier to trade.

To this end, manufacturers who can prove compliance with a harmonized European Standard will be entitled to affix a 'CE' mark to their product. In some ways, the CE mark is similar to the Kite Mark we are familiar with in Britain. There are, however, important

Performance and Durability of Bituminous Materials. Edited by J.G. Cabrera and J.R. Dixon. © E & FN Spon.
Published in 1996 by E & FN Spon, 2–6 Boundary Row, London SE1 8HN. ISBN 0 419 20540 3.

differences, partly in the way in which a CE mark is authorized, but more importantly in the legal weight that it carries in respect of the acceptability or fitness for purpose of the product to which it refers. In simple terms, an engineer rejecting a CE-marked product will not, as now, simply be being awkward, he will be in breach of EC (and hence UK) law.

It must, however, be appreciated that legal restraints will apply to both sides of the supply equation. The EC Regulations which deal with the enforcement of the CPD provide that it will be a criminal offence to place on the market construction products which are not fit for their intended use. Liability extends to a fine of £2,000 or three months imprisonment.

At the moment, although the CPD has been taken into UK law, there is no obvious effect. This is because the Regulations apply only to harmonized European Standards. In our sector, work is in hand to produce harmonized standards for aggregates, asphalt and concrete. This is carried out by CEN, the Comite Europeen de Normalisation, a grouping of the national standards bodies of all the European countries. The first harmonized Standards covering quarry products will probably be published in 2-3 years' time and then the CPD will take effect. Before the harmonized Standards are published, the EC will have taken a decision on the procedure to be used to demonstrate compliance for each group of products. Such procedures are known as 'Systems for attestation of conformity' and will be of great importance to producers of construction products.

ATTESTATION OF CONFORMITY

The EC considers this subject of such importance in removing barriers to trade that it has set up a committee solely to draft rules for the attestation of conformity of construction products. The proposed machinery is described in a series of Guidance Documents. These Guidance Documents provide a choice of systems varying in complexity, from which a choice has to be made taking into account the importance of the product.

The alternative attestation systems are made up from the following basic elements:

Factory Production Control

This means a quality system conforming to Guidance Document no. 7, which is derived from EN29002 (BS5750, Part 2). While the required system has much in common with EN29002, there are some major differences, for the following reasons:

1. EN29002 has a broad scope, to take account of all the interactions between a customer and a supplier. CPD is, however, solely concerned with demonstrating compliance of a product with a Standard for the purposes of CE marking. Sections of EN29002 relating to Contract Review - the checking of customers' order information - are, therefore, irrelevant and are left out of 'factory production control'.

2. EN29002 is not specific about the necessary frequencies of inspection and test but leaves these open to agreement between the producer, the customer and the third-party auditor. Because CPD is trying to ensure a 'level playing field'

without barriers to trade, it is necessary to set within the technical specifications the minimum requirements in this respect and these must be met within the 'factory production control'. The European Standards will, therefore, differ from British Standards, because they will include requirements for:-

- control of raw materials
- control of production process
- control of test equipment
- control of finished product

Approved Bodies

Some of the systems for attestation require surveillance or audit by a recognized independent third party. Such third parties are described as 'approved bodies' and will be authorized by the member states.

Guidance paper no. 6 sets out the requirements for an approved body. These are based on EN45000, the European Standard for Certification Bodies, but include certain qualifying statements which could be worrying.

Clearly, if the whole CPD procedure is going to work, countries need mutual respect for one another's CE mark. The UK already has a well developed idea of third-party QA and a body, the National Accreditation Council for Certification Bodies, which guarantees the strict independence and integrity of third parties. Some other EC countries are less well developed in this direction and may be expected to be more lax in their policing of CE marking. There is a risk of UK producers being penalized by the British tradition of playing by the rules in this respect and we shall need to proceed with caution.

The CPD recognizes three types of approved body:

1. Certification bodies to certify either factory production control or product conformity.

2. Testing laboratories.

3. Inspection bodies to carry out inspections as part of the certification process.

PRODUCT CERTIFICATION

Some of the proposed attestation systems require an element of testing by an approved body to certify the conformity of the product. There are three elements to such a scheme:

1. Initial type testing carried out by the approved body to confirm that the product meets the requirements of EN.

2. Surveillance of factory production control by the approved body.

3. Audit testing of normal output by the approved body to confirm continuing compliance.

The above elements are combined in the following ways to give four choices of attestation system for the CPD:

System 1: certification of the conformity of the product by an approved certification body.

System 2: declaration of conformity of the product by the manufacturer with initial type testing by an approved body and factory production control by an approved body.

System 3: declaration of conformity of the product by the manufacturer with initial type testing by an approved body and factory control by the manufacturer.

System 4: declaration of conformity of the product by the manufacturer with initial type testing and factory production control by the manufacturer.

The roles of manufacturer and approved body in each of these systems is shown in the simplified table (Table 1).

TABLE 1

System	Factory Production Control		Type Testing	Certification
	Operation	Surveillance		
1	M	A	A	A
2	M	A	M	M
3	M	M	A	M
4	M	M	A	M

M = manufacturer A = approved body

Clearly, as one progresses from System 1 to System 4, there is a reducing involvement of the third party, down to no involvement at all in System 4. Importantly, however, it must be appreciated that the manufacturer has essentially the same obligations throughout, particularly in respect of the 'factory production control'. It is also important to remember that these obligations are legal requirements stemming from the CPD.

CHOICE OF ATTESTATION SYSTEM

Before a harmonized product Standard is published, a choice has to be made as to which of the four attestation systems should apply to the products involved. The decision is ultimately taken by the Standing Committee for Construction of the European Commission, but first there is detailed consultation with all sides of industry in the member states. We are currently at that stage. The initial proposal for the form which attestation should take is made by the CEN Technical Committee responsible for the product Standard.

It would, therefore, seem that there is a strong likelihood of a third-party element being required in attestation systems for asphalt. The difficult factor to gauge is the 'least onerous route' requirement and how this will be interpreted by the European Commission. They are supposed to take into account the systems already in use in the member states. These will vary from the Product Certification Scheme by 'Quality Control Associations' operating in Germany to almost nothing at all in some southern European countries.

It is difficult what will best suit the UK without a better understanding of the legal basis of attestation and how far this will modify the current contractual arrangements. If, however, it is a legal obligation to carry out the various activities of factory production control and declaration of conformity, it may be more comfortable to do this with the backing of a third party than to be continually open to challenge. Clearly, however, the choice of attestation system will have cost implications for industry.

PROGRESS ON TC 227 WG1

WG1 has decided in principle that Attestation of Conformity should apply to the loose asphalt "in the lorry" and that the procedure should be conducted in two halves:-

(a) Type Testing

and

(b) Factory Production Control

Type Testing involves demonstrating that the mix, correctly made, is capable of meeting the performance related requirements of the specification. There is ongoing debate about whether this should be based on a laboratory "design procedure" or on tests on cores cut from a trial pavement. Whichever is finally chosen the result is a "Job Mix Formula" for the mix around which normal plant control is exercised.

Factory Production Control is the operational Quality System used to demonstrate consistent manufacture of the proven Job Mix Formula. A detailed Quality Schedule has been prepared covering the aspects of manufacture to be controlled. Agreement has still to be reached over tolerances, test frequencies and the extent of any performance testing of the finished product.

PART FOUR
LOW ENERGY CONSTRUCTION METHODS AND MATERIALS

22 THE BEST PRACTICE PROGRAMME IN THE UK ROADSTONE INDUSTRY

P. MacDONALD
Energy Technology Support Unit, Harwell, Oxfordshire, UK,
on behalf of the Energy Efficiency Office

ABSTRACT

Energy costs the road construction industry about £60 Million a year. In common with other sectors of industry, the Energy efficiency Office (EEO) is working with companies to help them reduce energy costs by lowering energy consumption. Energy use can be reduced by the introduction of new processes and modern plant, but in many cases significant energy savings accrue from the introduction of low cost measures involving little or no capital expenditure. Therefore, in addition to supporting projects that demonstrate the use of energy efficient technological developments the EEO also sponsor activities to promote the adoption of management systems that concentrate on the control and reduction of energy use.

For over fifteen years the EEO has funded projects in the minerals industry and several projects have direct relevance to companies involved in road construction. Road recycling, both hot and cold mix, have been promoted as examples of developments resulting in direct and indirect energy savings. The consequences of these and other projects have been to reduce energy consumption in the process itself and also by eliminating upstream energy-intensive processes.

The EEO, through its Best Practice programme, will continue to offer assistance to the road construction industry to reduce energy consumption. Prospects for energy savings are good; the introductions of new binders mean that hot mix processes can now be conducted cold and developments in coating plant technology ensure that less energy is used for each tonne of roadstone manufactured.

1 Introduction

The Energy Efficiency Office of the Department of Environment estimates that the non-metallic minerals industry currently has an annual energy bill of around £400M - almost 170 PJ of energy. The cement industry is the sub-sector that uses the greatest amount of energy - about half of the overall total, and 24 PJ/yr, worth about £60 million, is used for

Performance and Durability of Bituminous Materials. Edited by J.G. Cabrera and J.R. Dixon. © E & FN Spon.
Published in 1996 by E & FN Spon, 2–6 Boundary Row, London SE1 8HN. ISBN 0 419 20540 3.

road materials. Energy costs represent about 10% of total production costs for roadstone manufacturers, for drying, heating, mixing and motive power.

Some industries consuming large quantities of non-metallic minerals, such as producers of glass, bricks, pottery and steel, have been involved with Energy Efficiency Office initiatives for several years, through well-developed and co-ordinated programmes. The EEO Best Practice programme strategy for the UK non-metallic minerals industry contains plans for projects throughout the diverse sectors of the industry and a promotional programme to bring the information to end-users. Drawn up, in consultation with trade associations, it promotes awareness of energy efficiency amongst producers of primary minerals, aggregates and mineral products, including road materials.

2 EEO-funded Projects to Date

Previous projects have tended to concentrate on the promotion of the considerable benefits in both direct and indirect energy savings of roadstone recycling. Projects to date are summarised below:

Project	**Organisation**	**Publication No.**	**Publication Date**
BEST PRACTICE PROGRAMME Cold Mix Asphalt	Bardon Roadstone		Active Project
Cold-mix In-Situ Recycling of Roadstone	J Kennedy	GI 17	December 1993
Pulverised Fuel Ash as a Bituminous Filler	University of Leeds	RD 52	March 94
Pulverised Fuel Ash as a Bituminous Filler	University of Leeds	RD 52	March 94
Cold Road Recycling	Epsom & Ewell BC	NPFP 60	July 1992
ENERGY EFFICIENCY DEMONSTRATION SCHEME Road Reconstruction Using a Cold Recycling Process	Kent CC	EPP 343	March 1989
Energy Saving Potential of Roadstone Recycling	Kent CC	EPP 241	March 1989

Currently underway is a New Practice project to promote energy savings in the manufacture of road materials, using a cold-mix concept implemented by Bardon Roadstone. (Details of the installation and the process employed are contained in an abstract in this publication.

3 Future Plans

Recycling of material is only one method by which the energy input to road construction can be reduced. Other lower-energy options include:

- in-situ cement stabilisation of the foundation material.
- thin wearing courses.
- cold bituminous emulsion binders
- longer-life bituminous mixes

Over two to three years the following activities are planned:

- A pilot survey of energy in Northern Ireland, due to be published in Summer 1994, will be followed by a comprehensive survey of energy use across the UK. This date will be published in an Energy Consumption Guide which will identify the specific processes and types of operation that consume the largest amounts of energy. It will also show users how their energy consumption compares with that of others in the same industry. The survey will encompass all industries, including manufacturers of roadstone.

- **A Good Practice Guide** will be commissioned to describe the processing aspects and benefits of all low-energy road construction methods. Many new processes and systems for road construction have been imported following successful and widespread implementation in continental European countries (particularly France). The Good Practice Guide will include reference to technologies adopted, or under development, in other countries. The wider environmental benefits, in terms of upstream energy and natural resource consumption, will be emphasised.

- Organisations will be canvassed for projects suitable for **Good Practice Case Studies.** Suitable subjects include energy efficiency roadstone plants and energy management.

4 Conclusions

Best Practice programme assistance is available to promote energy efficiency and to help reduce energy bills at the many and widely-distributed sites producing minerals and mineral products. Previous initiatives, particularly in cement production and roadstone recycling, have resulted in savings of over £11 million/yr in energy costs.

The EEO estimate that the cost-effective energy efficiency measures identified in its latest strategic overview of the sector could save the road construction industry up to 12% of it's energy bill, worth over £7 million/yr, with a corresponding reduction in atmospheric emissions.

23 COLD MIX MACADAM PRODUCTION

J. CRICK
Bardon Roadstone Ltd, Coalville, Leicestershire, UK

1 Introduction

It appears to be one of life's maxims that adversity is the mother of invention. Bardon Roadstone have developed an innovative product in response to two specific problems:

(a) The environmental pressures of maximising recycling activity and minimising energy use. Both are opportunities not fully exploited by a hot mix installation.
(b) The demand for a permanent cold lay surfacing material (PCSM), which provides first time permanent trench reinstatements.

From a discussion of these business environment forces, the logic of the development route will become apparent.

2 Environmental Pressures

Current Government policy is directed at reducing aggregate demand from U.K. mainland quarries from the current 200 million tonnes per annum in 1991 down to 192 million in 2011. However, aggregate demand is anticipated to increase from 200 million to 240 million tonnes per annum in the same time period. As we have graphically illustrated a growing supply gap will become apparent. (See Fig. 1.)

This window of opportunity is open to two sources, being either:-

(a) Increasing market share of aggregate supply from either imports or from the coastal super quarries; or
(b) Increasing utilisation of recycled construction materials.

As a manufacturer of blacktop, Bardon Roadstone's chief area of experience has been the introduction of recycled asphalt into hot mix production. We have the ability to manufacture a satisfactory material, whose performance characteristics are at least as good as those incorporating only virgin aggregates. This has been via controlled trials in Staffordshire, Berkshire and the London Borough of Westminster, where rates of up to 35% of the finished product is recycled asphalt. Having noted no detrimental effect to end performance, Bardon Roadstone are content to add recycled material in

Performance and Durability of Bituminous Materials. Edited by J.G. Cabrera and J.R. Dixon. © E & FN Spon.
Published in 1996 by E & FN Spon, 2–6 Boundary Row, London SE1 8HN. ISBN 0 419 20540 3.

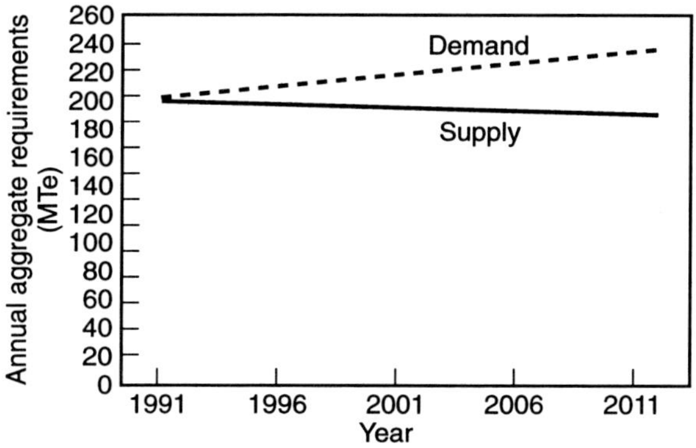

Fig. 1. Anticipated UK aggregate requirements. The graph illustrates the growing supply gap opportunity for recycled materials in construction products.

accordance with the Department of Transport's 7th Edition Specification for Highway Works Clause 902, where permitted.

It is also appropriate to note that as new road projects decline, the proportion of maintenance work will increase. The source of asphalt milling is principally planings of post design-life wearing course. Therefore a supply surplus to current demand for the recycled asphalt is likely in the longer term. This situation could be exacerbated by a decreasing utilisation in secondary construction work such as bridleways and farm access routes, the drop in greenfield applications such as sub-base, capping layer and structures backfills. Figure 2 gives figures as a guide to this relationship.

2.2 Europe

Many European countries possess lower quality aggregate resources with subsequent demand to import minerals. Furthermore, in countries

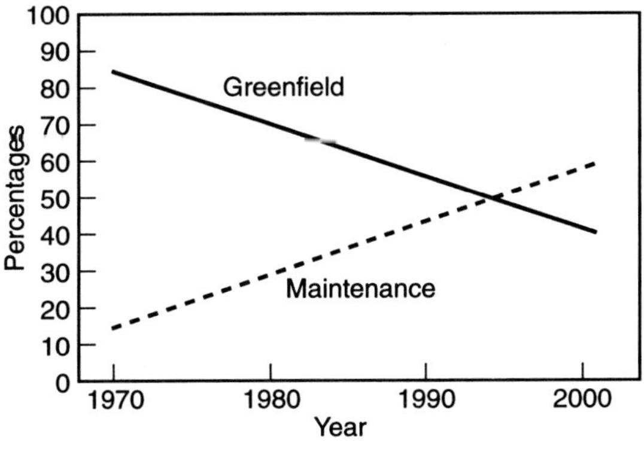

Fig. 2. Anticipated ratio of major works in the UK.

such as the Netherlands and Germany, environmental policies are more stringent. These two factors have led to pressure to maximise recycling activity. This is further enhanced by penal charges on the deposition and control of waste in landfill schemes. To recycle higher percentages in hot mix plants, a second drying drum has been added to facilitate this requirement. Bardon Roadstone's experience suggests that caution is exercised when considering hot mix recycling at rates exceeding the proven 35% level.

With lower total energy use and the incorporation of materials currently considered waste, recycling of asphalts has significant environmental benefits. Given the provision of a satisfactory performance product. However, carcinogenic tar-bound products cannot be hot-mixed, maintaining pressure for a safe method of disposal in continental Europe.

3 End Performance Specification

The way in which coated roadstone products are selected, designed and measured is changing. This is away from the empirical recipe specification, towards the determination of end performance engineering properties.

The first specification to utilise this change is the Highway Authority Utility Committee (H.A.U.C.) produced a code of practice, 'Specification for the Reinstatement of Openings in Highways', in response to Government legislation 'The New Roads and Street Works Act 1991'.

As an Appendix to this code, an approval procedure exists for permanent cold-lay surfacing materials (PCSM). This procedure for carriageway works incorporates elastic stiffness criteria. This table is reproduced as Table 1. At present the emulsion utilised evolve steam as are hot mixed during manufacture, thus a deferred binder must be utilised to ensure workability, storage longevity and ease of compaction. The development of load bearing strength therefore is delayed significantly. As such we are not aware of a permanent cold lay surfacing material which has the elastic stiffness

Table 1. Highway Authority Utilities Committee specification, showing minimum permanent cold-lay surfacing material requirements.

	Appendix 10: Minimum property requirement at 20°C for equivalence to:		
Permanent cold-lay surfacing material	50 pen hot laid Elastic stiffness (MPa)	100 pen hot laid Elastic stiffness (MPa)	200 pen hot laid Elastic stiffness (MPa)
20 mm nominal size basecourse	4600	2400	900
10 mm nominal size wearing course	3800	1900	800
6 mm nominal size wearing course	2800	1400	600

values as required by the governing specification for the carriageway works.

Typical stiffness values for conventional products conforming to British Standard Specifications are shown in Table 2. It is apparent from this data that the parameters specified for PCSMs possess stiffness value characteristics in excess of the standard/normal product performance values by a factor. The implications of this requirement are the enforcement of a discriminatory criteria and the potential embargo upon satisfactory alternative technology.

Table 2. Elastic stiffness values on materials tested for Bardon Roadstone to date.

Material size (mm)	Penetration	Description	Elastic stiffness (MPa)
14	100	Close grained macadam WC	960
14	50	30% 4–8kN HRA WC	2000
40	100	Dense macadam roadbase	3650
28	50	Heavy duty roadbase	2700
20	100	Dense basecourse sand fines	511
20	200	Permanent cold lay	1240
6	200	Permanent cold lay	630

4 Cold Mix Technology

Following a series of structural trials, Bardon Roadstone have developed a process which manufactures PCSM conforming to the HAUC specification. This process comprises two main elements:-

(a) The mixing plant.
(b) The material constituents.

4.1 Manufacturing Plant

The cold mix machine has been developed by Ammann to solve the safe deposition of tar bound millings problem. In continental Europe, the planings are hauled to a registered recycling site, crushed, screened and cold coated with either a bituminous emulsion or cement. This process produces a satisfactory coated roadstone or a lean mix concrete, dependant upon the customer requirements.

The Ammann control system ensures accurate increments of constituents are combined, with a sophisticated manufacturing process to ensure materials are thoroughly mixed without aggressive, emulsion breaking, stress developed.

4.2 Manufactured Material

The Ammann cold mix machine will be used in the U.K. by Bardon Roadstone in a similar mode, utilising the maximum percentage of quality controlled recycled asphalt materials. It will utilise the conglomerate particles as an aggregate and coat with a selected bituminous emulsion.

The water component in the cold mix emulsion acts as a viscosity-reducing agent, replacing the need for heat, providing lubrication to assist compaction. An optimum moisture content is therefore required to ensure maximum workability. The relatively high moisture content has enabled Bardon Roadstone to have the option of utilising as viscous as a 50 penetration base bitumen binder in a cold-mix cold-lay environment.

When assessing the Nottingham Asphalt Tester (N.A.T.) elastic stiffness results shown in Figure 3 the following observations are made:-

(a) A cold-mix cold-lay bituminous material can be manufactured in accordance with the HAUC specification as a permanent cold-lay carriageway surface material.
(b) A cold-mix cold-lay bituminous material can be manufactured whose performance is superior to a traditional hot-mix cold-lay material.
(c) A cold-mix cold-lay bituminous material can be manufactured whose performance is superior to a traditional hot-mix hot-lay material.

Our development has been achieved in conjunction with Nynas UK, taking advantage of their advance bitumen emulsion technology. During the trials product containing 100% virgin aggregate was compared with recycled asphalt materials. The fact that substantially increased proportions of emulsion are required, inevitably means increased volume of water to be forced out during compaction, or evaporated with time, to affect adequate curing. This substantial difference led to the conclusion that higher utilisation of recycled material leads to a quicker curing period. Therefore accelerating the development of load bearing capacity.

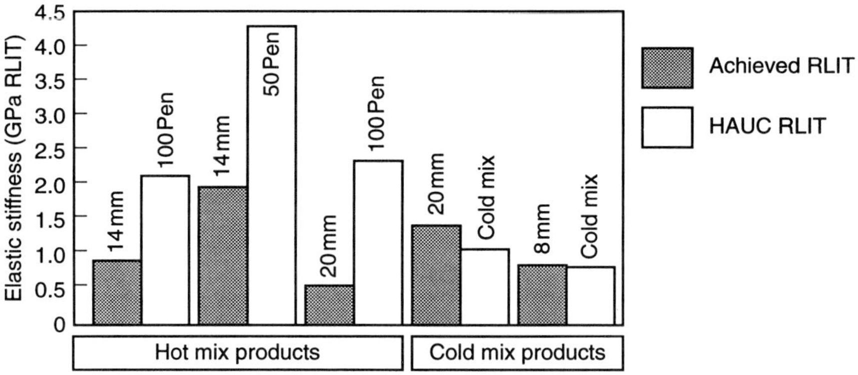

Fig. 3. NAT elastic stiffness summary of hot mix and cold mix products tested and HAUC specified.

5 Energy Efficiency

Where cold-mix materials can be utilised, there are energy-saving benefits in three categories: production, raw materials and end use.

(a) Production Efficiencies
- A cold mix machine needs no fuel to heat and dry the aggregate. A saving of 8 litres of fuel oil per tonne of manufactured material.
- The bitumen does not require heating. The physical operation of dryer drums, elevators, mixers and vehicle loading systems require additional energy. It is calculated that cold mixing required only 10% of the electricity required for a hot plant.

(b) Raw Materials
- The winning, processing and hauling of virgin aggregate to a remote manufacture plant requires more energy than recycling asphalt, yielding up stream energy savings.

(c) User Efficiencies
- To reinstate permanently first time using a PCSM provides less waste than either (a) a hot mix material which has to be used that day which inevitably leads to some wastage; (b) a cold lay deferred set material which is only useful as an interim reinstatement in carriageway works. This leads to two visits and consequential inefficiencies.

Inefficiency during reinstatement leads to inevitable traffic disruption. This traffic disruption leads to fuel waste by contractors and road users.

The energy saving aspects of the process have enabled the project to attract support from the Energy Efficiency Office, to promote the benefits across industry.

6 Conclusions

Bardon Roadstone are manufacturing cold-mix cold-lay product with the following benefits.

(a) Maximum utilisation of recyclable asphalt in a premium road application.
(b) Tho Ammann cold mix machine uses a state-of-the-art process to ensure consistent high quality manufacture.
(c) In collaboration with Nynas UK, development of a new cold emulsion meets the twin objectives of workability and end performance.
(d) Considerable fuel efficiencies compared to a traditional hot mix process.
(e) The Bardon Roadstone commitment to quality through Quality Assurance ensures the product will comply with end performance requirements first time, every time.

Through a structured development programme, a material has been produced with benefits to client, contractor and manufacturer alike.

At present, the product will meet the requirements of the H.A.U.C. specification, but results to date give confidence to continue the development of the product to meet a wide range of road construction applications.

24 A NEW DESIGN METHOD FOR DENSE COLD MIXTURES

A.F. NIKOLAIDES
Department of Civil Engineering, University of Thessaloniki, Thessaloniki, Greece

Abstract
Existing mix design methods are related primarily to open or semi-dense cold mixtures. This paper presents a new mix design method, exclusively geared, to dense graded cold mixtures having either continuous or gap graded gradation curve. The methodology uses the Marshall equipment and a modified marshall procedure. The design criteria are : Soaked stability, Retained stability, Water absorption, Total voids content, Degree of coating, Bitumen film thickness and Creep stiffness coefficient.
Keywords: Cold dense bituminous mixtures, Mix design, Bituminous emulsions.

1 Introduction

The use of dense graded cold bituminous mixtures was limited until few years ago due to their inherent problems (low degree of coating and inadequate early strength). Having solved these problems, arising primarily from the inefficiency of the existent bituminous emulsions, dense cold mixtures can be seen today as an attractive alternative to the corresponding hot mixtures in terms of energy consumption, protection of the environment from further air pollution and improvement of safety and health at work.

Originally, due to the intrinsic properties of the bituminous emulsions only open and semi-dense graded mixtures were used. Today, the improvement of the bitumen emulsions and the mixing techniques allow even dense graded type mixtures to be satisfactory produced and used in road maintenance and construction. Cold Dense Graded Mixtures (CDGM) with bituminous emulsions can be used in all layers of pavement, wearing course, base course and road base, provided that the volume of traffic is low or medium. However this restriction exists only when they are to be used as wearing courses. In such case and until further improvements are made on the mechanism of breaking of the emulsion in relation to cohesion and stability of the mix at the initial stage, CDGMs may also be used in heavy traffic areas provided that they are overlaid with at least 40mm hot mix layer.

Cold Dense Graded Mixture are defined as those mixtures consisting of continuously graded or gap graded aggregate

Performance and Durability of Bituminous Materials. Edited by J.G. Cabrera and J.R. Dixon. © E & FN Spon. Published in 1996 by E & FN Spon, 2–6 Boundary Row, London SE1 8HN. ISBN 0 419 20540 3.

gradation mixtures and bituminous emulsion as binding agent. These mixtures have low air permeability (lower than 10^{-8} cm^2), they are produced either in a stationary or in a moving plant and they are mixed and compacted at ambient temperatures.

Today there is no universally accepted mix design method for CDGM. Each authority or organization uses its own methodology derived from their own experience and laboratory investigations. A good overview of the available mix design procedure is given in reference (1).

The proposed mix design procedure for CDGM, presented in this paper, despite some similarities with the proposed by the Asphalt Institute (2) method is belied to be a more refined and integrated methodology for designing CDGM for optimum performance. It employs the use of Marshall equipment for the production and testing of the specimens as well as the static creep apparatus. The design criteria for optimum and long lasting performance are : soaked stability, retained stability, total void content, degree of coating, water absorption, bitumen film thickness and creep stiffness coefficient.

The proposed method and design criteria are applicable to base course and wearing course mixtures, with maximum aggregate size of 25.4mm, for low and medium traffic volume pavements.

2 Materials for CDGMs

2.1 Bitumen Emulsion

The bitumen emulsion could be of any type, preferably cationic, with or without modification to the bitumen, capable to coat satisfactory the given aggregate and in compliance with the current National or International specifications for bituminous emulsions. The type of bitumen used, after removal of water from the emulsion, should be similar to the type of bitumen used in equivalent hot mix. The percentage of bitumen in the emulsion should be preferably between 60 to 70% by weight.

2.2 Aggregates

The aggregates should be crushed, clean, durable and hard aggregates from natural parent rock or artificial aggregate or slag, suitable for paving mixtures and in accordance with the current specifications for equivalent hot mixtures.

The coarse and fine aggregate should have the required mechanical and physical properties as per equivalent hot mixtures specified in current national or international specifications.

The filler can be from the same parent rock or Portland cement or from any other material provided that they all comply with the current specifications as per hot mixtures. In case that Portland cement is used a maximum of 2% by weight of total aggregate is allowed.

The final gradation of the aggregate mix should be within the specified limits given in Table 1.

The Sand Equivalent and the Plasticity Index, determined from the required fractions of the aggregate mix, should be greater than 45% and NP respectively.

Table 1. Aggregate mix gradation limits for Cold Dense Graded Mixtures (with bitumen emulsion)

SIEVE SIZE	PERCENTAGE PASSING (% by weight)								
	TYPE I(*)	TYPE II	TYPE III	TYPE IV	TYPE V	TYPE VI	TYPE VII	TYPE VIII	TYPE IX
50 mm	100	-	-	-	-	-	-	-	-
38,1	90-100	-	-	-	-	-	100	-	-
25,4	-	90-100	100	-	-	-	90-100	100	100
19,1	60-80	-	90-100	100	-	-	65-98	97-100	97-100
12,7	-	60-80	-	90-100	100	100	50-75	70-100	77-98
9,52	-	-	60-80	-	90-100	-	40-60	56-80	67-85
4,76	20-55	25-60	35-65	45-70	60-80	75-100	35-50	50-60	60-70
2,36	10-40	15-45	20-50	25-55	35-65	60-80	30-48	45-60	55-70
0,6	-	-	-	-	-	-	10-45	15-60	18-70
0,030	4-15	5-18	5-20	5-20	5-25	15-65	-	-	-
0,074	0-5	0-5	3-8	3-8	3-11	5-20	3-6	3-8	3-8
Sand Equivalent	>45%	>45%	>45%	>45%	>45%	>45%	>45%	>45%	>45%
Plasticity Index	NP	NP	NP	NP	NP	NP	NP	NP	NP

(*) If aggregate size > 25.4mm, use nominal bitumen content as near optimum and adjust during construction using the criterion for proper coating

The water absorption of the final aggregate mix should not be greater than 1.7%. Aggregate mixes with grater than 1.7% water absorption may be used but after careful determination and investigation on the amount of bitumen absorbed by the given aggregate.

2.3 Added Water and Chemical additives

In all CDGM, when the aggregates are dry, a certain a small amount of water is necessary to be added to the aggregates in order to facilitate and improve the degree of coating with bitumen. The water should be clean and potable and the required amount should be determined with a procedure outlined in the design procedure paragraph.

Chemical additives may, sometimes, also be necessary to be added to the aggregate mix in order to decelerate or accelerate the braking time of the emulsion. In these occasions the type of the additive should be determined by the supplier of the emulsion, while the required amount should be determined by trial and error in the laboratory.

3 Types of Cold Dense Graded Mixtures

The CDGMs consisting of aggregate mix, bituminous emulsion, added water and chemical additive (if required) have lower air permeability and grater resistance to water attack in comparison to open or semi-dense graded mixtures. They include a wide range of aggregate gradation curves and hence a better use of available aggregate resources is obtained. The recommended limits of the aggregate gradations are as shown in Table 1. Generally, cold dense graded mixtures are preferred to the open or semi-dense mixtures for the same reasons as among hot mixtures.

3.1 CDGMs Type I to V (refer to Table 1)

The CDGMs Type I to V are similar to dense asphaltic concrete hot mixtures. Type I is the coarser of all and can be used as roadbase or base course mixture. Type V is the finer of all and can be used as wearing course mixture. The intermediate types of mixtures can be used as base course or wearing course mixtures depending on the circumstances.

3.2 CDGMs Type VI

This type of mixture is a cold sand-asphalt mixture and can be use almost exclusively as water proofing layer in low to medium traffic roads. In all cases this type of mixture should be covered with hot bituminous mixture.

3.3 CDGMs Type VII to IX

These type of mixtures are cold gap graded mixtures. They are distinguished by their lower air permeability, higher fine aggregate content and higher binder content than cold asphaltic concrete mixtures. Due to higher binder content these mixtures posses grater resistance to ageing of the bitumen and slightly grater stability.

Type VII is the coarser of all mixtures and normally is used as base course mixture. Type IX is the finer and can be used as wearing course. Type VIII can be used as either base course or wearing course material.

3.4 General Comment for All CDGMs

Covering the CDGMs, when they are used as wearing courses, with single surface dressing is not always necessary. It depends primarily on the type of road, traffic volume and season of construction. In case of surface dressing is decided, (for example when medium traffic volume and construction during autumn), it should be placed approximately a month after construction of the layer of cold mixture. Sanding of the surface is common for immediate opening to traffic in order to prevent possible pickup of the aggregates.

4 Design Procedure of CDGMs

4.1 General

The aim of the proposed methodology is to determine the optimum percentages of the ingredients of the cold mixture for optimum and long lasting behaviour. For this purpose the Marshall equipment is used as well as the static Creep tester. Given the suitability of the constituent materials the design methodology consists of two basic stages : The stage of examining the compatibility of the emulsion to the given aggregates and the stage of determining the properties of the mixture for optimum performance. The proposed method is applicable to either stationary plant or moving plant (mix in-place) produced mixtures.

4.2 Compatibility of Emulsion

The examination of compatibility of the emulsion is necessary since from the number of bitumen emulsions available, primarily slow to medium breaking, only few will produce a mixture with optimum properties, for a given aggregate type and gradation. The compatibility of the emulsion at this stage is judged by its ability to coat satisfactory, if possible 100%, all aggregates. This can be obtained by executing the Coating Test.

The Coating Test consists of visual inspection of the mix for obtainable degree of coating of the aggregates. The degree of coating is estimated and expressed as the percentage of coated area of the total aggregate surface area.

The coating ability of an emulsion is sensitive to the water content of the aggregate mix before mixing. For this reason the coating test is performed at varying water content of the aggregate mix prior to mixing. Hence the optimum moisture content for mixing is also determined at this stage. The moisture content for all types of CDGMs proposed should be between 1% to 4.0% by weight of dry aggregate.

The test is performed on batches of 500gr of representative sample of dried aggregate mix and by using the nominal percentage of bitumen emulsion given for each type of mixture, Table 2.

Table 2: Characteristic Properties of Cold Dense Graded Mixtures for Optimum Performance

PROPERTIES OF MIXTURES	LIMITS	COLD DENSE GRADED EMULSION MIXTURES								
		I	II	III	IV	V	VI	VII	VIII	IX
Nominal Bitumen content (%) (by w.t of total mix)	-	4.0	4.5	5.0	5.5	6.0	6.5	6.5	6.0	5.5
Soaked Stability (N)	MIN	1335	1335	1335	1335	1335	1110	1335	1335	1335
Retained Stability (%) (after 48h soaking)	MIN	50	50	50	50	50	50	50	50	50
Total Void Content* (%)	MIN MAX	6 12	6 12	6 12	6 12	6 12	6 12	6 12	6 12	6 12
Water Absorpt.(%) (after 48h soaking)	MAX	4	4	4	4	4	4	4	4	4
Bitumen Film Thickness (µm)	MIN	6	6	6	6	6	6	6	6	6
Degree of Aggr. Coating (%)	MIN	85	85	85	85	85	85	85	85	85
Permanent Deform. Performance	-	(Optimum bitumen content less than max. allowable determined by Creep test)								
Proposed layer Thickness (mm)	MIN MAX	80** 150**	50** 100**	40** 100**	30** 75**	25** 75**	25** 75**	30** 75**	30** 75**	30** 75**

* Total void content = air voids and voids filled with water
** It can be changed after compaction test trials on the job site

Mixtures with more than 75% coating are acceptable although 100% coating should always be aimed for. Mixing is carried out by hand and at room temperatures (±2 °C). More details for the coating test can be found in reference 3 and 4.

4.3 Properties' Determination of CDGMs

The stage of properties determination for optimum mix performance consists of determining the bitumen content for a given mixture so as to satisfy certain requirements. These requirements are shown in Table 2.

The properties: Soaked Stability, Retained Stability, Total Void Content and Water Absorption and Creep Coefficient criterium are referred to Marshall size specimens (100mm in diameter by approximately 62.5mm height) prepared by a modified Marshall procedure. The basic characteristics of this procedure is as follows :

> Mixing, compaction (50 Marshall blows) and testing takes place at room temperatures (±1°C). The specimens are tested, after one day curing in their moulds (in room temperatures) and one further day curing outside their moulds (in a 40 °C ventilated oven). The specimens, three per level of bitumen content, are compacted at moisture content for maximum or near maximum compaction, and then tested at room temperature for modified Marshall Dry Stability and flow.
>
> The moisture content for maximum compaction is determined at an initial stage using Marshall size specimens. Compaction takes place at various water content while the mixture has the nominal bitumen content (allow the mixture to rest for an hour and then force evaporate the water to the desired water content). A graph of bulk specific gravity versus water content is prepared, from which the water content for maximum bulk density is obtained.
>
> Three other specimens, after been cured as mentioned above are subjected to 48h capillary soaking and then tested for Soaked Marshall stability and flow, at room temperatures. The Retained Stability is then computed as the ratio of soaked stability over dry stability expressed as percentage. More details regarding the sample preparation and testing can be found elsewhere (3,4,5).

The proposed design method for CDGMs apart from the above mentioned criteria requires minimum Bitumen Film Thickness and maximum Water Absorption of the specimens after 48h soaking. These two criteria safeguard the mixtures from premature ageing of the bitumen and detrimental effect of water. Both parameters are of particular importance when CDGMs are going to be used as wearing course mixtures or in wet and cold climates respectively.

The results obtained for all the above mentioned parameters are plotted against percentage of binder; the typical curves are as shown in Figure 1. The optimum bitumen content is determined from these diagrams and is that for which all criteria requirements are satisfied in the best possible way.

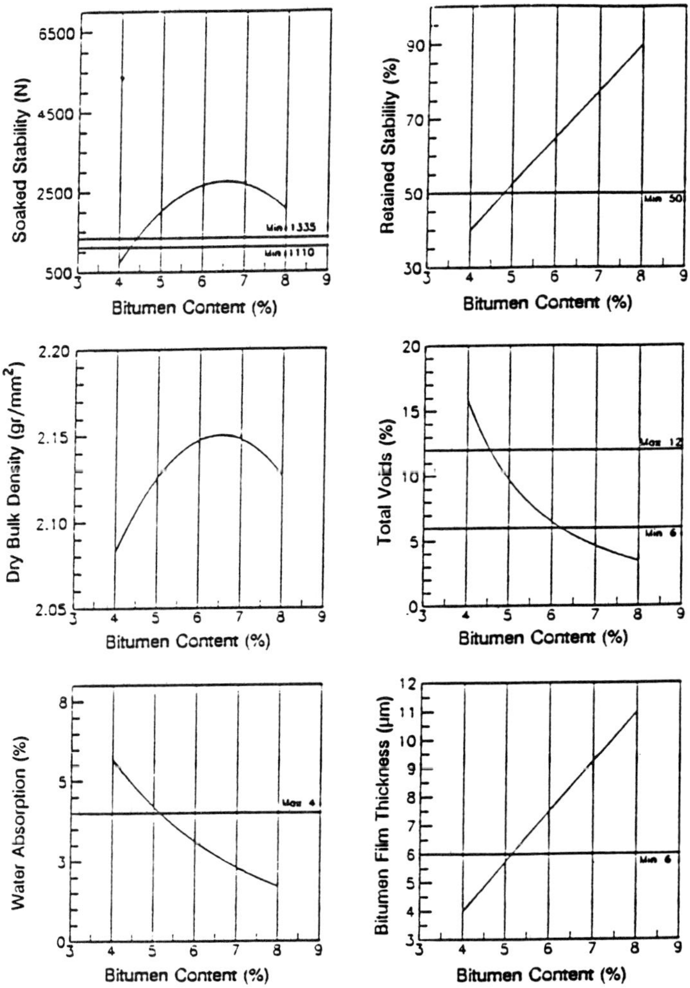

Figure 1 Typical diagrams for the determination of optimum binder content.

4.4 Compliance of the Design Mix with Permanent Deformation

In all cold mixtures, there is always the danger to use more than the actual required bitumen content since the ineffectiveness of an unsuitable emulsion to cover properly the given aggregate is sometimes corrected by increasing the amount of bitumen emulsion in the mix. The unnecessary increase in bitumen content may have a detrimental effect on the performance of the mixture in permanent deformation. This is more profound on gap graded mixtures rather than on the continuous graded cold dense mixtures (6). To avoid this the static creep test is recommended to be carried out, in addition, to all mixtures prepared for the modified Marshall design. The final design mixture is decided by taking into consideration its performance in static creep testing, which is expressed as the Creep Stiffness Coefficient. The procedure is as follows :

Two specimens per mixture are prepared as per modified Marshall test and tested immediately after curing (test conditions : 0.1 MN/m^2, 40 °C, 1h loading) using the CANIK creep tester (7) or a similar static creep device. From the creep strain measured the creep stiffness of the mixture, ($S_{mix,creep}$), is computed and plotted against corresponding stiffness of bitumen, (S_{bit}), in a double logarithmic scale (6). The resulted relationship is in the form :

$$\log Y = \log A + b \log X$$

where $Y = S_{mix,creep}$ and $X = S_{bit}$

This relationship was proved (7) to be independent of the operating parameters and therefore can be regarded as a genuine characteristic of the mix. Therefore, the slope, "b", of the straight line, named as Creep Stiffness Coefficient (CSC), indicates the sensitivity of the mixture to permanent deformation. The lower the CSC, the lesser the sensitivity of the mixture to permanent deformation is. The Creep Stiffness Coefficient for all mixtures with different bitumen content is plotted against bitumen content and a typical curve, as shown in Figure 2, is obtained. As it can be seen there is a sadden increase in CSC after a certain bitumen content is achieved. From this curve the maximum permissible bitumen content for acceptable behaviour in permanent deformation is determined. This is the binder content which corresponds to the interception of the two tangential straight lines to the curve, as shown in Figure 2.

4.5 Binder Content of the Optimum Design Mixture

The determined optimum bitumen content by the modified Marshall design should not be grater than the maximum permissible value of bitumen content determined in Figure 2. If the determined optimum binder content satisfies the above requirement this is the target value of the optimum design mixture. If this criterion is not satisfied, necessary adjustment of the bitumen content should be carried out provided that all other requirements of Table 2 are met. In case that this is impossible there is a strong indication that the bitumen emulsion is not the proper one for the given aggregate and even at this late stage, if no other modifications are possible to be made, it should be discarded.

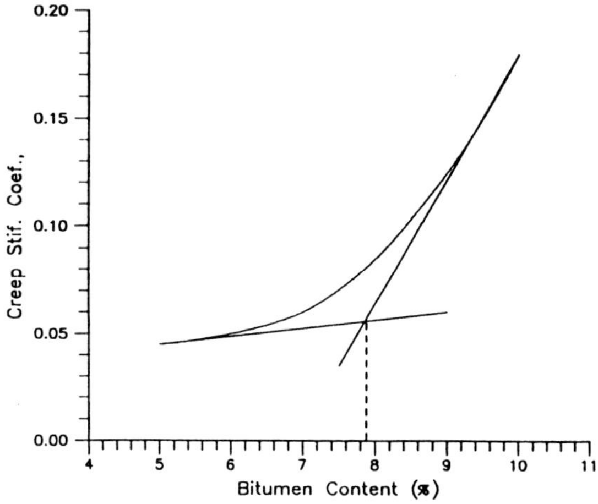

Figure 2 Determination of maximum allowable bitumen content for permanent deformation performance

5 Closing remarks

The proposed design method is believed to be a complete and integrated methodology for designing cold dense graded mixtures for optimum performance. It can be applied to either continuously or gap graded emulsion mixtures. It uses additional criteria, compared to other proposed design methods, which ensure the successful performance of CDGMs in pavements with low and medium traffic volume. It is the outcome of many years of research and application executed and supervised by the author in various countries. A similar version is currently in use in Indonesia and is been suggested to be used in Greece.

This methodology is going to be used in a large experimental study in which the long term behaviour of CDGMs is going to be monitored. At the end of this programme it is hoped to be linked with mix performance requirements so as to be incorporated in pavement design procedure using cold emulsion mixtures.

6 References

1 Walter H.F.(1980) Emulsion mix design methods: An overview. Transport Research Record 754, TRB, Washington.
2 Asphalt Institute (1989) Asphalt Cold Mix Manual. Manual Series No.14, Third Edition, Lexington, USA.
3 Nikolaides A.F. (1992) Dense Graded Cold Mixtures: Proposed design method. 1st National Conference on Bituminous Mixtures and Flexible Pavements, Thessaloniki, Greece, p.13.
4 Ministry of Public Works (1990) Paving Specifications Utilizing Bitumen Emulsions: Section 6.10-Dense Graded

Emulsion Mixtures. Directorate General of Highways, Jakarta.
5 Nikolaides A.F. (1983) Design of Cold Dense Graded Bituminous Emulsion Mixtures and Evaluation of their Engineering Properties. Ph.D Thesis, University of Leeds, England.
6 Cabrera J.C. and Nikolaides A.F. (1989) Creep Performance of Cold Dense Bituminous Mixtures. Journ. of the Institution of Highways and Transportation, No.10, Vol.35, p.7.
7 Cabrera J.C. and Nikolaides A.F. (1987) CANIK U.L.-A New Creep Testing Machine. Journ. of the Institution of Highways and Transportation, No.11, Vol.34, p.34.

An earlier version of this paper was presented at the First National Conference on Bituminous Mixtures and Flexible Pavements held at Thessloniki, Greece, in 1992.

25 PERFORMANCE OF COLD RECYCLED BITUMINOUS MATERIAL

S.J. BICZYSKO
Engineering Services Laboratory, Northamptonshire County Council, Northampton, UK

Abstract
Cold-mix bituminous material used as an alternative to hot-mix bituminous products should lead to overall energy savings. If the granulate incorporated in such mixtures is of a secondary utilisation then further energy savings should accrue.

The engineering equivalence and practical difficulties of using cold-mix formulations are not clearly defined.

A full scale trial has been constructed using cold recycled bituminous material as part of the rehabilitation of a relatively lightly trafficked rural road. The material was used as structural course in a haunch situation and extends over a length of one kilometre. A total bituminous construction method was adopted with the bituminous materials placed directly upon the subgrade soil.

The environmental aspects of placement and early life performance of the cold-mix formulation are discussed. Relative structural characteristics of the materials are described together with a transient deflection performance study over a one year service life in order to yield an indication of structural equivalence.

Keywords: Bituminous, Cold-mix, Equivalence, Fatigue, Hot-mix, Recycled, Stiffness, Trial.

1 Introduction

In the highway engineering field hot-mix bituminous materials dominate the UK construction market both for new projects and also in the maintenance of the established network of roads. Over the years a considerable depth and breath of experience has been accumulated with these hot-mix materials together with their tolerance to climatic effects at the time of placement and environmental influences during their service life.

From the standpoint of the Engineer it is important that any novel proposal or formulation should perform at least as well as the design requirements.

In order to yield equitable economic comparison between hot-mix bituminous material and alternatives such as cold-mix formulations it is necessary to consider 'whole energy

Performance and Durability of Bituminous Materials. Edited by J.G. Cabrera and J.R. Dixon. © E & FN Spon. Published in 1996 by E & FN Spon, 2–6 Boundary Row, London SE1 8HN. ISBN 0 419 20540 3.

costs', ie, the true cost to the economy and environment through the exploitation, production, haulage, placement and service life of these materials. The established hot-mix materials have the benefits of economy of scale and strategically placed production plants. At the centre of the economic considerations however, is the structural equivalence of cold-mix materials. It is critical in any economic evaluation comparing alternatives to ensure that they are equivalent in performance terms and it is therefore the design parameters, Figure 1, which require definition before the economics can be studied more closely.

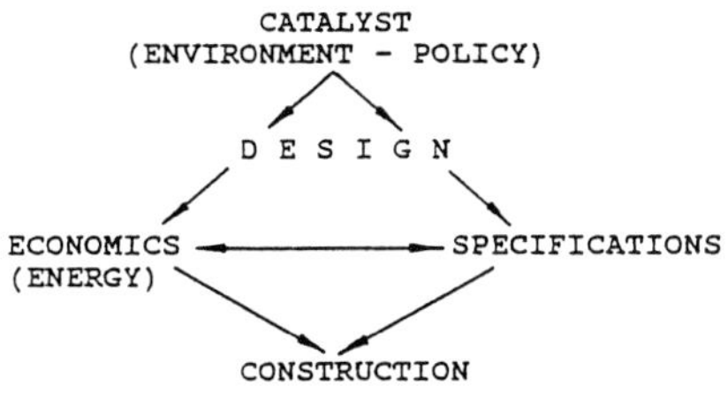

Fig.1. Design concept

A full scale trial was constructed in late 1992 to afford an opportunity for examination of a cold-mix bituminous formulation as compared with control sections of hot-mix material. The purpose of this trial, which extended over a length of one kilometre, was to study the placement characteristics, to establish the structural equivalence and to examine the in-service structural performance of the cold-mix material over an extended period of time.

2 Trial site

The trial site comprised a 500m length of unclassified single carriageway road which was subject to periodic improvement of carriageway standards by edge strengthening and widening. The road, which forms a link between the A5 and A43 to the south of Towcester, is not heavily trafficked but carries a regular pattern of commercial vehicles on a daily basis supplemented by substantially increased traffic use on a limited number of days per year. These increased flows arise from the traffic travelling to/from the nearby Silverstone motor racing circuit.

The site is virtually level with a consistent silty clay subgrade. The road was closed to traffic during the course of the works to facilitate unrestricted access and rapid completion of the construction.

3 Design considerations

The structural design of the edge strengthening works was prepared in accordance with published guides (1991a, 1991b). Two differing structural thicknesses were incorporated in the trial site: a 2.5msa design in the eastbound direction and a 10msa design in the westbound direction. Full depth construction, with the bituminous materials resting directly on the subgrade soil was chosen as the design solution (Fig.2). Within the trial two control sections of hot-mix materials were incorporated to provide a reference for the performance of the cold-mix material. The total bituminous structure was adopted since this afforded the most energy effective solution. Excessive depth of excavation and disposal of spoil is obviated. Additionally a granular foundation is avoided which can lead to serviceability problems owing to the lack of effective drainage systems in many rural road situations.

The intended surface for the trial site was application of surface dressing on the reconstructed areas. After completion of the trial however, a hot-mix macadam overlay was applied over the full carriageway width.

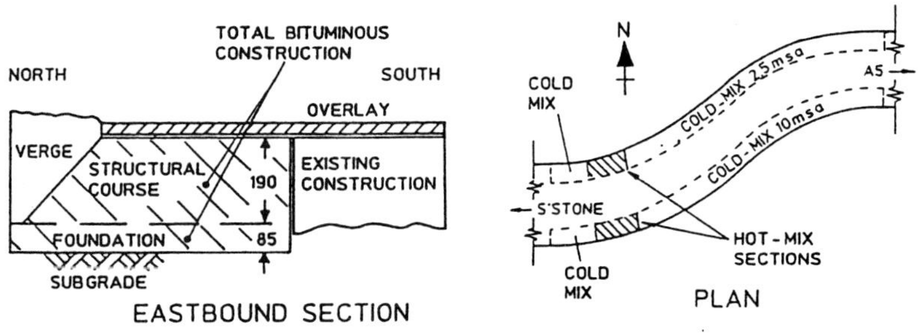

Fig. 2. Trial site details (2.5msa design)

4 Materials

The hot-mix used throughout the full depth of the control reference sections was a Macadam Basecourse material with 100 Pen grade bitumen binder (BSI,1993).

The cold-mix bituminous mixture was selected and incor porated into the trial on the basis of the following environmental criteria:-

a) The formulation should contain the greatest possible percentage of previously used materials (ie, granulate to have high recycled component).
b) The production arrangements to be of a type which could be operated as required on a regional centre

basis to process accumulated, previously used materials.

c) The production process to be as energy efficient as possible.

d) The material to remain in a usable and workable state for as long as possible.

The cold-mix used in the trial comprised processed reclaimed bituminous material (ie, a 'black' aggregate) together with added fine material, a foam extended bitumen binder and certain additives. The material was produced at an established mixing plant, transported to the site and placed in stockpile ready for use as required. The product was of a loose nature and remained in stockpile for some days prior to use. The cold-mix had the appearance of an unbound mixture in its stockpiled state.

5 Construction

The construction of the works was carried out swiftly by the use of side discharge plant operating in an unrestricted manner on the site.

After excavation of the margins of the existing road and verge and removal of the spoil the bituminous materials were placed directly upon the subgrade soil formation. The lower horizons (Foundation Platform) of the construction are less likely to be compacted to the same degree as succeeding layers (Structural Course) owing to the relatively low resilient support for compaction available from the subgrade clay soil.

During the process of construction a series of tests was carried out to yield information on the placement condition. (Table 1).

Both the cold-mix and the hot-mix bituminous materials were placed in the same way and with the same construction equipment.

		HOT MIX	COLD MIX
STRUCTURAL COURSES (Including Wearing Course)	Density (kg/m3)	2.36 - 2.38	-
	Equivalent Modulus (MPa)	-	50 - 110
	Air Voids (%)	7 - 9	-
	Thickness (mm)	190 (2.5msa)/250 (10msa)	
FOUNDATION PLATFORM	Density (kg/m3)	2.28 - 2.34	-
	Equivalent Modulus (MPa)	-	35 - 70
	Air Voids (%)	9 - 11	-
	Thickness (mm)	85	
SUBGRADE	CBR (%)	2 - 3	

Table 1. Material characteristics as installed.

During placement the cold-mix material behaved in much the same manner as an unbound aggregate mixture and immediately after compaction displayed a relatively low equivalent modulus value as determined by Plate Bearing Test.

6 Monitoring

The basis of evaluation of structural equivalence was determined from:-

a) A periodic sequence of transient deflection tests on site, and
b) A limited study of the engineering characteristics of the materials.

Transient deflection measurements were made on an approximate quarterly cycle on the trial site in order to enable comparison of the in-service deflections of the hot-mix control areas with the corresponding cold-mix material. The subsequent wearing course application over the whole site resulted in a structural pavement somewhat stiffer than intended. However, this effect is equal for both the hot-mix and cold-mix sections and therefore the basis of relative comparison of structural performance is still valid. The deflection measurements were carried out using a Deflection Beam (Kennedy et.al. 1978) and typical data is presented in Fig 3. The deflection characteristics of the pavement section may change further with time as the structure comes into equilibrium with its surroundings.

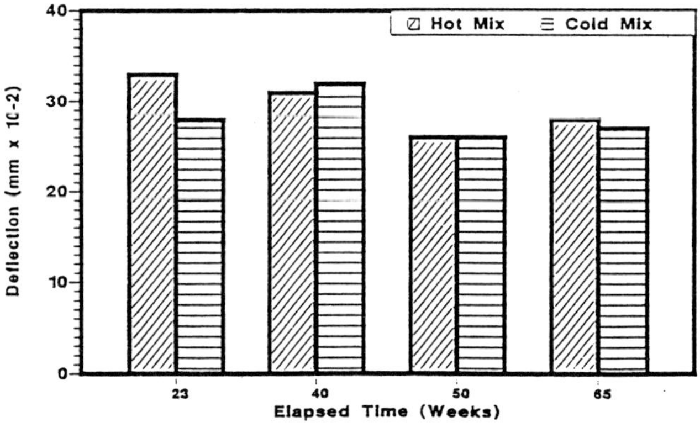

Fig.3. Transient Deflection (average values: eastbound)

For the study of the engineering characteristics of the materials it was intended to replicate test specimens in the laboratory manufactured to the same density as in the trial site. However, since there may have been some differences of particle orientation after compaction between laboratory prepared specimens and the material insitu it was considered more appropriate to carry out the assessment based on cores extracted from the site. Early attempts to obtain core specimens from the cold-mix materials in the trial were not successful and it was not until some time had elapsed that this could be achieved. The core specimens had to have sufficient integrity to allow them to be prepared for test.

Indirect Tensile Stiffness Modulus (BSI,1993a) was deter -mined on both the hot-mix and cold-mix core specimens. The cold-mix cores displayed a stiffness modulus of at least twice the value measured on the hot-mix control material. (Table 2).

		HOT MIX	COLD MIX
STRUCTURAL COURSES (Including Wearing Course)	Indirect Tensile Stiffness Modulus (MPa) at 20°C.	1900 - 2050	3700 - 4800
	Density (kg/m3)	2.36 - 2.38	2.02 - 2.07
	Thickness (mm)	190 (2.5msa)/250 (10msa)	
FOUNDATION PLATFORM	Indirect Tensile Stiffness Modulus (MPa) at 20°C.	600 - 1300	3100 - 4750
	Density (kg/m3)	2.28 - 2.34	1.99 - 2.04
	Thickness (mm)	85	

Table 2. Material characteristics (cores).

Determinations of fatigue resistance were also made on both hot-mix and cold-mix core specimens using the Indirect Tensile Fatigue Test (Univ. Nottingham, 1993b) and typical data is shown in Figure 4.

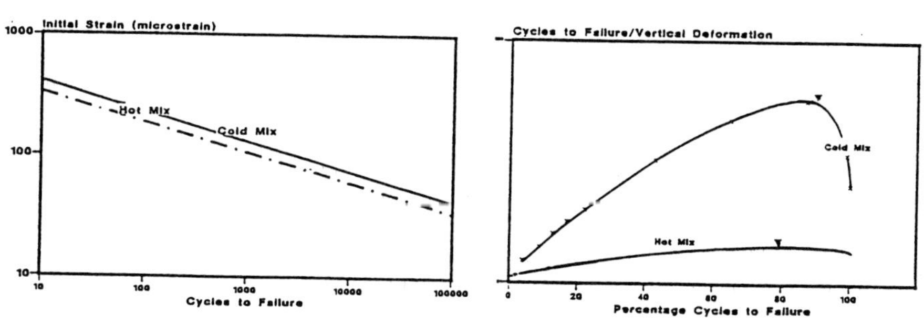

(a) Fatigue life (b) Crack propagation

Fig.4. Fatigue relationships.

7 Discussion

The placement and early life performance of the cold-mix material used in the trial was found to be not unlike an unbound mixture. The mechanics of transformation from an unbound mixture to a bituminous material are beyond the scope of this paper. Initially, however, it is considered that a hydraulic binder effect provides some structural strength at the early life stage. This is supplemented successively by the foam-mix binder and then the bituminous-/hydraulic binder combination. This process of 'strength' gain appears to yield a cold-mix material with stiffness properties better than those of a typical hot-mix macadam material placed in a road haunch situation. The hot-mix properties are however mobilised immediately after placement. With cold-mix where there is a period of time in its early life when the material is vulnerable to distress if time and environmental effects are not adequate to permit this process to proceed to a sufficient degree. The use of cold-mix material therefore needs to be carefully considered both in respect of traffic volume and climatic considerations at the time of placement to ensure that the stiffness gain occurs in an acceptable period. This can addressed by protection with either a hot-mix wearing course or, in appropriate conditions, by a thin veneer surface to isolate the material from the effects of erosion by traffic during its critical early life.

Stiffness modulus provides a measure of the load spreading ability of a material to reduce the stresses and strains developed in the subgrade through traffic loading. The stiffness modulus measured for the hot-mix material in the haunch trial is somewhat less than that which may be expected for similar material elsewhere (Powell et.al. 1984). There is also a stiffness modulus gradient between the lower, Foundation Platform, and the upper Structural Course layers in the hot-mix total bituminous construction. These effects may be due to the difficulties of placement and compaction in the confined working space of a haunch situation. However, it is assumed that the hot-mix material, which provides the control section of the trial site, is typical of performance which may be expected in this situation. On this basis the cold-mix material, which displays a greater and more uniform stiffness modulus (Table 2) would appear more efficient in terms of load spreading than the standard hot-mix material in this situation.

It is also important to assess a material in terms of its susceptibility to fatigue induced cracking under repeated loading. The fatigue life of both the hot-mix and cold-mix materials was found to be similar (Fig 4a). However the mechanism of crack propagation through the two materials appears to be significantly different (Fig 4b). The hot-mix material displays a relatively slow propagation of the crack between initiation and failure whilst the cold-mix material illustrates more rapid crack growth after the crack has been initiated.

The transient deflection study (Fig. 3) carried out on the trial site illustrates a similarity of performance between the hot-mix and the cold-mix materials for the first year of service life.

Consideration of the transient deflection performance of both materials in the trial site together with the laboratory determinations of stiffness modulus and fatigue characteristics would suggest that, at present, a structural equivalence of unity would be appropriate for design considerations in a haunch situation.

The initial laboratory determinations of the engineering characteristics are encouraging but further study is required in order to understand the cold-mix performance more fully. This needs to be set against a continuing monitoring of performance in service before a structural equivalence of less than unity in favour of the cold-mix formulation could be advocated.

8 Acknowledgements

The author is indebted to Mike Kendrick, Director of Planning and Transportation, Northamptonshire County Council, for permission to publish this paper. The contents and conclusions are those of the author and should in no way be attributed as policy of Northamptonshire County Council.

The trial site described is a facet of Northamptonshire Planning and Transportation Environmental Charter initiative.

9 References

British Standard Institution (1993), BS4987, Part 1 'Coated macadam for roads and other paved areas' Section 6.5.

British Standard Institution (1993), DD213 'Determination of the indirect Tensile Stiffness Modulus of bituminous mixtures.

County Surveyors Society (1991b) 'Practical Guide to Haunching' Report ENG/1/91.

Kennedy, C.K. et.al. (1978) 'Pavement Deflection : Equipment for Measurement', Transport and Road Research Laboratory, Report LR834.

Powell, W.D. et.al. (1984) 'The Structural Design of Bituminous Roads', Transport and Road Research Laboratory, Report LR1132.

Transport and Road Research Laboratory (1991a) 'Road Haunches : A Guide to maintenance practice' Report PA/SCR243.

University of Nottingham (1993) Test protocol 'Indirect Tensile Fatigue Test' (Unpublished) report of the LINK Bitutest Programme. Version 1.0).

26 CONSTRUCTION AND PERFORMANCE OF DENSE COLD BITUMINOUS MIXTURES AS STRENGTHENING OVERLAYER AND SURFACE LAYER

A.F. NIKOLAIDES
Department of Civil Engineering, University of Thessaloniki, Thessaloniki, Greece

Abstract
This paper presents construction details from two different projects where dense graded cold bituminous mixtures were selected to be used. The locations were Indonesia and Greece. At the first location, Indonesia, a dense cold mixture was used for the construction of a strengthening overlayer. At the second location, Greece, a dense cold mixture was used for the construction of surface layer in a newly constructed pavement. The performance of the above two applications after certain period of time is also reported in this paper.
Keywords : Cold dense bituminous mixtures, design, construction, performance

1 Introduction

Cold bitumen emulsion mixtures posses some advantages over conventional hot mixtures since they are produced and laid at ambient temperatures. They are "friendlier" to the environment, "cheaper" in energy requirements, "safer" to work with, "easier" to mix and handle and they "travel" longer distances. The ease in mixing arise also from the fact that the size and the installation (or re-installation) time of the mixing plant is much smaller and shorter, respectively, than of a hot mixing plant. Furthermore, cold mixtures can successfully produced by travel plants.

The above factors, specifically the last two, make cold mixtures particular attractive in areas where hot mixing plants are few and apart or there is complete absence of them. Such areas are remote areas with no dense road network or islands with small road network. In these areas the road network is either hot paved, in the past, at a high cost, or paved by blade mixing or in the extreme case is not paved at all. Maintenance works or upgrading the road network soon or later is required. Cold mixtures may provide the only available and cost effective solution.

Dense graded bituminous mixtures were decided to be used in two different locations, with completely different climatic conditions but for similar reasons.

Performance and Durability of Bituminous Materials. Edited by J.G. Cabrera and J.R. Dixon. © E & FN Spon. Published in 1996 by E & FN Spon, 2–6 Boundary Row, London SE1 8HN. ISBN 0 419 20540 3.

The fist location was situated in Java, Indonesia, at a latitude of 5°. The old paved road showed premature cracking and needed urgent strengthening. Premature cracking caused by two reasons: the early ageing of the bitumen due to oxidation and the vehicle overloading. The early oxidation of the bitumen in the tropics, due to the climatic conditions, is a well known problem. The acute problem of vehicle overloading has been studied and reported elsewhere (1). The site had no hot mix plant nearby. However, it must be emphasized that, generally, due to the physiology and size of the country (many islands in a vast area) and the sparse road network, the presence of a hot mixing plant close to the project site is always a problem. Hence, cold dense graded mixture was decided to be used for the construction of a strengthening overlayer. Mixing took place in a stationary plant and laying carried out by a conventional finisher.

The second location was in Naxos island, Greece, at a latitude of 37°. Naxos is situated in Cyclades in the Aegean sea and had no hot mix plant. All flexible pavements constructed or maintained in the past by mixing on the road by a grader motor followed by surface dressing. A cold dense bituminous mixture was decided to be used as a surface layer on top of a road base layer in a newly constructed pavement. The dense cold mixture designed according to a new proposed methodology (2).The mixing and laying of the mixture was carried out by a travelling plant.

For both projects, the use of cold dense mixtures was on a trial base. Successful application was going to solve the common problem of availability of asphaltic mixing plants. Additionally, was going to provide a more cost effective alternative to hot mixtures for maintenance or construction of new pavements in these areas.

2 The projects

2.1 Sites and pavement sections

The trial section in Indonesia was situated on the county road Jember to Situbondo in East Java and had a length of 2.5Km. The width of the pavement was 6m. The strengthening overlayer with dense cold mixture laid over a cracked old flexible pavement. The cracks caused by fatigue failure of the pavement. The thickness of the constructed layer was determined to be 67mm according to analytical calculation carried out using overlayer theory. Due to the peculiarities of the site (high humidity, frequent and heavy seasonal rain and high oxidation of the bitumen) some sections were decided to be covered by a single or double surface dressing. Analytically these sections are shown in Figure 1. The surface dressings carried out one month after completion of the construction of the overlayer. The estimated daily traffic was approximately 3000 commercial vehicles (typical range of total vehicle load 2.5 to 12tons). Construction took place February 1990.

The project site in Naxos island, the largest of the Cyclades islands complex in the Aegean sea, was 12.5Km away from the capital Chora on the county road Sangri to Agiassos. The total

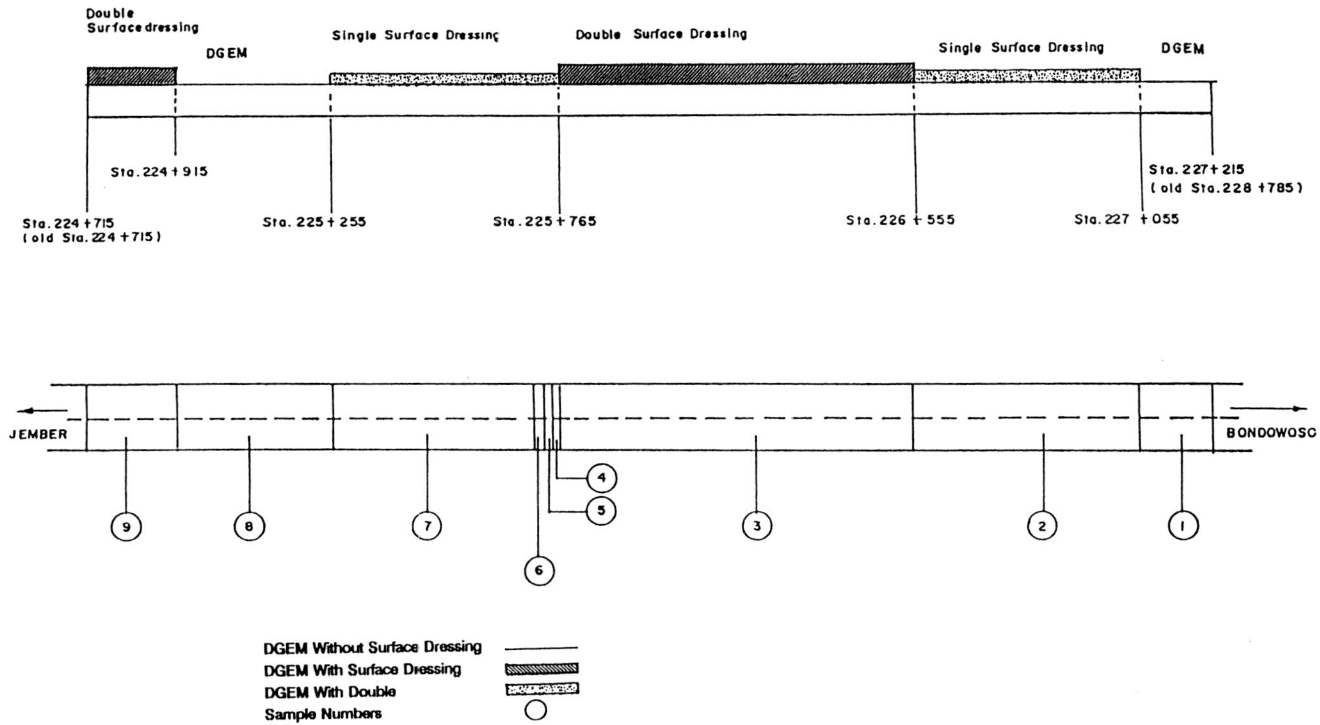

Figure 1 Trial sections of strengthening overlayer with dense graded cold mixtures.

length of the project was 3.4Km. The old unpaved road, with original width approx. 5m, was widened to an average 6.8m width prior to the construction of the pavement. The typical cross section of the newly constructed pavement was : a granural sub-base layer of variable thickness, to act as regulating and correction layer, 100mm granural road base, and 50mm cold dense bituminous mixture as surface layer. No surface treatment of any kind was applied on the surface layer. The traffic volume on this section, at the time of construction, was extremely low, less than 100 commercial vehicle (c.v.) per day (typical range of total vehicle load 2.5 to 12tons). Traffic was estimated to increase after two years to an estimated volume of 1000 c.v. per day (typical range of total vehicle load 2.5 to 12tons). Construction took place in December 1993.

2.2 Material used

a) Aggregates

All aggregate used in the fist project was local crushed igneous rock. The aggregate was delivered in three sizes, coarse (nom. size 25.4mm), medium (nom. size 14.0mm) and fine (nom. size 9.5mm). The properties of the aggregates are as shown in Table 1. The three fractions of aggregates blended such as to produce a gradation falling within the specified limits shown in Table 2.

The aggregate used in the second project in Naxos and for all layers was crushed limestone from a local quarry. The properties of the aggregate are as shown in Table 1. The limestone aggregate was delivered in two sizes, medium (nom. size 12.5mm) and sand (nom. size 5.0mm). The two fractions blended such as to produce a gradation falling within the specified limits shown in Table 2.

b) Bitumen emulsion

The bitumen emulsion in both projects were cationic type slow setting. The properties of both emulsions used are as shown in Table 3.

2.3 Mix design

The mix design for the project in Indonesia carried out, originally, by the contractor using the design methodogy for grave-emulsion. The recommended bitumen content was 4.03% by wt. of total mix. Other properties of the mixture given were : Marshall wet stability 1998N and total void content 8.6% (on specimens compacted by 50 blows both sides).

Another mix design carried out by the author on behalf of the consultant office C.P. Corne & Associates and the Ministry of Public Works of Indonesia, using the methodology given in the National specifications(3). The recommended bitumen content derived was 6.0% by wt. of total mix. Other properties of the mixture for optimum performance are as shown in Table 4. It must be noted that the mix design followed was proposed by the author and is very similar to the one followed in the project in Greece.

The optimum mixture, for project in Greece, designed in accordance to the proposed new methodology (2). The results obtained were : optimum bitumen content 5.0% by wt. of total mixture, added water to aggregate prior mixing 2.5% -3.0% by wt.

of dry aggregate and water content of the total mixture for optimum compaction 5.4% by wt. of dry mix. The other properties of the mixture are as shown on Table 4.

Table 1. Properties of aggregates

Properties	Crushed igneous	Crushed limestone
Apparent Spec.Grav. - coarse - medium - fine	 2.643 gr/cm^3 2.660 2.660	 - 2.680 gr/cm^3 2.680
Water absorption - coarse - medium - fine	 1.6 % 1.7 1.7	 - 0.32 % -
Sand equivalent	45 - 57 %	55 %
Plasticity	N.P.	N.P.
Los Angeles	18.3 %	30 %

Table 2. Specified aggregate gradation limits

Sieve size	Project 1 Java	Project 2 Naxos
	Percent passing (%)	
25.4mm	100	-
19.1	90-100	-
12.5	-	100
9.52	60-80	90-100
4.76	35-65	60-80
2.36	20-50	35-65
0.3	5-20	5-25
0.074	3-8	3-11

3 Construction details

The construction of the overlayer in East Java started in February 1990. The mixing plant and the material depot was approximately 30Km away from the project site. Aggregates stockpiled on a prepared clean area in enough quantities for two to three days work. Aeration of the aggregates, specially the fine fraction, was sometimes necessary since the storms were causing an increase of the desired moisture content of the aggregate. The environmental temperature throughout the construction was 30 to 32 °C and the relative humidity very high, approx. 90%.

Prior laying the strengthening overlayer the old pavement surface was checked for cleanliness (free of mud etc.) and then tack coated, using diluted cationic emulsion at a rate of 250-350 gr/m^2.

Table 3. Properties of cationic bituminous emulsions

Properties	Project 1 Java	Project 2 Naxos
Bitumen content (%)	58-60	59-61%
pH	-	3.5
Viscosity		
- Cannon at 25°C (cP)	17.3	-
- Saybold at 25°C (sec)	-	68
Storage Stability, 24h (%)	0.59	0.6
Sieve test (%)	-	0.02
Coating ability (%)	95-100	95-100
Tests on the residue		
Penetration, 25°C (pen)	105	87
Ring & Ball (°C)	46	48
Ductility, 25°C (cm)	60	>100

Table 4. Properties of dense cold mixtures

Mixture properties	Project 1 Java	Project 2 Naxos
Bitumen content (%)	6.00	5.0
Soaked stability, 48h (kN)	2.23	5.5
Retained stability (%)	78	99
Total voids (%)	9.2	10.9
Water absorption, 48h (%)	2.6	3.3
Bitumen film thickness (μm)	7.8	6.9
Degree of coating (%)	95-100	95-100
Max. permissible bit. cont. for acceptable permanent deformation (%)	-	5.7

The works in Naxos for the widening of the road and the construction of sub-base and road base started in July 1993. The works for the construction of the cold mixture surface layer started beginning of December and finished the 12th of the same month. The local limestone aggregate was stockpiled on a site adjacent to the project. Due to the low daily production of the quarry, 500 to 600 ton of aggregate material was collected before the commencing of the works. The aggregate most of the time had natural moisture 2 to 3%. The small quantity of extra water, if required, added by spraying to the conveyor belt prior mixing.

The air temperature during construction of the cold mix surface layer was varied from 10°C to 19°C. Most of the work was carried at an average temperature of 14°C. There was no rain during the days of construction. The humidity was normal to low, approx. 50%.

Prior laying the surface course layer the surface of the road base was prime coated with diluted cationic emulsion at an approximate rate of 1.0 kg/m^2.

3.1 Mixing, laying and compaction

a) East Java's project

Mixing of the dense cold mixture, in Indonesia's project, carried out in a stationary cold mixing plant. The plant was consisting of three aggregate storage hoppers, a volumetric aggregate dosage system, conveyor belt, a double paddle mixer and a delivering system for the emulsion and the water, if required. The maximum production capacity per hour was 60 tons.

The produced mixture directly loaded to awaiting trucks, having been inspected for cleanliness (left soil, oils etc.). Due to frequent storms (construction took place during raining session) the loaded trucks were covered with plastic cover in order to protect the cold mixture from "washing off".

The cold mixture arrived on site laid by a conventional finisher. The heating system of the screed was switched off. Compaction took place almost immediately after laying using a vibratory steel roller (8ton) and a kneumatic tire roller (8-10ton). Vibration was only used at the longitudinal joints. Compacted density determinations were carried out at frequent intervals with the use of nuclear density meter. The compacted layer opened to the traffic as soon as possible.

As it was mentioned earlier, some sections covered with single or double surface dressing, for details see figure 1. The binder used was bituminous cationic emulsion. The surface dressing carried out one month after completion of works.

The main problem occurred during construction was only the frequent storms, during which every activity had to be stopped. It must be reported, however, that extremely premature ravelling occurred (after two days) on the sections laid the first two days. This was due to laying a mix with low bitumen content (4.2%, according to contractor's design). Immediate modification of the mixture, based on the mix design carried out by the author, solved the problem. Those ravelled sections covered with double surface dressing.

b) Naxos' project

The mixing and laying of the cold dense graded bitumen mixture carried out by a travelling plant, hopper type-Midland Paver machine. The aggregate was fed at the front of the machine from a tipped lorry and transported to the mixer by a conveyor belt. The aggregate, prior mixing, was wetted with the required amount of water at a point just before the entrance to the mixer. The bituminous emulsion was pumped to the mixer at the required quantity from the 5ton bitumen tank of the machine. The temperature of the emulsion during mixing was varied between 10°C to even 60°C. The best results in terms of coating obtained when the emulsion temperature was around and below 30°C. At the low temperature of 10°C, coating was still excellent but the workability of the mix was decreased which affected also the proper performance of the machine (mix too stiff). At the high temperature of 60°C, coating of the aggregates was drastically decreased (coating as low as 50 to 60%). The situation was improved with a slight increase in the amount of added water to the aggregate (a further increase by approximately 0.3%).

Laying of the cold bituminous mixture was carried out immediately by the hydraulically adjusted screed unit. The screed unit had a vibrating screed plate, electronic sensors for automatic thickness adjustment and no heating unit. The absence of no heating system on the screed plate may have caused the frequent stops for cleaning the surface of the screed plate. The screed was capable to lay from 3.1m up to 3.8m width of lane.

Compaction took place, almost immediately, 5 to 10 min after laying, using a 12ton steel-tired roller and 8ton pneumatic-tired roller. The steel-tired roller was used for breakdown rolling and the pneumatic-tired roller for the intermediate and final rolling. For the type of mixture used, the required thickness of 50mm thick compacted layer and the conditions faced on site, three to four passes of the steel roller together of no less than ten passes of the pneumatic roller were found sufficient for proper compaction. Compaction, always, started from the outside edge of the lane being paved towards the other edge near to the central line of the road. The steel roller started to compact approximately 100mm away from the edge in order to avoid lateral displacement of the mix. The outer edges were compacted after the breakdown rolling with the pneumatic roller. The longitudinal joints were first compacted with the steel roller.

The paved lane of 3.1m width was given to traffic immediately after compaction finished. No surface treatment of any kind was applied to the finished surface. However, instructions were given to all construction vehicles to avoid heavy breäking and spot U-turns on the freshly laid mat.

4 Quality control during construction

Thorough and systematic checking was carried out, during constructions, on the following : the water content of the stockpiled aggregate, the aggregate gradation and the bitumen content in the mixture. Simple coating test and bitumen content determination on the emulsion was performed on every new delivery of the emulsion. The air temperature and the temperature of the bitumen emulsion in the tank was also recorded.

The water content of the stockpiled aggregate was varying, during the day and from one day to another. This variation was found to be in the first project from 0.5% to 7.0% and in the second from 1.5 to 3.0% by wt. of dry aggregate. Hence, it was always necessary to make the appropriate changes in the added water or to aerate the aggregate.

The variation of the aggregate gradation of the mixtures,

determined after bitumen extraction, was within the specified limits for both projects. The variation of bitumen content in the mixtures was also within the specified limits in Naxos' project. Typical results obtained in Naxos project are as given to Table 5. In East Java's project a wide range of binder content was observed mainly at the begging of the project.

Table 5 Laboratory results during construction from Naxos' project

Sieve size (mm)	**Target curve[1]**	**Variation range**	**Specified limits**
	Percentage passing (by weight)		
12.7	100	100	100
9.52	98	99-97	90-100
4.76	68	66-74	60-80
2.36	38	37-47	35-56
0.30	14	13-17	5-25
0.074	9	8-11	3-11
Binder Content (% by wt of total mix)	**Target Value**	**Variation range**	**Specified limits**
	5.0	4.76-5.35	4.7-5.3

(1) After combination of available local aggregates

Measurements were also made for the required thickness of the layer and it was found that the compacted thickness varied within acceptable limits in both project.

5 Performance evaluation of constructed pavements

Twelve months after construction, pavement evaluation carried out in East Java's project. This consisted of visual inspection, core sampling, density measurements and straight edge beam measurements.

The condition of the pavement in all section, with or without surface dressing, was found to be in a very good condition. No cracking or ravelling or rutting was observed. In some areas, however, and on the sections not covered with surface dressing, "reverse" macro-texture was observed. This was due to the departure of some coarse aggregates took place during the first few days of construction and had nothing to do with the long term deterioration of the pavement.

The surface dressed sections showed better macro-texture to that of the uncovered sections. Double surface dressed areas had better macro-texture than singe surface dressed areas.

The laboratory tests on bitumen content and aggregate gradations confirmed the laboratory results obtained during construction.

Interesting results obtained from the recovered bitumen. The penetration and R&B values decreased and increased respectively, particularly when the mix samples taken from the undressed sections. The original penetration of 105pen decreased after only one year to an average value of 50pen. Similarly, the R&B value of 46°C increased to 48°C. On bitumen recovered from surface dressed sections the corresponding changes were not significant. The results obtained were on average 90pen and 46°C. The above finding proves the severe oxidation of bitumen occurred in the tropics and the importance of having as much bitumen film thickness as possible.

Three years after construction, further visual inspection was carried out and it was reported that all sections were in good condition with no signs of serious defects.

For Naxos' project, three months after construction, detailed visual inspection carried out. All paved area was in excellent condition with no signs of any failure. The only remark was that construction joints had to be improved in relation to levelling. Frequent inspections will be continued.

6 Conclusion

Evaluating the results of the two projects executed it is concluded that dense cold bituminous mixtures can, successfully, be used in maintenance, as strengthening overlayer, and in new constructions, as surface layer, of the flexible pavements.

The climatic conditions do not restrict their application provided that the suitable materials are used, the right mix design is applied and followed, the right equipment is used, there is consistency in the produced mixture and the construction is executed according to specified rules and conditions.

The need to cover for protection the dense cold mixture, by surface dressing, proved to be unnecessary, even under tropic conditions. The traffic volume, however, was low or medium.

Stationary or travel cold mixing plants produce equally good mixtures provided the emulsion used is a suitable one. Travelling plants are more flexible to move from a site to a site and they minimise the transport cost of the mixture.

Subject to the long term behaviour, Indonesia, Greece and many other countries, which have remote areas or islands with no facility of a hot mixing plant, have an attractive and more cost effective alternative to that of hot mixtures for the construction or maintenance of the flexible pavements. Additionally, cold dense mixtures may also contribute to the efforts of investing the taxpayer's money in a better way and keep the environment clean.

Can the cold mixtures substitute completely the hot mixtures? The answer at the moment is no. Will they do so in the future? The answer must be why not. One should never be afraid of the heat or the water.

7 References

1 Sepang P.,Aryato B.E.,Scott M.L., Corn C.P. (1990) The catastrophic impact of vehicle overloading on Indonesia's toll roads. 4th Annual Conference on Road Engineering, Jakarta.

2 Nikolaides A.F. (1993) Proposed Design Method for Cold Dense Graded Bituminous Mixtures. 5th Eurobitume Congress, Stockholm, Vol.3, p.615.

3 Ministry of Public Works (1990), Paving specifications utilizing bitumen emulsion, Section 6.10-Dense graded emulsion mixtures. Directorate of Planning, Jakarta

4 Nikolaides A.F. (1990) Site investigation of Jember-Situbondo project where dense graded emulsion mixtures used as strengthening overlayer, Part I : Inspection during construction. C.P. Corn & Assocs, Report to the Department of maintenance and rehabilitation, Ministry of Public Works.

5 Nikolaides A.F. (1990) Site investigation of Jember-Situbondo project where dense graded emulsion mixtures used as strengthening overlayer, Part II : Inspection after one year. C.P. Corn & Assocs, Report to the Department of maintenance and rehabilitation, Ministry of Public Works.

8 Acknowledgement

The author wishes to thank the Ministry of Public Works of Indonesia and the consultant office C.P. Corne & Assocs for the co-operation and the provision of laboratory personnel and testing facilities. He also wishes to thank AKTIS S.A. for giving him the opportunity to design and supervise the first full scale project in Greece on cold dense graded mixtures. He also thanks BITOUMINA S.A. and her laboratory team for allowing him to develop together the most suitable bituminous emulsion for the dense limestone mixture.

27 DESIGN OF LOW ENERGY HOT ROLLED ASPHALT

J.G. CABRERA and S.F. ZOOROB
Civil Engineering Materials Unit, Department of Civil Engineering, University of Leeds, Leeds, UK

Abstract
A laboratory study on the effect of pulverised fuel ash on the properties and performance of hot rolled asphalt is presented. The aims of the study were to assess the effect of pulverised fuel ash on the mixing, handling and compaction of hot rolled asphalt (HRA), and to quantify the influence of temperature of mixing and compaction on the engineering properties and performance of HRA.

Eight HRA mixes were studied, these consisted of four aggregate-sand combinations and two fillers. The aggregate-sand combinations were selected as representative of the materials used in the north of England and Scotland to produce HRA. The fillers were a conventional limestone powder and pulverised fuel ash from four thermal power stations from the geographical areas where the aggregate-sand combinations were obtained.

The study was carried out following the Leeds Design Method which includes the assessment of engineering properties, performance parameters and quantification of workability using the Workability Index. The temperature of mixing was varied from 140°C to 110°C, and the temperature of compaction from 125°C to 85°C.

The data obtained in the laboratory show conclusively that pulverised fuel ash improves the workability of hot rolled asphalt. Hot rolled asphalt made with pulverised fuel ash filler can be mixed at 110°C and compacted at 85°C without affecting its engineering and performance properties. The reduction of energy requirements for production and placement of HRA are of considerable magnitude and warrant a field trial.

The advantages of the new hot rolled asphalt containing pulverised fuel ash are:

a) Reduction of direct expendable energy,
b) Improved mix workability resulting in greater ease in achieving design densities in the field,
c) Reduction of material rejection due to loss of temperature,
d) The incorporation of a waste product in an engineering structure.

1 Introduction

One of the most important properties of bituminous mixtures is that their compaction characteristics in the laboratory can be repeated during construction in the field. Poor performance of bituminous mixtures in road pavements is in many cases attributed to poor mixing and inadequate compaction. Thus mixes that can be mixed, handled and placed

Performance and Durability of Bituminous Materials. Edited by J.G. Cabrera and J.R. Dixon. © E & FN Spon.
Published in 1996 by E & FN Spon, 2–6 Boundary Row, London SE1 8HN. ISBN 0 419 20540 3.

without difficulties are said to be workable. Workability is a parameter which indicates these attributes in a bituminous mix.

Most bituminous mixes can be made workable if a high enough temperature of compaction is maintained during the process, this is obtained by heating the mineral aggregates, filler and bitumen to relatively high temperatures, and transporting and laying the mixes in short periods to avoid loss of temperature. Many mixes become unworkable when they reach temperatures of approximately 120 °C.

The production and placement of bituminous mixtures like hot rolled asphalt is an intensive energy process and therefore designing mixtures which can be produced and placed at lower temperatures is part of intensive work being carried out by many investigators.

The Civil Engineering Materials Unit (CEMU) of the University of Leeds has for many years investigated and developed methods for the design of high performance materials and the assessment of their durability. In 1977, it started projects on the use of pulverised fuel ash (PFA) in bituminous composites. The project reported in this paper is an outcome of part of the work started in 1983, when it was already suggested that PFA could be used to design bituminous composites of high workability.

The project on the design of low energy hot rolled asphalt using PFA was supported by the Energy Efficiency Office of the Department of the Environment, Cleveland County Council, National Power and Tilcon Limited.

The main objectives of the study reported in this paper were :

- To assess the effect of PFA on the engineering and performance properties of hot rolled asphalt.
- To assess the influence of changes on the temperature of mixing and compaction in conventional and PFA hot rolled asphalt.
- To validate any findings using a wide range of mineral aggregates and fillers.

2 Materials

The mineral aggregates and limestone fillers were selected and supplied by Tilcon Limited and the pulverised fuel ash (PFA) fillers were supplied by National Power, as being representative materials covering a wide range of the UK's aggregate and filler supply industry.

2.1 Mineral Aggregates and fillers

Four coarse aggregates and four sands were used in the study. The origin and code given to these materials are shown in Table 2.1

Table 2.1 Coarse Aggregates and Sands used in the study

Coarse Aggregates		Sands	
Code	Name	Code	Name
C1	Cairneyhill Basalt	B1	Douglasmuir asphalt sand
C2	Northumberland Basalt	B2	Blaydon asphalt sand
C3	Mancetter Basalt	B3	Meriden fine asphalt sand
C4	Yorkshire Gritstone	B4	Tyrham Hall asphalt sand

The fillers selected were limestones from Tilcon Limited quarries and PFA fillers from four Power Stations belonging to National Power. Their code and name are given in Table 2.2.

Table 2.2 Fillers used in the study

Code	Name	Code	Name
A1	Drax PFA	A5	Ballidon Filler
A2	Blyth PFA	A6	Marsden Limestone
A3	Thorpe Marsh PFA	A7	Scottish Limestone
A4	West Burton PFA		

2.2 Binder

For all the mixes investigated, the Binder used was supplied by Tilcon Limited and consisted of a straight run nominal 50 pen grade bitumen. Table 2.10 gives the bitumen properties.

2.3 Properties of the Mineral Aggregates

2.3.1 Coarse Aggregates and Sands

The properties to characterise the coarse aggregates and sands were :

a) Relative densities.
b) Water absorption.
c) Particle size distribution.

These are shown in Tables 2.3, 2.4 and 2.5.

Table 2.3 Sand and Aggregate Properties

Material	Relative Density on Oven dried basis	Relative Density on a saturated and Surface dried basis.	Apparent Relative Density.	Water Absorption (% of dry mass)
B1	2.489	2.494	2.502	0.210
B2	2.383	2.402	2.430	0.808
B3	2.534	2.547	2.568	0.512
B4	2.579	2.595	2.621	0.611
C1	2.861	2.910	3.009	1.719
C2	2.893	2.923	2.982	1.032
C3	2.728	2.756	2.807	1.030
C4	2.753	2.763	2.780	0.353

Table 2.4 Coarse Aggregate Sieve Analysis Values

Sieve Size	Mancetter Basalt	North-umberland Basalt	Cairneyhill Basalt	Yorkshire Gritstone
	Cumulative Percentage Passing			
20 mm	100	100	100	100
14 mm	96.09	94.08	98.31	86.8
10 mm	30.75	19.05	31.51	9.79
8 mm	5.06	2.13	4.41	0.99
6.3 mm	1.21	1.03	0.43	0.74
5 mm	0.16	0.81	0.23	0.51

Table 2.5 Fine Aggregate Sieve Analysis (% Passing)

Sieve Size	Meriden	Blaydon	Douglasmuir	Tyrham
5 mm	99.56	100	100	99.78
3.35 mm	95.74	99.83	99.70	94.88
2.36 mm	93.97	99.58	98.68	93.30
1.18 mm	92.56	97.41	95.04	92.41
600 um	90.71	85.26	85.84	90.68
425 um	83.89	73.34	73.75	88.12
300 um	61.67	63.86	57.19	68.99
75 um	3.512	3.60	2.27	0.86

2.3.2 Fillers

The properties of fillers which were measured were :

a) Specific Gravity.
b) Bulk density in Toluene.
c) Voids in dry compacted filler.
d) Particle size distribution using laser diffraction.
e) Surface area.

These properties are presented in Tables 2.6 to 2.9 and Figures 1 and 2.

Table 2.6. Filler Properties

Filler	Relative Density	Voids of Dry Compacted filler	Bulk Density in Toluene (g/ml)	% retained on 75um sieve
A1	2.179	0.211	0.450	14.64
A2	2.412	0.275	0.638	12.77
A3	2.249	0.281	0.344	9.16
A4	2.384	0.235	0.422	11.81
A5	2.773	0.255	0.612	3.30
A6	2.824	0.243	0.625	5.72
A7	2.725	0.248	0.638	8.48
Specific limits	-	-	0.5 - 0.9	< 15.00

Table 2.7 Particle Size Distribution (Microns); PFA fillers

	Size Band		Cumulative weight below = % passing			
Upper	**Lower**	**Mid Range**	**W. Burton**	**Blyth**	**T. Marsh**	**Drax**
188	87.2	137.6	96.5	95.4	93.5	87.3
87.2	53.5	70.35	83.6	77.1	76.5	63.8
53.5	37.6	45.55	75	67.1	65.4	55.8
37.6	28.1	32.85	66.7	57.6	55.5	48.3
28.1	21.5	24.8	59.1	50.9	47.5	43.1
21.5	16.7	19.1	54.7	45.4	41.4	39.8
16.7	13	14.85	48.4	38.5	34.2	35.2
13	10.1	11.55	39.7	30.6	26.6	28.6
10.1	7.9	9	33.2	24.1	20.8	22.2
7.9	6.2	7.05	25	15.8	13.9	15.1
6.2	4.8	5.5	18	9.1	8.7	9.7
4.8	3.8	4.3	14.1	6.3	5.9	7.1
3.8	3	3.4	12.4	5.2	4.5	5.9
3	2.4	2.7	11.3	4.5	3.6	4.9
2.4	1.9	2.15	9.6	3.2	2.4	3.5

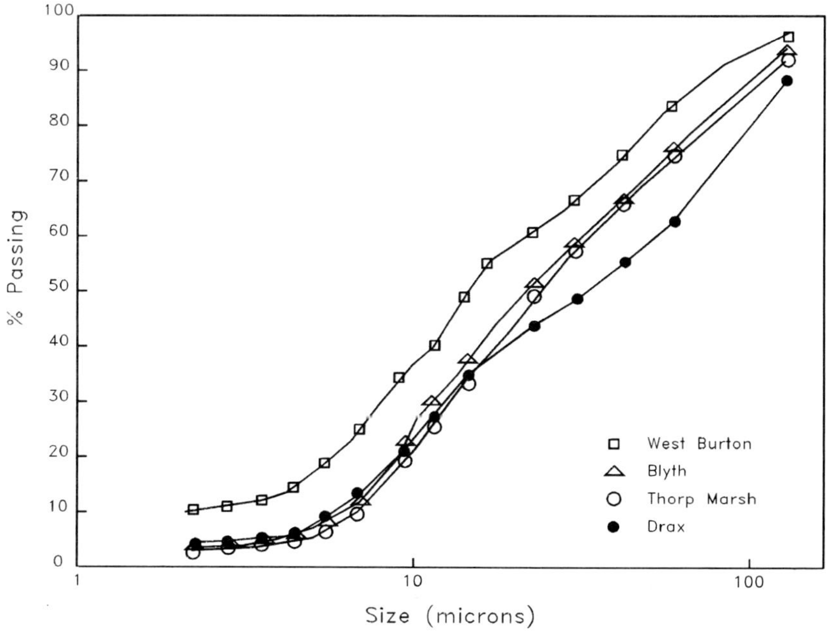

Figure 1. Particle size distribution of PFA fillers used in the investigation.

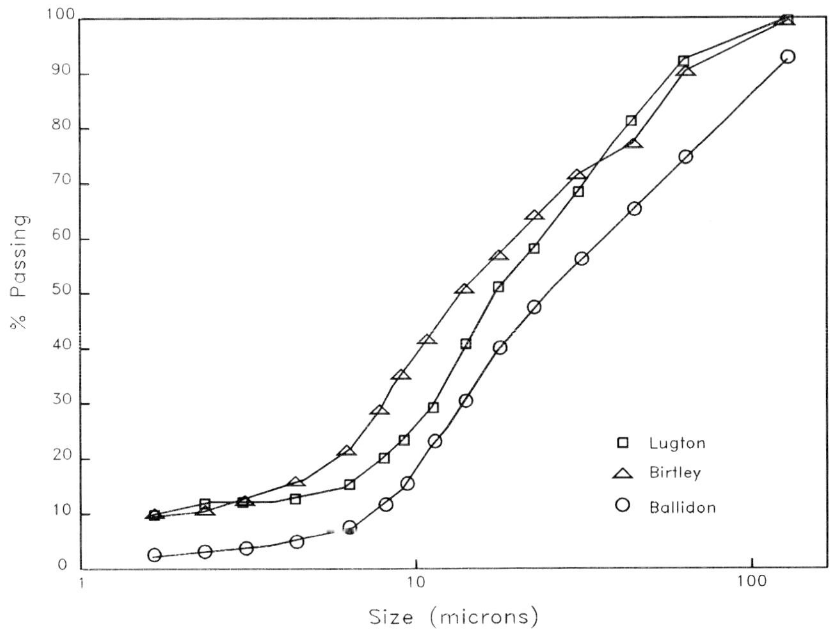

Figure 2. Particle size distribution of limestone fillers used in the investigation.

Table 2.8 Particle Size Distribution (Microns); Limestone Fillers

Size Band			Cumulative Wt. below = % Passing		
Upper	**Lower**	**Mid Range**	**Scottish**	**Marsden**	**Ballidon**
188	87.2	137.6	93.5	100	100
87.2	53.5	70.35	76.5	93	91.8
53.5	37.6	45.55	65.4	81.8	78
37.6	28.1	32.85	55.5	69.3	71.1
28.1	21.5	24.8	47.5	58.7	63.6
21.5	16.7	19.1	41.4	50.7	56.9
16.7	13	14.85	34.2	40.7	51.1
13	10.1	11.55	26.6	31.2	42.7
10.1	7.9	9	20.8	26.6	35.9
7.9	6.2	7.05	13.9	19.7	27.5
6.2	4.8	5.5	8.7	14.2	19.6
4.8	3.8	4.3	5.9	13	14.7
3.8	3	3.4	4.5	12.8	12.7
3	2.4	2.7	3.6	12.4	11.4
2.4	1.9	2.15	2.4	10.2	9.7

Table 2.9 Filler Diameters, Specific Gravities and Surface Areas

Filler Type	Mean Diameter (micron)	Specific Gravity g/cc	Surface Area (m^2/g)	Calculated Surface Area (m^2/g)	Shape Factor
Drax PFA	14	2.179	0.196	0.215	2
Blyth PFA	10	2.412	0.248	0.213	3
Thorpe Marsh PFA	11	2.249	0.242	0.218	3
West Burton PFA	9	2.384	0.279	0.233	3
Ballidon Filler	7	2.773	0.309	0.217	3
Marsden Limestone	7.4	2.824	0.287	0.171	3
Scottish Limestone	5.4	2.725	0.407	0.180	3

2.3.3 Binder

The bitumen properties measured are shown in Table 2.10.

Table 2.10 Properties of Bitumen

Penetration 25 °C (×0.1mm)	52
Softening Point (R&B) °C	53
Relative Density	1.029
Penetration Index	-0.20

2.3.4 Observations on the Properties of the Materials Tested

There is a marked difference between the two types of filler. Limestone fillers are on average finer than PFA fillers and have a higher specific gravity. The shape factor number (1) which is a measure of the sphericity of a particle shows that pfa is predominantly spherical in shape while limestone is not. This characteristic of PFA allows it to function as a filler in a solid-liquid or solid-plastic composite without unduly increasing the viscosity of the composite.

The particle size distribution and measured surface areas of the PFA fillers show that Drax PFA is the coarsest of the fillers and that it exhibits a low measured surface area. This result indicates that the Drax PFA is less effective in terms of enhancement of the workability properties of the HRA. Scottish limestone filler on the other hand is the finest filler in terms of measured surface area but it's calculated surface area is the lowest of all fillers. This is an indication of the irregular nature of the particles which can give problems regarding the effect on the workability of the filler-bitumen system.

The properties of the coarse aggregates and sands are within the expected range for materials used in the manufacture of HRA. The bitumen properties are similarly those corresponding to a typical bitumen within the specification range for the manufacture of HRA.

2.4 Preparation of Aggregate Blends

The proportions of coarse aggregate, sand and filler required to produce size distributions within the specifications given in **BS 594: Part 1:1985**, were :

Coarse Aggregate	:	Fine Aggregate	:	Filler
34%	:	56%	:	10%

An example of the resultant particle size distribution for mix 2 is shown in Figure 3 and the quantitative values of particle size distribution for all the combinations shown in Table 2.11 are given in Table 2.12. These combinations labelled M1 to M8 are the aggregate frameworks used for the preparation of the eight HRA mixes of this project.

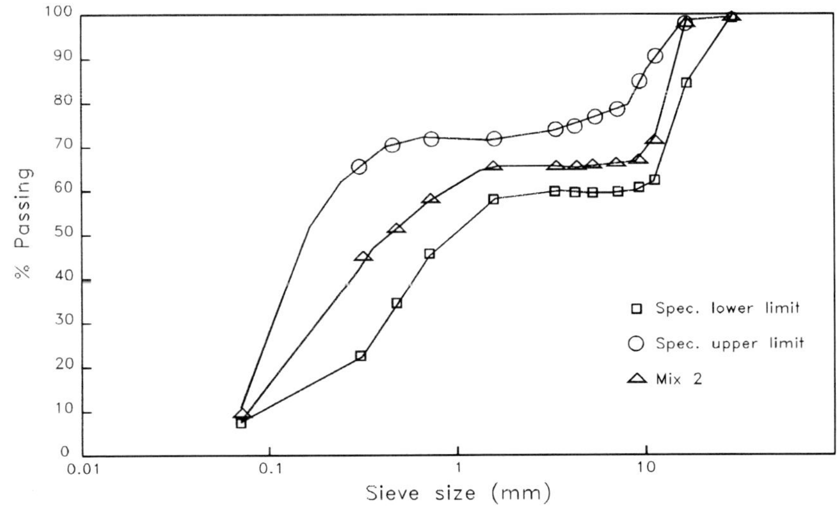

Figure 3. Particle size distribution for mix.2.
Coarsc and finc aggregates = Birtley, filler = Birtley limestone.

Table 2.11 Code and Combination of Mineral Aggregates to produce the blends for the investigation

Mix Code	Coarse Aggregate	Sand	Filler
M1	Northumberland Basalt	Blaydon asphalt sand	Blyth PFA
M2	Northumberland Basalt	Blaydon asphalt sand	Marsden Limestone
M3	Cairneyhill Basalt	Douglasmuir asphalt sand	Drax PFA
M4	Cairneyhill Basalt	Douglasmuir asphalt sand	Scottish Limestone
M5	Mancetter Basalt	Meriden Fine asphalt sand	West Burton PFA
M6	Mancetter Basalt	Meriden Fine asphalt sand	Ballidon Limestone
M7	Yorkshire Gritstone	Tyrham Hall asphalt sand	Thorpe Marsh PFA
M8	Yorkshire Gritstone	Tyrham Hall asphalt sand	Ballidon Limestone

The combination of coarse aggregate, fine aggregate and filler for all mixes resulted in particle size distributions slightly different but within the Specification Limits in shown Table 2.12.

Table 2.12 Particle Size distribution of the eight mixes used in the study

Sieve Size	Mixes 1 & 2	Mixes 3 & 4	Mixes 5 & 6	Mixes 7 & 8	Specification Limits
			Percentage Passing		
20 mm	100	100	100	100	100
14 mm	97.98	99.42	98.67	95.51	85 to 100
10 mm	72.47	76.71	76.45	69.32	60 to 90
8 mm	66.72	67.50	67.72	66.33	57 to 86
6.3 mm	66.35	66.14	66.41	66.25	57 to 81
5 mm	66.27	66.07	65.81	66.05	57 to 77
3.35 mm	66.18	65.91	63.67	63.31	58 to 73
2.36 mm	66.04	65.34	62.68	62.42	60 to 72
1.18 mm	64.82	63.27	61.85	61.88	57 to 73
600 μm	58.02	58.12	60.82	60.92	45 to 72
425 μm	51.34	51.35	57.00	59.48	34 to 68
300 μm	46.03	42.08	44.55	48.77	23 to 60
pfa 75 μm	8.92	8.00	9.18	7.88	8 to 12
Limestone 75 μm	10.29	7.95	9.88	8.02	8 to 12

3 Method of Testing

3.1 Sample Preparation

The dried aggregate blend was placed in containers and heated in an oven to the required mixing temperature. The mixing and compaction temperatures used in the project are given in Table 3.1.

The dry hot mineral aggregate blend was then mixed mechanically in a preheated mixer at the required mixing temperature for approximately 1.5 minutes, the hot fluid bitumen was subsequently added and mixing continued until all aggregate were fully coated with the bitumen and there was no visible signs of uncoated particles.

Table 3.1 Mixing and Compaction Temperatures

Code	Mixing Temperature °C	Compaction Temperature °C
T1	140	125
T2	140	115
T3	130	115
T4	130	105
T5	120	105
T6	120	95
T7	110	95
T8	110	85

3.2 Mixing and Compaction

The samples were compacted in the laboratory, using the Gyratory Testing Machine (GTM) (3). The main characteristic of the GTM compactor is that it facilitates the application of an axial static pressure at the same time that the specimen is subjected to a dynamic shear 'kneading motion'. The compactors that can apply a combination of static-dynamic energy are probably the ones that have most resemblance to the mode of energy applied in the field by construction plant.

For each combination of mixing-compaction temperature, four samples were prepared using the GTM for both compaction and measurement of the Workability Index. The compaction conditions in the GTM were :

Vertical pressure	0.7 MPa	Angle of Gyration	1°
No. of revolutions	30.		

These conditions give an energy of compaction of the same order as the energy of compaction applied by 50 blows of a Marshall hammer.

3.3 Testing of Specimens

3.3.1 Workability

The problem of measuring the workability of bituminous mixes is not a recent one. Marvillet and Bougault (6), for example proposed a method based on the measurement of the torque required to mix the mineral aggregates with bitumen. More recently Fordice (4) developed a procedure based on the parameters obtained from the triaxial test of bituminous mixtures. In this study the method used to assess workability is the one developed by Cabrera (5) which briefly consists of the following steps :

a) Monitor the specimen height reduction during the compaction process by reading the height control gauge of the GTM at 5 revolution intervals.

b) Use the heights registered to calculate the volume of the specimen and its porosity at 5, 10,15, 20, 25, and 30 revolutions. Use the following formulae .

$$V_i = \frac{10.12^2}{4} \times \pi \times h_i$$

where V_i = Volume of specimen at i revolutions (cm^3).
h_i = Height of specimen at i revolutions (cm).

$$D_i = \frac{W_a}{V_i}$$

D_i = Density at i revolutions (g/cm^3).
W_a = Weight of the specimen in air (g).

$$P_i = \left(1 - \frac{D_i}{SG}\right) \times 100$$

Pi = Total porosity at i revolutions (%).
SG = Specific gravity of specimen.

The SG for each specimen is obtained using the following formula :

$$SG = \frac{100}{\frac{P_{wa}}{SG_a} + \frac{P_{ws}}{SG_s} + \frac{P_{wf}}{SG_f} + \frac{P_{wb}}{SG_b}}$$

Pw = percentage weight in mix.
a = coarse aggregate
s = sand
f = filler
b = bitumen.

A graph is plotted relating Pi with the Log_{10}[number of revolutions (i)]. The experimental plot should approximate a linear relation of the form :

$$P_i = A - B \log_{10}(i)$$

where : A, B are constants,
A = intercept with the y axis.
B = slope of the line.
i = number of revolutions.

The Workability expressed by the "Workability Index" (W.I.) is defined as the inverse of the constant A, i.e. the porosity at zero revolutions.

$$W.I. = \left(\frac{1}{A}\right) \times 100$$

3.3.2 Density

Densities were obtained according to BS 598 (2).

3.3.3 Marshall Stability and Flow

Stabilities and Flows were obtained according to BS 598 (2).

3.3.4 Permeability:

Permeability measuring techniques of bituminous materials are mostly based on ideas originally developed for measuring the permeability of mortars and concretes using differential pressure techniques, (7). Permeability was measured in this study using the Leeds Air Permeameter. This is a non-destructive test which allows the determination of permeability in a very short period (8)

The apparatus consists of :

a) A steel mould.
b) A water container of sufficient volume.
c) A manometer to control differential pressure.
d) A graduated cylinder of 50ml capacity.
e) Two rubber membranes of constant internal diameter (100mm).
f) A Stopwatch.
g) Silicon grease for sealing.

In the laboratory, the two rubber membranes are placed inside the mould and folded over the top and bottom edges of the mould. This is to ensure sealing between the specimen, the mould and the cup. The air trapped between the membranes and the mould is removed by suction through a pipe fixed to the middle of the mould. A very thin layer of silicon grease is applied around the cup, one membrane is folded over the cup and the other over the mould to secure sealing the specimen. The mould assembly is then placed over a perforated base. A schematic diagram of the Leeds Permeameter is presented in Figure 4

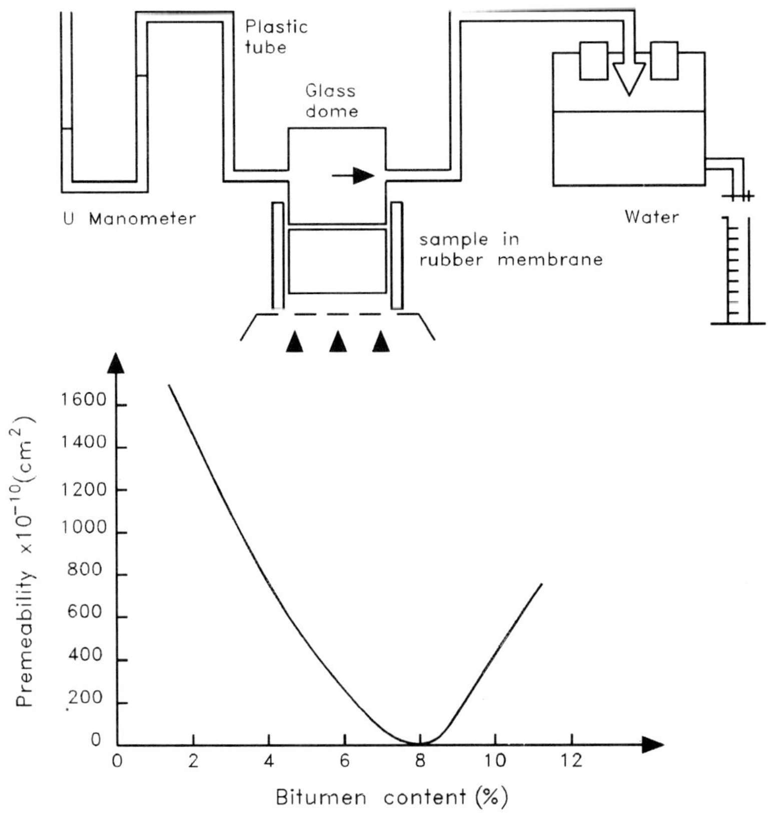

Figure 4. Permeability measurement

A pressure difference is obtained by opening the valve of the water container. After the pressure has stabilised, the time taken for 50ml of water flowing from the container into a graduated cylinder is registered. This procedure is repeated three times or more to obtain a representative flowing time.

The results are given in units of intrinsic permeability K in cm^2.

$$K = \frac{\mu \cdot V \cdot L}{A \cdot \delta P \cdot \delta t}$$

where A = Cross sectional area of the sample.
K = Permeability (cm^2).
μ = Viscosity of fluid air (poise).
V = Volume of fluid passing (cm^3).
L = Length of porous medium (cm).
δP = Differential pressure (dynes per cm^2).
δt = Time taken for fluid volume V to pass through the porous medium (in seconds).

For specimen with 101.2 mm diameter, air viscosity = 1.85×10^{-4} poise at 20°C, and volume of fluid of 50 cm^3, the equation is :

$$K = \frac{1.143 \times 10^{-4} \times L}{\delta P \times \delta t}$$

The intrinsic permeability value can be corrected to coefficient of permeability by multiplying K by a constant :

$$k = K \times 10^5 \text{ where k is in cm/sec.}$$

3.3.5 Creep

The creep test is carried on duplicate specimens at 40°C using the "Canik" static creep testing machine developed at Leeds University (9). The test lasts two hours, and gives results which allow the characterisation of the mixes in terms of their long term deformation behaviour.

The conditions under which the test is carried out are :
a) Test temperature 40°C.
b) Preloading for 2 minutes at 0.001 MPa.
c) Constant stress during test equal to 0.1 MPa.
d) Duration of test : 1 hour loading and 1 hour unloading.

4 Determination of the optimum bitumen content (o.b.c.)

The Leeds Design Method (3), recommends that the optimum binder content should be obtained by averaging the binder contents corresponding to the following parameters :

1 Maximum Stability.
2 Maximum Density.
3 Minimum Voids in the Mineral Aggregate.
4 Maximum Compacted Aggregate Density.
5 Minimum Permeability.
6 Maximum Stiffness.

The optimum value obtained should lie within 3 - 5% porosity and below 4 mm Flow.

In theory, parameters 2,3 and 4 should give the same o.b.c., therefore in this study, the o.b.c. for all mixes was obtained by averaging the o.b.c. for parameters 1,4,5, and 6. The o.b.c's for the mixes prepared at different temperatures of mixing and compaction were then averaged and the results used as the o.b.c. for each mix combination independently of the temperature of mixing and compaction. An example of the values of o.b.c.'s for mixes M3 and M4 are shown in Table 4.1. It can be seen that the values are on average equal to 7% for all temperatures.

Mixes 1 and 2 were selected for a field trial as a second phase of this project. This was reported by Rockliff (10).

Table 4.1 Example of the procedure to find the optimum binder content

Mix 3	T1	T2	T3	T4	T5	T6	T7	T8	
C.A.D.	7.0	7.5	6.5	7.0	7.5	7.5	7.0	6.5	
Perm.	6.0	6.0	6.5	7.0	6.0	6.5	6.5	7.5	
Stab.	6.5	7.0	7.0	7.5	7.5	7.5	6.5	6.5	
1h. Stiff.	7.0	7.0	7.5	7.5	7.5	7.5	7.0	7.5	
Average	6.5	7.0	7.0	7.0	7.0	7.5	7.0	7.0	o.b.c.= 7%
Mix 4	**T1**	**T2**	**T3**	**T4**	**T5**	**T6**	**T7**	**T8**	
C.A.D.	7.0	7.5	7.5	7.0	6.5	7.0	7.0	7.5	
Perm.	6.5	6.5	6.5	7.0	6.5	7.5	6.5	7.0	
Stab.	7.0	7.0	7.0	7.0	7.5	7.5	7.5	7.5	
1h. Stiff.	7.0	7.5	7.0	7.0	7.0	7.5	7.0	7.0	
Average	7.0	7.0	7.0	7.0	7.0	7.5	7.0	7.5	o.b.c.= 7%

5 Analysis of Results

5.1 Stability Values

In general, Stability values decrease as mixing and compacting temperatures decrease. In all cases, the Stability of the conventional Hot Rolled Asphalt mixes were only slightly higher than their counterpart PFA mixes. Nevertheless, the stability of all mixes satisfy the criteria for roads with medium traffic (up to 6000 vehicles/lane/day) See Tables 5.1 and 5.2 for the stability and flow design parameters.

Also M1 (Northumberland Basalt, and Blyth PFA) see Figure 5, and M7 (Yorkshire Gritstone, Thorpe Marsh PFA) satisfy the requirement for heavy traffic at almost all temperatures of mixing and compacting.

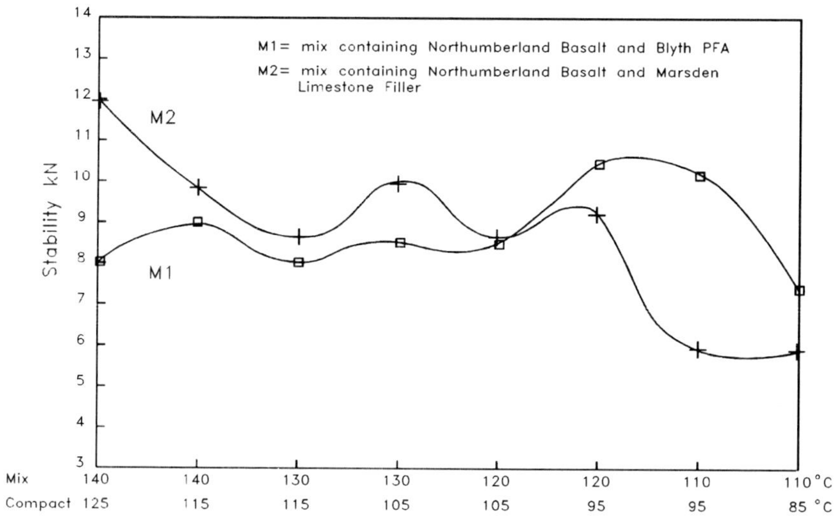

Figure 5. Stability values versus mixing and compacting temperatures.

Table 5.1 Criteria for the Stability of laboratory designed asphalt. BS 594 : Part 1 : 1985

Traffic (commercial vehicles per lane per day)	Marshall Stability of complete mix (kN).
Less than 1500	2 to 8
1500 to 6000	4 to 8
Over 6000	6 to 10

Table 5.2 Asphalt Institute Design Criteria

Traffic	Light	Medium	Heavy
Compaction	2 × 35	2 × 50	2 × 75
Stability (kN)	3.33	5.33	8.00
Flow, (0.25mm)	8 - 18	8 - 16	8 - 14
Porosity (%)	3 - 5	3 - 5	3 - 5

Note; The American Asphalt Institute imposes much higher Stability values than the British Standards. This could be attributed to the fact that the properties of the denser continuously graded American Asphaltic Concrete allow higher strengths to be used. The British gap graded HRA on the other hand does not use the aggregate-aggregate interlock to carry the traffic, it uses the stiffness of the bitumen-filler interface to perform the same job. Hot Rolled Asphalt does have the advantage of better fatigue performance and the flexibility for controlling the surface texture of the wearing course.

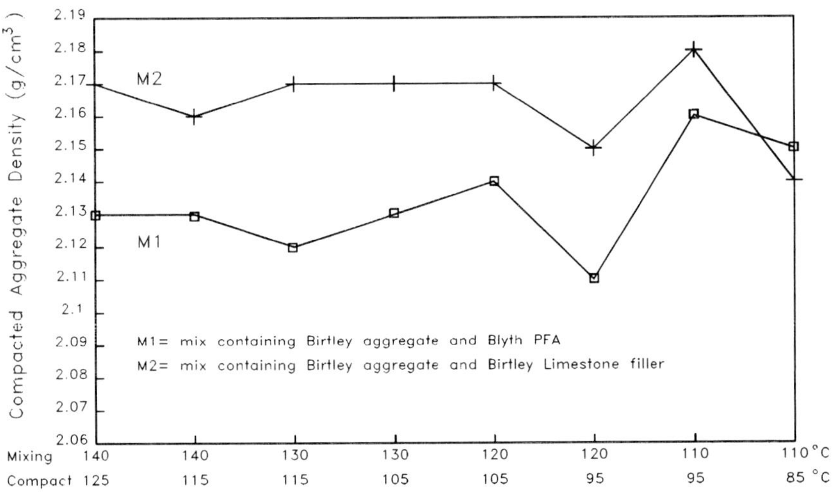

Figure 6. Compacted aggregate density (CAD) versus mixing and compacting temperatures.

5.2 Flow values

None of the flow values measured exceed 4 mm except mix 2 (Northumberland Basalt, Marsden Limestone) at temperature T6 (mixing 120°C, compacting 95°C) and T7 (mixing 110°C, compacting 95°C).

PFA mixes exhibit consistently lower flow values than their counter part conventional mixes at all mixing and compacting temperatures.

5.3 Compacted Aggregate Density (CAD)

The values of CAD for all PFA mixes at all mixing and compacting temperatures were slightly lower than the values of the conventional mixes, see Figure 6. One exception being mix 5 (Mancetter Basalt, West Burton PFA).

PFA is a naturally lighter material than limestone, hence it occupies more volume per unit weight, therefore causing lower CAD values.

5.4 Voids in Mineral Aggregate (VMA)

There is no marked change in VMA values as temperature of mixing and compaction decrease. Also for each aggregate type, both conventional and PFA mixes do not exhibit a great change in VMA values within the range of testing temperature, see Figure 7.

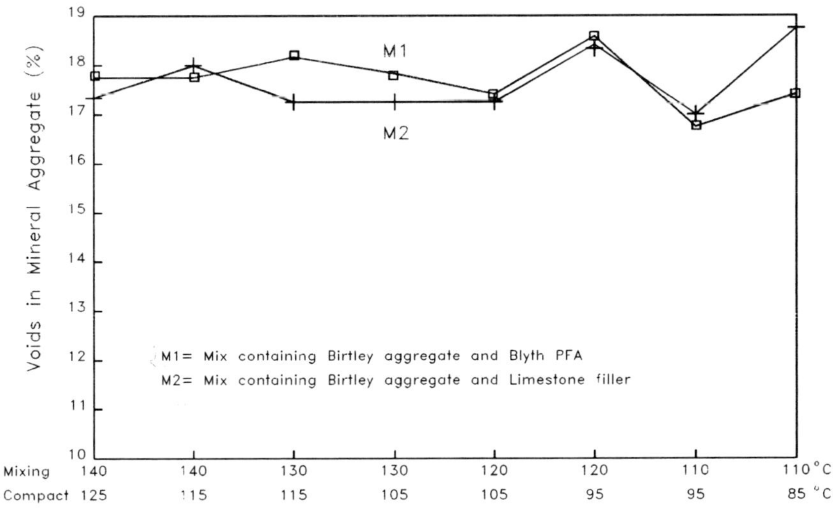

Figure 7. Voids in mineral aggregate (VMA) versus mixing and compacting temperatures.

5.5 Porosity

Except for mix 1 (Northumberland Basalt, Blyth PFA), all mixes containing PFA filler showed lower porosity values than their conventional filler mix counterpart. In general porosity values were low, (< 6%), see Figure 8.

5.6 Workability Index Values (W.I.)

As expected W.I. values decrease as temperature decreases due to the increase in bitumen viscosity as the softening point is approached. Most importantly, all PFA mixes at all mixing and compacting temperatures exhibit higher W.I. values than conventional HRA mixes, example Figure 9. This can be attributed to the fact the PFA particles have more rounded and less angular texture aiding workability. In some cases the gap is quite large. Example mix 5 (Mancetter Basalt, Westburton PFA) and mix 6 (Mancetter Basalt, Ballidon Limestone). This is clear evidence that PFA mixes will compact better than limestone even at the lowest temperature of compaction used in the laboratory.

5.7 Creep Stiffness

Analysis of creep test results carried out on the hot rolled asphalt containing PFA filler show the normal variation in stiffness of mix values with respect to stiffness of bitumen. The stiffness of the bitumen was obtained from a Van der Poel nomograph. The nomograph gives values of stiffness as a function of the time of loading, the temperature difference between test conditions, the Softening Point temperature, and the Penetration Index.

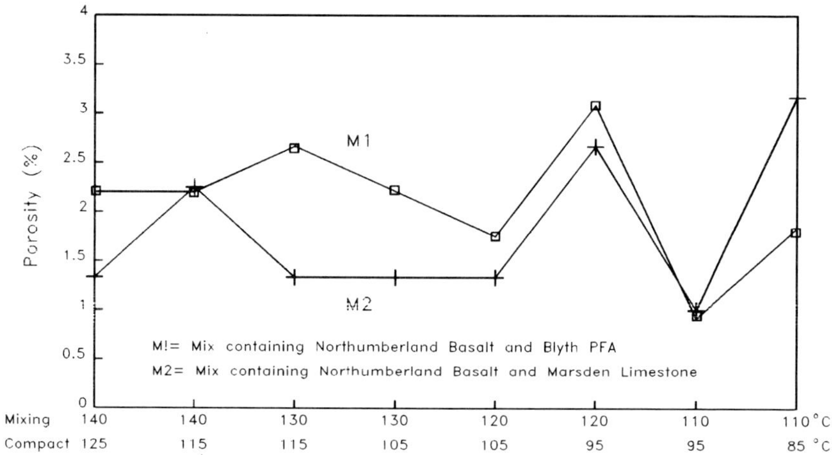

Figure 8. Porosity values versus mixing and compacting temperatures.

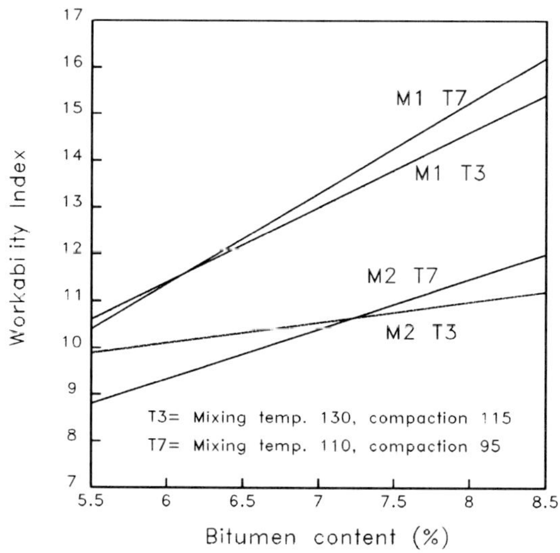

Figure 9. Workability index values versus bitumen content for mixes M1 and M2 at temperatures T3 and T7.

From the S_{mix} - S_{bit} experimental values, regression lines were obtained.

These regression equations are of the form :

$$\text{Log Smix} = \text{X Log Sbit} + \text{C}$$

They were used to obtain the S_{mix} values at one hour loading time. Figure 10 shows an example for mixes 1 and 2.

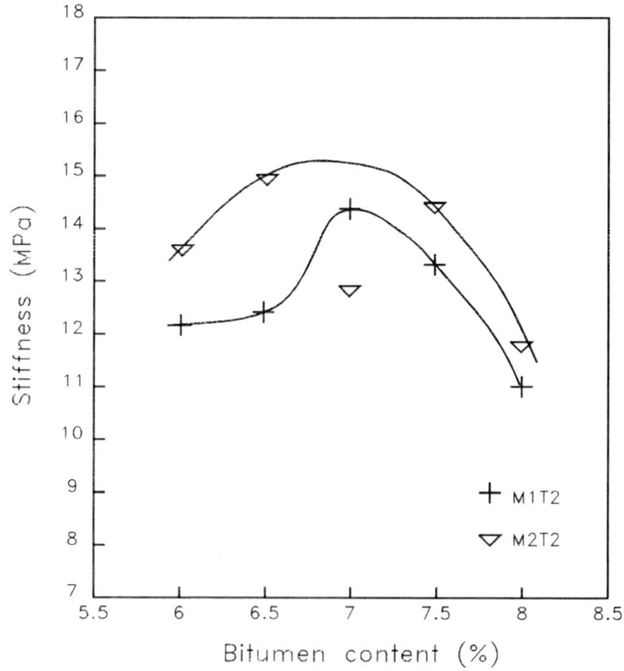

Figure 10. Stiffness of mix (MPa) versus bitumen content.

7 Conclusions

From the results obtained in the Laboratory and reported in this study the following conclusions are offered :

1 PFA filler hot rolled asphalt has far higher workability index than conventional hot rolled asphalt for any of the aggregate combinations used. This finding implies that hot rolled asphalt containing PFA can be mixed and compacted at temperatures as low as 110°C - 85°C respectively without impairing its engineering and performance properties.

2 The savings in energy input are considerable and thus PFA-HRA can be classed as a low energy material.

3 Replacement of limestone filler with PFA does not affect the optimum bitumen content of Hot Rolled Asphalt.

4 The Stability and Flow of the PFA mixes satisfy the criteria for medium traffic (up to 6000 CVd), laid down by the Ministry of Transport and the Asphalt Institute of U.S.A. for the range of temperatures tested.

5 All other properties measured show that at lower temperatures the PFA mixes exhibit marginally better properties.

6 The successful outcome of this study allowed us to proceed to a field trial which is reported in another paper during this conference.

References

1 Cabrera J.G and Hopkins C.J. 'The influence of PFA shape on the properties of concrete. Ash Tech 84, Second International Conference on Ash Technology, pp. 393-398, London.

2 BS 598: Part 107: 1990, Sampling and examination of bituminous mixtures for roads and other paved areas.

3 Cabrera J.G. 'Hot bituminous mixtures: Design for performance'. 1st National Conference on Bituminous Mixtures and Flexible Pavements. University of Thessaloniki, Greece, pp 1-12, 1992.

4 Fordice D. and Al Nageim H.E. 'Triaxial test procedure to predict the voidage within wearing coarse HRA'. Proc. Conf. on Durability and Performance of Bituminous Highway Materials. Hatfield Polytechnic, April 1989.

5 Cabrera J.G. 'A new method for the assessment of the workability of bituminous mixtures'. Highway and Transportation No. 11, pp 17 to 23, 1991.

6 Marvillet J. and Bougault P. 'Workability of bituminous mixes. Development of a workability meter'. Proc. Ass. of Asphalt Paving Technologists Vol. 48, pp 91 -110, 1979.

7 Cabrera J.G. and Lynsdale C.J. 'A new Gas Permeameter for measuring the Permeability of Mortar and Concrete'. Magazine of Concrete Research, Vol. 40, No. 144, Sept. 1988.

8 Cabrera J.G. and Hassan T.Q.M., 'Quality Control During Construction of Bituminous Mixtures using a simple air permeability test', 1st National Conference on Bituminous Mixtures and Flexible Pavements. University of Thessaloniki, Greece, pp 191-200.

9 Cabrera J.G. and Nicholaides A.F., 'Creep performance of cold dense bituminous mixtures'. Highways and Transportation, Vol. 35, no.10, pp. 7-15, 1988.

10 Rockliff D. 'The use of pulverised fuel ash as a filler in hot rolled asphalt mixtures-practical aspects'. Proc. Performance and Durability of Bituminous materials Leeds, March 1994.

28 THE USE OF PULVERISED FUEL ASH AS A FILLER IN HOT ROLLED ASPHALT MIXTURES - PRACTICAL ASPECTS

D. ROCKLIFF
Tilcon Limited, Scotton, Knaresborough, UK

Abstract
Tilcon Limited have supported a research project carried out by the Civil Engineering Materials Unit at the University of Leeds. The primary aim of research is to assess the energy savings which may result from the use of pulverised fuel ash as a filler in hot rolled asphalt wearing course mixtures. Laboratory evaluation of four typical "design type" hot rolled asphalt mixtures established the validity of the proposal for the range of aggregate types used in four coating plants from the English Midlands to Central Scotland. The production stage of the project involved the production and laying of material on a Principal Road controlled by Cleveland County Council. The material was divided between a control mixture with ground limestone filler, and the trial mixture. The paper gives details of the energy monitoring which took place during production, suggesting a potential reduction in energy use of around 10%. Details of the evaluation of the trial site are given, including delivery and rolling temperatures, rate of spread of chippings and measured texture depth. The paper concludes that hot rolled asphalts with PFA filler are worthy of further consideration.
Keywords: Aggregates, Energy, Hot rolled asphalt, Limestone filler, Pulverised fuel ash, Rolling temperature, Texture depth.

1 Introduction

Energy use and its associated costs are closely monitored by all well managed companies. The manufacture of bituminous materials is an energy intensive process, and constant efforts are made to control the consequent costs.

Typically, energy accounts for about seven percent of the cost of hot rolled asphalt wearing course. An aggregate temperature of up to 180°C is used to ensure that moisture is removed from the sand and coarse aggregate, and to ensure that the 50 pen. grade bitumen is at a viscosity suitable for mixing.

The temperature of manufacture also affects the length of time available for satisfactory placing of the material at the laying site. Daines (1985) identified the following factors as affecting the time available for a mixture to be properly laid and compacted:

Performance and Durability of Bituminous Materials. Edited by J.G. Cabrera and J.R. Dixon.
Published in 1996 by E & FN Spon, 2-6 Boundary Row, London SE1 8HN. ISBN 0 419 20540 3.

Weather - wind speed and air temperature.
Thickness of the layer.
Initial laying temperature - at the paver screed.
Minimum compaction temperature - when rolling is complete.

Daines also indicates that compaction time will be reduced by 40% if the initial laying temperature falls from 160°C to 140°C; but it is increased by 30% if the temperature above which compaction can be substantially completed is lowered from 100°C to 90°C.

As a producer of bituminous mixtures must recognise the needs of the paving crew and the ultimate customer, there is little scope for reducing production temperatures unless a comparable material can be found with better compactability at lower temperatures.

Cabrera (1991) suggested that the use of Pulverised Fuel Ash (PFA) in place of ground limestone filler could form the basis of a hot rolled asphalt wearing course mixture with improved workability. The concept was considered worthy of further research, if the improved workability could provide handling characteristics equivalent to those of conventional mixtures, but at lower temperatures.

A research project was established by the Civil Engineering Material Unit at the University of Leeds to investigate mixtures with PFA as a filler. The investigation is reported by Cabrera and Zoorob (1994). The work has been supported by the Energy Efficiency Office of the Department of Environment, National Power plc, Cleveland County Council and Tilcon Limited.

This paper sets out the practical aspects of the manufacture of a trial quantity of the mixture, and its subsequent laying and compaction on a Principal Road in the County of Cleveland.

2 Laboratory investigation

Laboratory evaluations of PFA as a filler in asphalt mixtures have been carried out a number of times. National Power plc, hold unpublished work carried out by the Road Research Laboratory in 1953, which investigated the use of bottom ash as a substitute for mineral aggregate. This prompted other work, which led to a road trial in Nottinghamshire in 1960. The primary aim of this work was to use power station ashes as a direct substitute for natural aggregates at conventional temperatures.

Between 1977 and 1984, the Civil Engineering Materials Unit at the University of Leeds carried out studies into the effects of PFA on the properties of bituminous materials. This work established that the main benefit of using PFA was that it improves the workability characteristics of bituminous mixtures.

The recent project extends this work. The first stage established that the concept of improved workability at lower temperatures was valid over a range of typical hot rolled asphalt wearing course mixtures, using samples taken from four Tilcon coating plants, as set out in Table 1.

The evaluation used the Leeds Design Method for bituminous mixtures (1990), over a range of mixing and compaction temperatures. Full details of the work are given by Cabrera and Zoorob (1994).

The work indicates that PFA filler mixtures have an optimum binder content which is about 0.5% less than the equivalent limestone filler mixture, and that the Stability of the PFA mixtures is less sensitive to the mixing and compaction temperatures used.

Table 1. Aggregates and fillers used in the laboratory evaluation

Coarse aggregate	Fine aggregate	L'st filler	PFA filler
Midlands basalt	Midlands sand	Ballidon	Thorpe Marsh
Yorks. gritstone	Doncaster sand	Ballidon	Drax
N'land whinstone	Tyne & Wear sand	Marsden	Blyth
Scottish basalt	Strathclyde sand	Lugton	Blyth

3 Production of the trial site materials

A trial batch of seventy tonnes of Designation 30/14 design type hot rolled asphalt wearing course with PFA filler was manufactured on Thursday 11th April 1991, at the Tilcon Coating Plant in Blaydon, Tyne and Wear.

Blaydon Plant is a modern conventional asphalt plant fitted with computer based controllers. A separate dryer feeds "hot" screens and a batch mixer with a nominal capacity of two tonnes.

The trial batch was mixed as part of the deliveries to the road trial site in Cleveland. The mix components were chosen to match the mixture tested in the laboratory, as follows:

Coarse aggregate - Northumberland basalt (whinstone).
Fine aggregate - Blaydon asphalt sand.
Filler - Control wearing course - Marsden dolomitic limestone;
- PFA wearing course - Blyth Power Station PFA.

The binder contents chosen were as follows:

Bitumen - Shell Bitumen, 50 pen grade. - Control - 7.7 %
PFA - 7.2 %

The control wearing course is an established Designation 30/14 design type asphalt, used widely in the North East of England. The binder contents chosen are slightly higher than those suggested by the laboratory work, but reflect the local clients' desire to ensure long term durability.

The actual production sequence was as follows:

Control HRA at about 160°C - 30 tonnes
PFA filler (11%) HRA at about 130°C - 30 tonnes
PFA filler (9%) HRA at about 140°C - 40 tonnes
Control HRA at about 170°C - 50 tonnes

During the first load of PFA wearing course, problems were encountered with the filler screw-feed system. Manual rodding had to be used to maintain the flow of PFA. This phenomena is not unknown on PFA silos on ready mixed concrete plants and can be overcome by simple mechanical modifications.

Attempts to match the feed rate of the sand with the slow feed rate of the filler, and the need to ensure that the sand was properly dry, resulted in the dryer burner controller becoming unstable below about 135°C. To ease the consequent production problems - the plant had a full order book on the day of the trial - the dryer temperature was raised by 10°C and the filler content dropped by 2% after the first 30 tonnes. The second 40 tonnes were manufactured at a more satisfactory rate.

4 Mixture composition analysis

Two samples were taken from both the Control and the PFA mixtures, for analysis using the binder direct/filler by difference method described in Clause 4.2 of British Standard BS 598 : Part 102 (1989).

The analysis of all four samples confirmed compliance with the specification for Designation 30/14 mixtures set out in BS 594 (1985). Variability between individual results was well within that expected for a modern coating plant.

The results are summarised in Table 2., using the average of each pair.

Table 2. Mixture composition analysis results

% passing sieve	Control	PFA	Specification
20 mm.	100	100	100
14	99	98	85 - 100
10	73	73	60 - 90
2.36	66	64	60 - 72
600μm	59	60	45 - 72
212	30	27	15 - 60
% binder content	7.8		7.1 - 8.3
		7.3	6.6 - 7.8
% filler content	9.5	9.5 *(1)*	8.0 - 12.0
		7.7 *(2)*	8.0 - 12.0

For the PFA mixtures, the binder and filler contents are slightly in error because a proportion of PFA particles have a particle density which is lighter than that of methylene chloride. The problem can be overcome by using one of the filler direct (pressure filter) methods of analysis described in Clauses 4.1 and 4.2 of the Standard.

5 Energy use evaluation

An energy use evaluation was carried out under the supervison of Tilcon's Group Engineer, who controls a database of energy use information for all of the Company's coating plants.

Gas meter readings were recorded at five minute intervals to give details of the energy used in drying and heating the aggregate. The electrical power consumption of the mixer was monitored using a recording ammeter, which sampled the power load at about twenty second intervals.

The gas use evaluation was inconclusive. Long term energy monitoring suggests that a saving of 8% could be expected to result from the lower mixing temperature - equivalent to about 12p per tonne. However, the burner instability caused by the problems with the low feed rate of PFA masked any beneficial effects.

The evaluation of the power consumption of the mixer showed the PFA mixture to require about 7% more power - equivalent to about 0.5p per tonne. Further study showed a close agreement between the mixer power consumption for the PFA mixtures and a conventional mixture at the same temperature. This indicates that the viscosity of the bitumen is the dominant factor in power consumption during the mixing stage.

6 The road trial site

The road trial site chosen by Cleveland County Council is the eastbound slow lane of Principal Road A689 between Wolviston and Billingham in Cleveland. This dual carriageway road links the Motorway A1(M) with the Trunk Road A19 and the industrial complexes of Teesside. Each weekday, the road carries around 4,500 commercial vehicles in each direction. Many of the commercial vehicles are five or six axle articulated units.

Stockton Borough Council, acting as agents for Cleveland County Council, let a maintenance contract for the cold planing and replacement of the wearing course over about a 1.5 km. of the dual carriageway. The road trial represented one morning of about two weeks' work.

The hot rolled asphalt wearing course was laid at a nominal thickness of 40 mm. The morning of the trial was a dry, warm but breezy spring day with average wind speeds in the range 5 to 7 m/s. The paving train consisted of a conventional Blaw Knox tyred paver, a Bristows chipper fed with 20 mm. nominal size precoated chippings, and two Aveling Barford three point deadweight rollers.

Temperatures at all stages in the laying procedure were monitored using thermocouple-type electronic thermometers. The rate of spread of the precoated chippings was also measured, using the procedure described in British Standard BS 598 : Part 108 (1990). The results of the monitoring are summarised in Table 3. The measurement of texture depth was carried out using a laser texture meter as described in British Standard BS 598 : Part 105 (1990), during the initial evaluation of in-service performance (Section 7 of this paper).

British Standard BS 594 : Part 2 (1987) recommends that rolling should be substantially completed at a temperature above 85°C. To comply with this requirement, common practice accepts a minimum paver delivery temperature of 130°C, and an minimum initial rolling temperature of 125°C.

Table 3. shows that the PFA mixtures were still receptive to chippings at much lower temperatures, and that some slight excess rolling may have occurred. The values of Sensor Measured Texture Depth (SMTD) are close to the minimum value of 1.03, although it should be noted that this method is only now used as a quality control tool because of its uncertain precision.

Table 3. Monitoring of the road trial

	Control.1	PFA(11%)	PFA(9%)	Control.2
Temperature, °C				
- Delivery	155	125	135	160
- Paver screw	145	120	125	140
- Initial rolling	130	110	105	130
PCC spread, kg/m²	13	14	13.5	14
Average SMTD	0.98	0.92	0.84	1.00

A short length of the PFA(9%) material was deliberately left to cool before rolling commenced. At an initial rolling temperature of 88°C the mixture was still receptive to chippings. However, the section was subsequently found to have a SMTD of 1.07. A core taken from the section also had a porosity which was about three percent higher than adjacent cores. This area has greater potential for failure due to chipping loss and fretting of the matrix - although there is no evidence so far.

The evidence supports a realistic minimum initial rolling temperature of 95°C, which is 30°C lower than the accepted minimum for limestone filler mixtures.

7 In-service performance

In-service performance of the trial section has been carried out jointly by the Cleveland County Council Materials Laboratory, under the control of Mr G Poad, and the Civil Engineering Materials Unit at the University of Leeds.

The evaluation methods have included:

Taking of cores for evaluation of porosity, stability and flow.
Surface texture measurements.
Rut depth measurement.
Dynamic deflection measurement (deflectograph)

The results of the evaluation may be the subject of future papers. After nearly three years of service the PFA section appears to be performing as well as the adjacent control sections.

8 Conclusions

The practical aspects of the trial were satisfactory. Much to the surprise of the many experienced contracting and client staff who witnessed the road trial, the PFA mixtures could be laid and compacted at reduced working temperatures without changes to working practice and adverse effects on the appearance of the finished wearing course. Energy savings and the potential for reduced amounts of unworkable material can be expected.

To utilise the advantages of PFA filler mixtures to the full, the current specification limits for delivery and rolling temperatures must include an evaluation of the workability of the mixture rather than a consideration of the viscosity of the bitumen in isolation.

PFA filler hot rolled asphalt wearing courses are worthy of further consideration. It is suggested that they could be of particular use where long workability times are advantageous, such the reinstatement of highway openings.

9 Acknowledgements

The author would like to acknowledge the support of the senior management of Tilcon Limited throughout the project, and to thank the staff of Blaydon Coating Plant and Eastern Area Contracting for their efforts during the trial. Special thanks are due to Arthur Ridley and the Eastern Area Technical Team.

10 References

Daines, M.E. (1985) Transport and Road Research Laboratory, Research Report 4, Cooling of bituminous layers and time available for their compaction.

Cabrera, J.G. (1991) Highways and Transportation No. 11, pp 17 - 23. Assessment of the workability of bituminous mixtures.

Cabrera, J.G. and Zoorob, S.E. (1994) Design of low energy hot rolled asphalt.

British Standards Institution, British Standard BS 594, Hot rolled asphalt for roads and other paved areas.
- Part 1 (1985), Specification for constituent materials and asphalt mixtures.
- Part 2 (1985), Specification for the transport, laying and compaction of rolled asphalt.
- (Both now superseded by a 1992 edition)

British Standards Institution, British Standard BS 598, Sampling and examination of bituminous mixtures for roads and other paved areas.
- Part 102 (1989), Analytical test methods.
- Part 105 (1990), Methods of test for the determination of texture depth.
- Part 108 (1990), Methods for measurement of the rate of spread of coated chippings.

29 NEW DEVELOPMENTS IN WEARING COURSES

A. CHRISTIE
Alfred McAlpine Quarry Products Ltd, Billinge, Wigan, Lancashire, UK

Summary

New materials are now being made available to the Road Surfacing Industry for surfacing in all situations from High Speed Motorways to Minor Country roads, and this paper describes the changes taking place and the alternative materials becoming available to the design engineer.

1 Introduction

During the past few years, through the influence of the European Economic Community and an exchange of ideas and technology between the nations, the road construction industry in the United Kingdom has been examining materials used in continental Europe in the construction of the European road network.

The development and introduction of new materials into the U.K takes time, and in the past, for some reason, we in the United Kingdom have had to re-invent the wheel every time we develop new products before they are used, but following a programme of research, trials and tests plus the various environmental pressures being applied to the industry, 1993 saw the application of porous asphalt on our motorway network under a full-scale contract.

For well over 25 years, hot rolled asphalt wearing course, with the addition of pre-coated chippings, has been the traditional material used as a wearing course on all of our high speed motorways and trunk routes, and also a considerable percentage of our minor roads.

In Europe , little or no hot rolled asphalt wearing course is used, and over the same period of time, asphaltic concrete of various qualities has been the traditional material used. However, with improvements in the properties of the binders developed by the oil companies, and in particular the addition of polymers to the bitumen with the resultant enhancement of the properties of the bitumen, European technology has produced a series of new and exciting wearing courses.

Performance and Durability of Bituminous Materials. Edited by J.G. Cabrera and J.R. Dixon. © E & FN Spon. Published in 1996 by E & FN Spon, 2–6 Boundary Row, London SE1 8HN. ISBN 0 419 20540 3.

2 Porous Asphalt

It is therefore appropriate that we start by discussing porous asphalt. Porous asphalt , as with the other materials in this paper, will rely heavily on European experience. The Permanent International Association of Road Congresses (P.I.A.R.C.) recently published a document on porous asphalt, and Table 1 below lists the working group members, with the United Kingdom a notable absentee.

PERMANENT INTERNATIONAL ASSOCIATION OF ROAD CONGRESSES 1993
WORKING GROUP MEMBERS
AUSTRIA
BELGIUM
FRANCE
GERMANY
ITALY
NETHERLANDS
SWITZERLAND

Table 1 - P.I.A.R.C. Working Group Members

This in itself is not surprising considering the minimal production of porous asphalt in the U.K. - see Table 2 - and it was no coincidence either that a Belgian representative was Chairman of the working group.

COUNTRY	% POROUS OF ALL HOT MIXES
BELGIUM	9
NETHERLANDS	6
SWITZERLAND	6
AUSTRIA	5
ITALY	5
FRANCE	4
GERMANY	1
UNITED KINGDOM	MINIMAL

Table 2 - Production of Porous Asphalt (1990)

Porous asphalt is now being laid in the United Kingdom, and its relative merits have been discussed on numerous occasions and in many technical papers, but with the current environmental pressures, it is interesting to see a comparison of the relative noise levels of a number of wearing courses, and to note just how quiet a well-designed porous asphalt can be - see Table 3

Surface	dB (A) External
Sett Paving	82
Concrete	76 - 85
Surface Dressing	74 - 81
Asphaltic Concrete	72 - 79
Porous Asphalt	69 - 76
80 km/hr Dry	
P.I.A.R.C P.120	

Table 3 - Tyre / Road Contact Noise Levels

So why has porous asphalt been slow to develop in the United Kingdom? Table 4 perhaps provides part of the answer. In respect of the United Kingdom figures, the price of the modified binder for porous asphalt may well reduce in the future, but the price levels of the mixed materials may reduce only slightly because of the United Kingdom's insistence on the use of very high P.S.V. aggregates.

COUNTRY	Price / Tonne			
	Modified Binder (1)	Dense Mix with Modified Binder (2)	Porous Asphalt with Normal Bitumen (2)	Porous Asphalt with Modified Binder (2)
Belgium	2.3	-	1.2	1.5
Netherlands	1.7	1.3	1.0	1.2
Switzerland	2.2	1.5	1.1	1.6
Austria	1.5 - 2.0	1.1	1.0	1.2 - 2.0
Italy	7.1	1.5	-	2.0
France	1.7 - 2.0	1.2 - 1.3	1.0	1.2 - 1.3
Germany	1.8 - 5.7	1.1 - 1.4	1.3 - 1.5	2.4 - 2.7
United Kingdom	3.2	3.0	2.0	3.3

(1) Cost ratio in comparison with normal bitumen

(2) Cost ratio in comparison with dense mix with normal bitumen P.I.A.R.C. P.149

Table 4 - Porous Asphalt Mix Cost Comparison

Porous asphalt is also considered to give only approximately half the strength life to any pavement structure in comparison with hot rolled asphalt. Its design life is said to be less than traditional surfacings, whilst its service life may be affected by the blocking of the pores by traffic grit. There is also a necessity to alter the de-icing salt regime, as porous asphalt tends to freeze more quickly than traditional surfacings. These disadvantages need to be mentioned as a balance to the benefits of noise reduction and the reduction of spray.

United Kingdom authorities (and some of the European authorities) are not convinced that there are any overall reductions in accidents where porous asphalt is used compared with traditional surfacings. There is no evidence to support this statement - it is simply that motorists feel safer driving on porous asphalt because it is so quiet, and as a result tend to drive much closer to the vehicle in front.

3 Proprietary Materials

In addition to porous asphalt, a number of new and innovative materials are slowly being introduced into the Industry. Table 5 illustrates the difference in thickness of some of the new surfacing techniques, and each can be discussed.

	THIN SURFACING	ULTRA THIN SURFACING	MICRO SURFACING
DEFINITION	Hot mix Generally laid by conventional paver in Nominal layer 20mm - 25mm thick	Hot mix Laid by specialist plant in Nominal layer 12mm - 18mm thick	Cold mix Laid by specialist plant in Nominal layer 6mm - 12mm thick
FEATURES	Traditional surfacing technique with regulating ability	High output machines, thick tack coat, some regulating ability	Improved slurry surfacing operations, regulating by 2nd layer method

Table 5 - Thicknesses and Features of New Propriotary Materials

3.1 Hot Mixed Thin Surfacings

Hot mixed thin surfacings were developed in France in the 1980's. They are typically laid by conventional paving equipment with nominal layer thicknesses, around 20mm. Generally they are proprietary systems, and usually the mix will contain a polymer-modified binder. The hot mix can be produced in any typical coating plant equipped with the necesssary quality control. In the

United Kingdom, thin surfacings use hardstone aggregates with a PSV dependent upon the location.

In France, thin surfacings have been laid on the road network since 1985; they currently have a market share which exceeds 10 million square metres per year. Hot mix thin surfacings are used and are very popular on the French motorway system. They are chosen by toll road operators and by national authorities on the primary road system and also by local authorities for the secondary road network.

The application of a thin surfacing has a number of advantages, particularly in the traffic management aspect, and also in the speed of operation compared with the more traditional wearing course materials. Conventional paving equipment is used, typically one paver and two rollers. There is no need to feed chippings into a chipping machine, thus eliminating one of the major problems from the traffic management angle when laying a traditional hot rolled asphalt wearing course, and whilst the laying rates in tonnage terms are very similar to hot rolled asphalt, in terms of area they are virtually doubled, thus speeding up the laying operation considerably.

The thin surfacing technique has been introduced into the United Kingdom and a number of contracts have been carried out.

3.2 Hot Mix Ultra Thin Surfacings

Hot mix ultra thin surfacings were again developed in France during the 1980's. These are in the main proprietary systems generally laid by specialist machinery with a nominal thickness of 15mm and these also have been introduced into the U.K. The application is carried out by a specialist paving machine which also spreads a tack-coat immediately prior to laying the hot mix.

An alternative system uses modified traditional paving equipment where the tack-coat is applied ahead of the paver followed by an application of a thin layer of chippings to enable the paver and delivery lorries to run on the tack-coat without damaging it.

3.3 Micro Surfacings

Micro surfacings are mainly cold mix materials laid by specialist plant with a nominal layer thickness of between 6mm and 12mm. Again developed in Europe, they use sophisticated emulsion-based technology and have been developed from the old slurry seal methods.

These ultra and micro thin surfacings have again been the result of the advantages and advances in technology of polymer-modified binders, in place of cementitious binders, Some of the micro surfacings contain fibres, and machines have been developed with fibre-dosing apparatus contained within the paving machine. Most of these types of systems are again proprietary, and in many cases are patented due to the very high research and development costs the companies have had to bear. One recently developed machine takes an aggregate to the front and can therefore work on a continuous basis and appears to offer an alternative to surface dressing.

4 The Future

In addition to the improvements to and acceptance of innovative materials for wearing course in the United Kingdom, there are also moves to produce high-strength road bases which could be said to be a flexible concrete approach. In the United Kingdom, the dense bituminous macadams have been developed into heavy-duty macadams, and the next step is to produce a material which is known in France as EME (Enrobé Module Élevé) or high modulus blacktop. This carefully designed hot mix uses bitumen with a penetration of 20 or less, and the aim, with these materials, is to design blacktop pavements with a design life of 40 years in order to compete with cementicious concrete.

The Road Construction Industry, particularly with regard to surfacings, is going through some exciting and most interesting times. It is also pleasing to see the acceptance of these innovative materials by the Department of Transport, and the encouragement being given by the Department to the development and the introduction of new and fresh ideas into the industry.

5 References

1. T.R.J Fabb, " The Case for the Use of Porous Asphalt in the U.K." Refined Bitumen Association, London, June 1992

2. J.C.Nicholls and M.E.Davies, 1992 "Cleanig of Porous Asphalt" Department of transport TRL working paper WP/MC/49 (unpublished)

3. J.Bellanger, Y.Brosseaud, JL.Gourdon, "Thinner and Thinner Asphalt Layers for the Maintenance of French Roads" proceedings of the Transportation Research Board, 71st annual meeting, Washington D.C. 1992

4. International Slurry Sealing Association. "Micro surfacing - Pavement Resurfacing" 1101, Connecticut Avenue, Washigton D.C.

Author index

Subject index

This index is compiled from the keywords assigned to the papers, edited and extended as appropriate. The page references are to the first page of the relevant paper.